C0-ATR-924

THE FUTURE OF AMAZONIA

Also by David Goodman

FROM FARMING TO BIOTECHNOLOGY: A Theory of Agro-Industrial Development (*with Bernardo Sorj and John Wilkinson)*

FROM PEASANT TO PROLETARIAN (*with Michael Redclift)*

THE INTERNATIONAL FARM CRISIS (*co-editor with Michael Redclift)*

Also by Anthony Hall

DEVELOPING AMAZONIA: Deforestation and Social Conflict in Brazil's Carajás Programme

DEVELOPMENT POLICIES: Sociological Perspectives (*with J. Midgley*)

COMMUNITY PARTICIPATION, SOCIAL DEVELOPMENT AND THE STATE (*with J. Midgley et al.*)

DROUGHT AND IRRIGATION IN NORTH-EAST BRAZIL

The Future of Amazonia

Destruction or Sustainable Development?

Edited by

David Goodman
Senior Lecturer in Economics
University College, London

and

Anthony Hall
Lecturer in Social Planning in Developing Countries
London School of Economics and Political Science

St. Martin's Press New York

Scholarly and Reference Division,
St. Martin's Press, Inc., 175 Fifth Avenue,
New York, N.Y. 10010

First published in the United States of America in 1990

Printed in Hong Kong

ISBN 0–312–04914–5

Library of Congress Cataloging-in-Publication Data
The Future of Amazonia: destruction or sustainable development?/
edited by David Goodman and Anthony Hall.
p. cm.
Includes index.
ISBN 0–312–04914–5
1. Environmental policy—Amazon River Region. 2. Environmental policy—Brazil. 3. Amazon River Region—Economic policy. 4. Land settlement—Amazon River Region. 5. Ecology—Amazon River Region.
I. Goodman, David, 1938– . II. Hall, Anthony L., 1947– .
HC188.A5F88 1990
363.7′0981′1—dc20 90–35991
CIP

To the memory of Francisco ('Chico') Mendes and other victims of rural violence in Amazonia

To the memory of Francisco (Chico) Mendes and other victims of rural violence in Amazonia

Contents

List of Illustrations

List of Tables

Acknowledgements

The editors have been fortunate to receive the wholehearted cooperation of contributors in the preparation of chapters for publication in this volume. Institutional support has been provided by the London School of Economics and Political Science and University College, London. The secretarial assistance of Janet Roddy at UCL is gratefully acknowledged.

DAVID GOODMAN
ANTHONY HALL

List of Acronyms

ABA	Associação Brasileira de Antropologia/Brazilian Association of Anthropologists
ALBRAS	Alumínio do Brasil/Aluminium Company of Brazil
ALUMAR	Alumínio do Maranhão/Aluminium Company of Maranhão
ALUNORTE	Alumínio do Norte/Aluminium Company of the North
BANACRE	Banco do Estado de Acre SA/Bank of the State of Acre
BASA	Banco da Amazônia SA/Bank of Amazonia
CAPEMI	Caixa de Pecúlio dos Militares/Military Pension Fund
CEDI	Centro Ecumênico de Documentação e Informação/Ecumenical Centre for Documentation and Information
CEDOP	Centro de Documentação e Pesquisa na Amazônia/Centre for Documentation and Research in Amazonia
CGT	Confederação Geral do Trabalho/General Workers' Congress
CIMI	Conselho Indigenísta Missionário/Indian Missionary Council
CNBB	Conferença Nacional dos Bispos Brasileiros/National Conference of Brazilian Bishops
CNS	Conselho Nacional dos Seringueiros/National Council of Rubber-Tappers
COBAL	Companhia Brasileira de Alimentos/Brazilian Food Company
COLONACRE	Companhia de Colonização do Acre/Acre Colonisation Company
CONTAG	Confederação Nacional dos Trabalhadores na Agricultura/National Confederation of Agricultural Workers
CPT	Comissão Pastoral da Terra/Pastoral Land Commission

CSN	Conselho de Segurança Nacional/National Security Council
CUT	Central Única do Trabalho/Unified Labour Centre
CVRD	Companhia Vale do Rio Doce/Rio Doce Valley Company
DNER	Departamento Nacional de Estradas e Rodagem/National Department of Roads
DNPM	Departamento Nacional de Pesquisas Minerais/National Mineral Research Department
EMATER	Empresa de Assistência Técnica e Extensão Rural/Technical Assistance and Rural Extension Enterprise
EMBRAPA	Empresa Brasileira de Pesquisa Agropecuária/Brazilian Enterprise for Agricultural Research
ELETROBRAS	Centrais Elétricas do Brazil SA/Brazilian Central Electricity Board
ELETRONORTE	Centrais Elétricas do Norte do Brasil SA/Northern Brazil Electricity Board
FUNAI	Fundaçao Nacional do Indio/National Indian Foundation
GEBAM	Grupo Executivo da Baixa Amazonas/Executive Group for the Lower Amazon
GETAT	Grupo Executivo das Terras do Araguaia–Tocantins/Executive Group for the Lands of the Araguaia–Tocantins
IBDF	Instituto Brasileiro de Desenvolvimento Florestal/Brazilian Institute of Forestry Development
IBRA	Instituto Brasileiro de Reforma Agrária/Brazilian Institute of Agrarian Reform
IBGE	Instituto Brasileiro de Geografia e Estatística/Brazilian Institute of Geography and Statistics
INCRA	Instituto Nacional de Colonização e Reforma Agrária/National Institute for Colonisation and Agrarian Reform
MEAF	Ministério Extraordinario para Assuntos Fundiários/Extraordinary Ministry for Land Tenure Affairs
MINTER	Ministério do Interior/Ministry of the Interior

MIRAD	Ministério da Reforma e do Desenvolvimento Agrário/Ministry of Agrarian Reform and Rural Development
NGO	Non-government organisation
PAD	Projeto de Assentamento Dirigido/Directed Settlement Project
PDA	Programa de Desenvolvimento da Amazonia/Development Programme for Amazonia
PGC	Programa Grande Carajás/Greater Carajás Programme
PIC	Projeto Integrado de Colonização/Integrated Colonisation Project
PIN	Programa de Integração Nacional/Plan for National Integration
PMDB	Partido do Movimento Democrático Brasileiro/Party of the Brazilian Democratic Movement
PND	Plano Nacional de Desenvolvimento/National Development Plan
PNRA	Programa Nacional de Reforma Agrária/National Agrarian Reform Plan
POLAMAZONIA	Programa de Polos Agropecuários e Agromineirais da Amazonia/Programme of Agricultural and Agro–Mineral Poles of Amazonia
POLONOROESTE	Programa de Desenvolvimento Integrado do Noroeste do Brasil/Northwest Brazil Integrated Development Programme
PROBOR	Programa de Incentivo à Produção da Borracha Natural/Natural Rubber Production Incentive Programme
PRODIAT	Projeto de Desenvolvimento Integrado do Araguaia–Tocantins/The Araguaia–Tocantins Integrated Development Project
PROTERRA	Programa de Redistribuição de Terras e Estímulos à Agroindustria do Norte e Nordeste/Programme of Land Redistribution and Stimuli to Agro-Industry in the North and North-East
PT	Partido dos Trabalhadores/Workers' Party
SEMA	Secretaria Especial do Meio Ambiente/Special Secretariat for the Environment
SUCAM	Superintendência da Campanha de Saude

	Pública/Superintendency for the Public Health Campaign
SUDAM	Superintendência de Desenvolvimento da Amazônia/Superintendency for the Development of Amazonia
SUDHEVEA	Superintendência da Borracha/Superintendency for Rubber
SUFRAMA	Superintendência da Zona Franca de Manaus/Superintendency of the Free Port of Manaus
STR	Sindicato dos Trabalhadores Rurais/Rural Workers' Union
UDR	União Democrática Ruralista/Rural Democratic Union

Glossary

caboclo	native farmer of Amazonia; backwoodsman
cerrados	savanna region of the central Brazilian plateau
empate	(=literally 'draw' or 'stand-off') a form of collective, face-to-face resistance developed by rubber-tappers against rainforest destruction and land-grabbing
Escola Superior de Guerra	War College
garimpo	mining prospector
grilagem	fraudulent land dealings, including the acquisition of land by force
grileiro	landgrabber
jagunço	hired gunman
latifundiário	large estate owner
latifúndio	large estate with extensive areas of under-utilised land
minifundiário	operator of a smallholding below the size considered adequate to support a family
minifúndio	smallholding
Nova República	New Republic, name applied to the transitional civilian government established in 1985, often derisively
posseiro	settler without legal land title
posse	right to usufruct acquired by occupancy
roça, roçada	smallholding, cultivated plot
seringueiro	rubber-tapper
soldados da borracha	wartime rubber-tappers
terra devoluta	unoccupied public land

Notes on the Editors and Contributors

Alfredo Wagner Berno de Almeida is an anthropologist who until February 1987 was Director of the Land Conflicts Office of the Ministry of Agrarian Reform and Development (MIRAD), Brazil.

Keith Bakx is a sociologist and a recent Research Associate, Department of Sociology, University of Liverpool.

Christopher Barrow lectures on environmental issues at the Centre for Development Studies, University College of Swansea, University of Wales.

José Alberto Colares is an economist in the Planning Secretariat, State of Pará and a researcher at the Nucleus for Higher Amazonian Studies (NAEA), Belém, Brazil.

Philip M. Fearnside is an ecologist and Research Professor in the Department of Ecology, National Institute for Research in the Amazon (INPA), Manaus, Brazil.

Peter A. Furley is a soil scientist who teaches in the Department of Geography, University of Edinburgh.

David Goodman is an economist at University College London and the Institute of Latin American Studies, University of London.

Anthony Hall is a sociologist who lectures on social policy and planning in developing countries in the Department of Social Administration, London School of Economics and Political Science, University of London.

João Hébette is a sociologist teaching at the Nucleus for Higher Amazonian Studies (NAEA), Federal University of Pará, Belém, Brazil.

George Martine is a sociologist and demographer based in Brasília, who is Director of the UNDP research programme on the social impact of development in Brazil.

José de Souza Martins is a sociologist at the University of São Paulo and an advisor to the Pastoral Land Commission (CPT) of the National Conference of Brazilian Bishops (CNBB).

Emilio F. Moran teaches anthropology at the University of Indiana, Bloomington, Indiana, USA.

Luc J. A. Mougeot is a geographer in the Nucleus for Higher Amazonian Studies (NAEA), Federal University of Pará, Belém, Brazil.

Norman Myers is a consultant in ecology and development.

Fred T. Neto received his PhD from the Department of Geography, London School of Economics and Political Science, University of London in 1988.

João Pacheco de Oliveira Filho teaches anthropology in the Postgraduate Programme in Social Anthropology, National Museum, Federal University of Rio de Janeiro, Brazil.

David Pearce is an economist teaching in the Department of Economics, University College London, University of London and is Director of the London Environmental Economics Centre.

David Treece lectures in Brazilian literature and cultural history in the Department of Portuguese and Brazilian Studies, King's College, University of London and is a consultant to the human rights organisation Survival International.

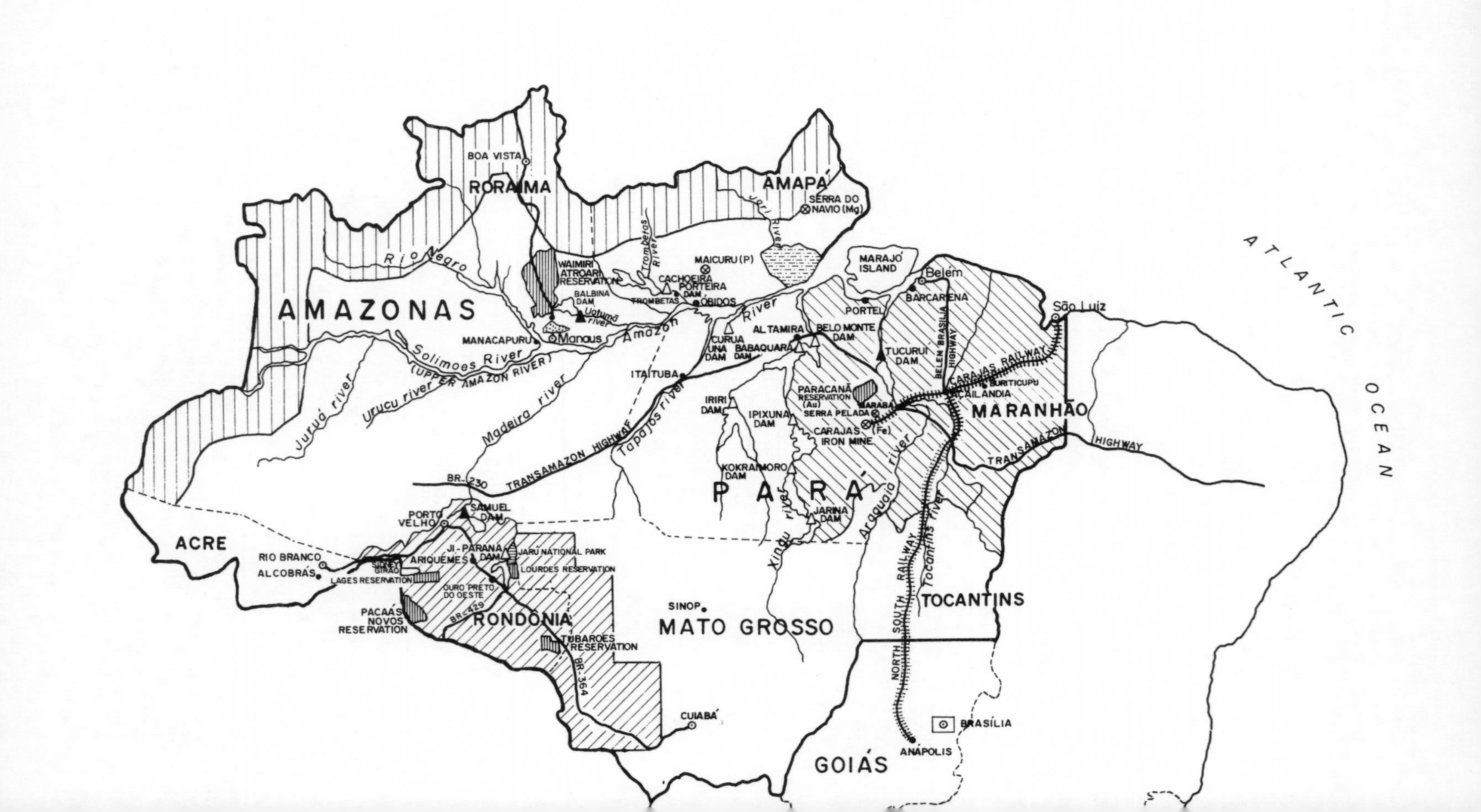
ATLANTIC OCEAN
AMAZONAS
RORAIMA
AMAPÁ
PARÁ
MARANHÃO
TOCANTINS
MATO GROSSO
GOIÁS
RONDÔNIA
ACRE
BOA VISTA
SERRA DO NAVIO (Mg)
Jari River
MARAJÓ ISLAND
Belém
BARCARENA
São Luiz
MAICURU (P)
WAIMIRI ATROARI RESERVATION
BALBINA DAM
Uatumã river
CACHOEIRA PORTEIRA DAM
TROMBETAS
ÓBIDOS
Trombetas River
Rio Negro
Manaus
MANACAPURU
Amazon River
Solimoes River
(UPPER AMAZON RIVER)
Urucu river
Juruá river
Madeira river
Tapajós river
Xingu river
Araguaia river
Tocantins river
PORTEL
ALTAMIRA
BELO MONTE DAM
CURUA UNA DAM
BABAQUARA DAM
TUCURUÍ DAM
BELEM BRASILIA HIGHWAY
CARAJÁS RAILWAY
BURITICUPU
AÇAILÂNDIA
PARACANÃ RESERVATION
MARABÁ
(Au) SERRA PELADA
CARAJÁS (Fe) IRON MINE
ITAITUBA
IRIRI DAM
IPIXUNA DAM
KOKRAIMORO DAM
JARINA DAM
TRANSAMAZON HIGHWAY
BR-230
HIGHWAY
NORTH SOUTH RAILWAY
BRASÍLIA
ANÁPOLIS
SINOP
CUIABÁ
PORTO VELHO
SAMUEL DAM
JI-PARANÁ DAM
ARIQUEMES
JARU NATIONAL PARK
LOURDES RESERVATION
OURO PRETO DO OESTE
BR-429
TUBARÕES RESERVATION
BR-364
PACAÁS NOVOS RESERVATION
LAGES RESERVATION
RIO BRANCO
ALCOBRÁS

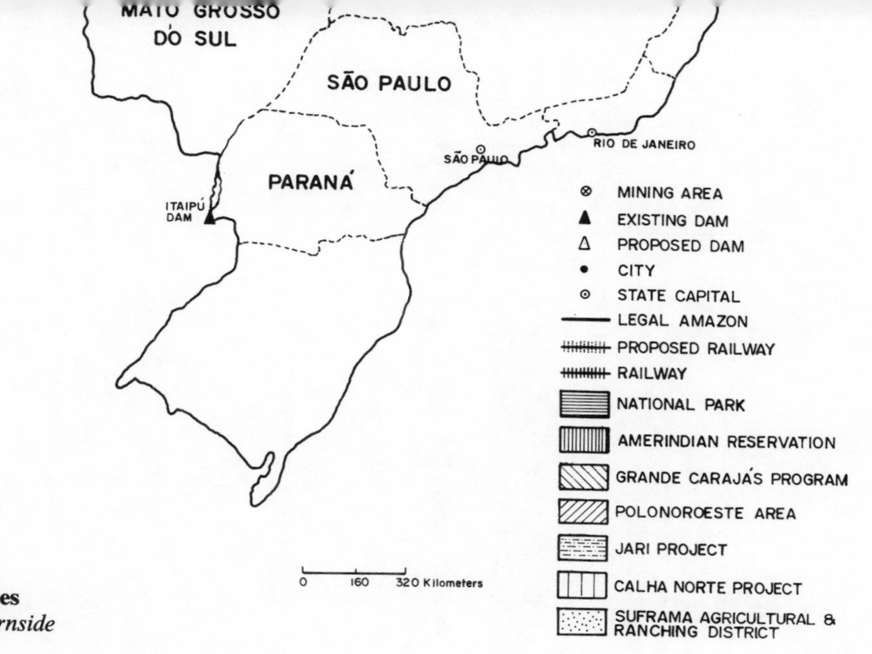

Brazilian Amazonia: Major Development Programmes
Reproduced by kind permission of Dr Philip M. Fearnside

1 Introduction

David Goodman and Anthony Hall

The 1990s will be decisive for Amazonia. Before the dawning of the third millenium, planners and policy-makers must decide whether the world's largest remaining area of tropical rainforest will follow much of Africa and South-East Asia down the path of irreversible destruction, or whether the resources of this vast region will be harnessed for the benefit of Brazilian society and the world as a whole. The present volume is concerned primarily with the Brazilian portion of the Amazon Basin, which occupies two-thirds of the region's 4.2 million square kilometres. It has been estimated that up to 12 per cent of Brazilian Amazonia had been deforested by the end of 1988, or some 350,000 square kilometres. Yet while this figure seems insignificant when compared with the 72 per cent and 63 per cent loss rates recorded for West Africa and Southern Asia respectively (WRI, 1986), there is certainly no justification for complacency. As several contributors to this volume point out, deforestation rates are proceeding exponentially in large parts of eastern Amazonia and Rondônia. On average about 20,000 square kilometres are destroyed every year; in 1988 alone, 12,000 square miles of rainforest disappeared, an area the size of Belgium. In 1989 a region the size of West Germany will experience a similar fate. If present trends continue, by the middle of the next century the Brazilian rainforest will have ceased to exist as a sustainable ecosystem.

The purpose of this book is twofold. First, it aims to describe and analyse current strategies of economic development and frontier integration in Brazilian Amazonia, highlighting their major environmental (that is, social and ecological) impacts upon the region. Secondly, this analysis goes beyond mere criticism of existing policies and contributes to a growing body of opinion, both within and outside of Brazil, which insists that urgent action is required on the part of decision-makers to halt the progressive destruction currently under way. In addition, it is argued that more sustainable development options do exist for Amazonia than those currently prioritised by official and private investors, and that these must be explored. Without such a major commitment by both government and private enterprise, at domestic and international levels, there is little likelihood that the pace of deforestation, with all its negative consequences, will diminish in the forseeable future.

In proclaiming the need for prompt attention to these issues, the contributors to this book are not basing their arguments on the purist notion of preserving the environment merely for its own sake. They are, on the contrary, motivated by a common belief that the minimisation of natural capital loss and the introduction of more sustainable forms of land-use in Amazonia, whatever their intrinsic merit in environmental terms are, more importantly, essential prerequisites for balanced growth. That is, a pattern of economic development which distributes the benefits of rainforest exploitation in a way which permits the majority of participants to benefit, while incurring a minimum of ecological damage. Those responsible for the formulation and implementation of Amazonia's current predatory development policies have so far, however, been content to encourage the pursuit of immediate gains through relatively destructive activities, ignoring the longer-term consequences for Amazonia, for Brazil and the world.

There is now an abundance of research evidence which unequivocally demonstrates both the actual and potential importance of Amazonia's fragile rainforest ecosystem in generating significant benefits for humanity. In terms of their biodiversity, tropical rainforests are unrivalled. Covering only 6 per cent of the world's land area, rainforests contain at least 50 per cent, and perhaps 90 per cent, of all known plant and animal species. Amazonia's tropical moist forest (TMF) is, in fact, the richest single region, harbouring at least 20 per cent of all higher plant varieties, some 30,000, as well as one-fifth of all bird species (Wilson, 1989; Brundtland Report, 1987; WRI, 1986; Wetterberg, 1981; Myers, 1985). Rainforests are a vital source of medications, with over 40 per cent of all modern prescriptions containing drugs of natural origin; destruction of the TMF could obstruct researchers in their efforts to find a cure for many diseases such as cancer and even AIDS. In addition, of course, the TMF provides a wide range of industrial materials such as dyes, resins, fibres, oils and tannins, and also forms an invaluable reservoir of genetic material which scientists, at the dawn of the biotechnological revolution, have hardly started to tap. Given that barely 20 per cent of rainforest species have so far been catalogued, there is a very real danger that many will be permanently lost as a result of deforestation before their usefulness can be assessed. By the late twenty-first century, according to one estimate (Ledec, 1985), up to 50 per cent of all animal and plant species, mostly of rainforest origin, will have vanished in this way.

As well as destroying an irreplaceable source of valuable raw materials, the widespread deforestation which results from current development practices in Amazonia could also lead to substantial ecological disruption. Environmental services provided by the TMF include watershed protection, soil stabilisation and the maintenance of regional, national and possibly global climatic stability. Although the rainforest occupies 6 per cent of the Earth's surface, it receives almost 50 per cent of total rainfall, soaking it up like a sponge and releasing it gradually. As has been witnessed already in many areas of Amazonia, loss of forest cover greatly increases soil erosion, reducing agricultural yields and leading to the siltation of rivers and reservoirs. Increased run-off leads to greater flooding, disrupting transport, fishing and floodplain-based farming. Since at least half of Amazonia's rainfall is self-generated through evapotranspiration, forest loss has been linked to the onset of drought conditions. This phenomenon has been recorded in Mexico, Costa Rica, India and Malaysia (Myers, 1985; Ledec, 1985). There is mounting evidence that a similar process is taking place within localised parts of Amazonia such as Manaus and southern Pará and, indeed, further afield in southern Brazil, as a direct result of rapid TMF loss (Salati and Vose, 1984; Sioli, 1985). Thus, not only is an unknown but surely large number of species being destroyed; the productive potential of Amazonia is being severely and permanently curtailed by the scale and pace of deforestation.

Yet the repercussions of rainforest destruction are felt beyond Amazonia and even Brazil or South America. It is now generally agreed that the burning of biomass such as tropical vegetation has contributed significantly to the so-called 'greenhouse effect', through which an accumulation of carbon-dioxide and other gases in the atmosphere will lead to an estimated increase in global mean temperatures of between 2 and 12 degrees Centigrade (WRI, 1986). As the polar ice-caps melt in response, vast lowland areas will be submerged all over the world, and crop production patterns significantly modified. In Brazil there has been recent speculation, based on US satellite data, that Amazonian deforestation has helped to enlarge the hole in the ozone layer above Antarctica, a gas which protects the Earth from the Sun's harmful ultra-violet rays (*The Guardian*, 20 April 1988).

There are, therefore, sound ecological reasons why urgent action is necessary to slow down the rate of deforestation in Amazonia. Yet environmental concerns must also embrace the vital social role of Amazonia in providing a means of livelihood to millions of small-scale producers and their families, both indigenous populations and migrants. While Brazilian policy-makers have possibly been guilty in the past of using Amazonia in a careless manner for 'syphoning off' and resettling land-hungry peasant farmers from other regions, there can be no doubt that, as small-scale producers are expelled from more fertile lands in the Centre–South by the spread of large-scale commercial farming and agribusiness, the rainforest will, despite recurring problems of rural violence and land reconcentration, have to absorb further waves of migrants for a good few years to come. Around the world some 200 million people depend upon rainforests for their survival. In Brazilian Amazonia this figure could amount to over 3 million individuals at present. The preservation and sustainable development of this region is thus of paramount importance not simply in terms of maintaining natural capital stocks or species for future use. More immediately, it is essential to ensure that the TMF ecosystem is settled in an environmentally sound manner so that its benefits accrue more equitably to those groups which have little alternative for economic survival.

It is not the aim of this volume to repeat in detail the history of Amazonian development but, for the purpose of providing a background against which current official priorities may be viewed, it is worth briefly sketching its major landmarks. There is no single explanation for the drive to occupy and develop Amazonia, which has become so intense during the past 20 years in particular. Rather, a variety of reasons may be cited which, in the short to medium term at least, appear to be mutually supportive of each other even if, in the long run, contradictions become apparent. It is difficult to disaggregate these factors and to assess, with any degree of precision, their relative importance. Yet if we consider the major thrusts of policy-making and implementation for the region a number of themes recur, especially during the post-1964 period of more concerted and aggressive Amazonian development pursued both by successive military governments and, indeed, the civilian 'New Republic' after 1985.

An overriding goal of the Brazilian State has undoubtedly been a joint concern to promote 'national integration', in both an economic and political sense, combined with efforts to occupy and secure Brazil's frontiers against the perceived threat, real or imagined, of foreign incursions. Indeed, there is a long history of nationalist fears over the country's territorial integrity in Amazonia. Before its 1964 intervention the military had, through the powerful *Escola Superior de Guerra*, officially endorsed a policy of permanent occupation of all Brazilian lands (Branford and Glock, 1985). Such ideas were also reflected in subsequent publications which condemned the coveting of Amazonia's natural resources by foreign interests (Reis, 1982). Thus, the setting up of the Superintendency for the Development of Amazonia (SUDAM) in 1966 and the unleashing of 'Operation Amazonia' at the same time, reflected the growing concern of the new military regime that Amazonia should be developed more systematically.

Specifically geopolitical aims, however, became more evident from 1970 with the 'Plan for National Integration' (PIN) and development plans for both Brazil (PND I) and Amazonia (PDA I) for the period 1971–4. Although there were other social, political and economic aims being pursued simultaneously, as will be discussed below, the slogan which accompanied these plans captures the underlying nationalistic goals; '*integrar para não entregar*' (integrate in order not to forfeit). A massive highway-building programme was started, adding to the roads already built in the late 1950s and early 1960s (Belém–Brasília and Cuiabá–Santarém). During the 1970s major highways such as the Transamazon (5,000 km), *Perimetral Norte*, skirting the northern borders of Brazil with Venezuela, Guyana and Surinam (2,700 km) and Cuiabá–Porto Velho–Rio Branco, opening up Rondônia and Acre, were commenced. These major highways and the network of secondary feeder roads to which they gave rise permitted the rapid physical settlement of Amazonia. In addition to broad geopolitical goals, the early 1970s saw another justification for accelerating the pace of human settlement and State–military control in the region, that of counter-insurgency. Such was the perceived threat posed by a small band of Maoist guerrillas in the Araguaia region of eastern Amazonia that 20,000 troops were mobilised in three campaigns before the movement was finally crushed in 1975. Almost unknown within Brazil at the time due to heavy censorship, it was only later that the scale of the military intervention against the insurgents and the local population in what was labelled the 'Araguaia guerrilla war' became public knowledge (Bourne, 1978; Doria *et al.*, 1978).

If preoccupations over national sovereignty were evident during the 1970s they were no less important during the following decade. The expansion of State influence over the process of Amazonian development has continued unabated. While some observers (e.g., Souza Martins, 1984) see the spread of huge government projects and programmes such as POLONOROESTE and the Greater Carajás Programme (including special bodies such as GETAT and GEBAM) as essentially a military phenomenon, others take the broader view that (whatever the rhetoric of free-market economics) it reflects a more general attempt by the State to exert more centralised control over the economy and society (Grindle, 1986; Hall, 1989a). Indeed, as both Pacheco de Oliveira Filho and Treece clearly demonstrate (Chapters 7 and 11 in this volume), the recent Calha Norte programme embodies the continuing importance of military influence within the civilian government and its national security doctrine of securing Brazil's borders in Amazonia.

In addition to being the object of nationalistic concerns over the issue of territorial sovereignty, Amazonia has been viewed increasingly as a source of natural resources to fuel Brazil's economic development. Having laid virtually dormant during four centuries, the Amazon Basin's potential (for success and for failure) in this respect became apparent with the ill-fated rubber boom of 1870–1912, an enterprise which Getúlio Vargas tried to revive during the 1940s to boost production for the war effort with the 'Batalha da Borracha' programme. The inconsequential Superintendency for the Valorisation of Amazonia (SPVEA), established in 1953 as Amazonia's first regional development agency, was superseded in 1966 by SUDAM which, with the aid of generous subsidies in the form of tax and other incentives, set about encouraging land acquisition by southern businessmen, with pasture formation and cattle ranching considered to be the most appropriate development strategy for the region. From a handful in the mid-1960s, these ranches numbered 162 by 1969, benefitting from about $150 million in tax incentives, increasing sharply during the 1970s under PIN and POLAMAZONIA, to reach a figure of 630 by 1985, enjoying over $1 billion in subsidies in real terms (Branford and Glock, 1985; Hecht *et al.*, 1988; Gasques and Yokomizo, 1985; Mahar, 1988). The ranches average 24,000 hectares in size but can be vast, the largest covering half a million hectares or more, such as the Liquigas company's Suiá Missú and the Codeara enterprises.

While it is difficult to attribute responsibility for deforestation to different activities, there is general agreement that livestock development and pasture formation are the major culprits, with estimates of their contribution ranging from 40–75 per cent (Davis, 1977; Mahar, 1988; Pearce and Myers, Chapter 15 in this volume). This process has also been fuelled by high inflation encouraging property speculation based on substantial annual increases in real values, as well as laws which require a proportion of land claimed to be cleared of forest cover as a legal proof of occupation and subsequent ownership. The environmental consequences of these policies have been quite disastrous. The chapters in this volume by Fearnside (Chapter 8) and by Pearce and Myers (Chapter 15) discuss the negative ecological impacts of extensive deforestation and pasture formation under these circumstances: rapid fertility decline, soil erosion and degraded pastures, leading to poor economic returns.

Despite the intrinsically uneconomic nature of cattle ranching in Amazonia, however, government subsidies have continued relatively unabated. Neither is the ecological damage and rural violence engendered through growing land conflict as a result of land reconcentration on the frontier sufficient to inspire a change of policy. Although some minor restrictions were imposed in 1979, both SUDAM and the Carajás Programme continue providing fiscal incentives for livestock production. The political importance to the government of continuing to support wealthy business groups through such measures has been a recurrent feature of Amazonian development, the environmental price for which is now being paid. During the 1970s highway construction companies and other commercial interests exerted strong pressure on the Médici and Geisel governments to switch funding from directed 'social' colonisation for small farmers, discussed in greater detail below, to large-scale, corporate agribusiness and land acquisition (Bourne, 1978; Pompermeyer, 1984). This concern appears to persist and was perhaps reflected in President Sarney's declaration in July 1987 that he did not 'even want to hear' about the possibility of abolishing these subsidies (*Isto É*, 15 July 1987). In a gesture towards an increasingly vociferous Brazilian and international environmental lobby, however, some fiscal incentives were suspended on 12 October 1988, initially for a period of 90 days.

While the political importance of continued State support for livestock and pasture formation was recognised, however, so too was its limited economic return. During the late 1970s and 1980s, therefore, although livestock projects continued to absorb scarce public funds, mining replaced cattle ranching as the major investment priority. Mineral extraction and processing activities had been highlighted in the Second National Development Plan (PND II, 1975–9) and POLAMAZONIA proposals, which focussed on the Trombetas bauxite reserves and Carajás iron-ore deposits, the world's largest, discovered in 1967. These were systematised under the Greater Carajás Programme (PGC), instituted in 1980 under the Figueiredo government by presidential decree, and administered by an interministerial council with charge of a special fiscal incentive programme for investors within its 900,000-square kilometre area of jurisdiction, covering 11 per cent of Brazil and a quarter of the Amazon Basin itself. Its plans for integrated aluminium production, iron-ore extraction, pig-iron smelting and steel production along the 890-kilometre railway corridor between Carajás and São Luís, as well as vast infrastructural and associated agrolivestock and lumbering activities, make it the largest economic development programme for a rainforest area anywhere in the world (Hall, 1987, 1989a).

Budgeted originally at no less than $62 billion, foreign funding to the tune of $1.7 billion was immediately forthcoming. Some came in the form of multilateral aid from the World Bank ($405 million), but most assistance was bilateral between Brazil and foreign governments anxious to guarantee supplies of cheap raw materials or gain access to heavily subsidised investment opportunities; the EEC agreed to lend $600 million and Japan $500 million. Japanese aid advisors were also heavily involved in designing this regional development package, which has been condemned by some critics either as crude foreign interference (IBASE, 1983; Cota, 1984) or as constituting a distortion of the regional planning process primarily to support the global strategies of foreign transnationals rather than Brazilian development *per se*, as argued by Neto in Chapter 6 in this volume.

The PGC, as with other government initiatives in Amazonia, has been subjected to mounting criticism because of its apparent failure to take account of the environmental destruction which it is inducing, both directly and indirectly. The PGC interministerial council, for example, has approved plans for some twenty charcoal-fired, pig-iron smelters to be set up along the railway, mainly by steel companies from the South seeking to minimise production costs by utilising virgin rainforest as cheap fuel and taking advantage of financial incentives offered by the programme. According to recent estimates by the CVRD's environmental division, up to 600,000 hectares of forest will be indiscriminately destroyed every year in this fashion, in addition to which no serious plans have been made either for reforesting cleared areas or obtaining future charcoal supplies from plantations, as in fact dictated by law. The PGC has also, of course, speeded up deforestation indirectly in eastern Amazonia as a result of a rapid escalation in real estate values induced by the expansion of infrastructure such as highways and dams, which has attracted wealthier landed and commercial interests in search of investment opportunities. Pressure on land is further exacerbated by the attraction of workers to large construction projects within the PGC many of whom, upon completion of these schemes, revert to small-scale farming alongside thousands of other migrants, within a context of growing concentration of property ownership. This combination of settler farming, cattle ranching and property speculation in the Carajás area has led, as elsewhere in Amazonia, not only to ecological degradation but to severe rural violence caused by competition over access to land (Hall, 1987, 1989a).

If there has been consistent attention paid in development policy-making for Amazonia to broad geopolitical and economic objectives, the State has also shown no hesitation in simultaneously using such occupation strategies to serve internal, socio-political purposes. By using Amazonia to absorb displaced rural populations from the North-East and Centre–South of the country, the structural problems of growing land concentration, landlessness and the pressing need for agrarian reforms could be conveniently side-stepped by planners and politicians alike. Amazonia has been carelessly and cynically used to absorb growing numbers of small farmers squeezed out of their traditional areas; from the North-East by a combination of drought, land concentration and rural stagnation, and from the South of Brazil by the capitalisation and mechanisation of agriculture for soybean and wheat farming, linked to polarisation in landownership, growing landlessness and discriminatory agricultural policies (see, for example, Goodman, 1988).

A third major justification for accelerated development in Amazonia, complementing geopolitical and economic reasons, is thus the so-called 'safety-valve' function. In common with rainforest areas in other countries, such as Indonesia and Malaysia, the vast, sparsely populated lands of Brazilian Amazonia have been viewed as a solution for resettling 'excess' or displaced rural inhabitants from other areas of the country. This tradition goes back to the nineteenth century when thousands of drought refugees or 'flagelados' were transported at government expense from the North-East to work as rubber tappers. Throughout the 1950s and 1960s Amazonia received a small but steady flow of so-called 'spontaneous' migrants as agriculture and livestock activities moved westwards into Maranhão and northwards into Goiás and Mato Grosso, replacing the southern states of Paraná and Rio Grande do Sul as the new frontier (Goodman, 1978; Foweraker, 1981). The influx of small farmers into the Amazon Basin rose dramatically in the early 1970s, however, under PIN which was popularly thought to have been inspired by President Médici's despair at the human suffering caused by the North-East drought of 1970.

The huge road-building programme then undertaken, whatever the commercial pressures behind such a policy was intended, in the catchphrase of the day, 'to unite men without land to land without men'. Some $2.3 billion dollars was invested in building the Transamazon highway; its planned communities (the *agrovila*, *agrópolis* and *rurópolis*) were expected to absorb 100,000 families under the auspices of INCRA (Smith, 1982; Rich, 1985). The failure of public colonisation along the *Transamazônica* is well-documented, with only 8 per cent of the resettlement target achieved. Much has been written about the reasons for this shortfall, and its causes are varied; no feasibility studies were undertaken, what little planning took place was extremely poor, on top of which there was a low level of institutional support at all levels for colonists (Moran, 1981; Bunker, 1985). Moran's chapter in this volume (Chapter 4) underlines these problems which, as he demonstrates, have recurred even with private colonisation schemes supposedly in possession of more resources and expertise.

Above all and underpinning these factors was the short-lived official commitment to directed public resettlement. While businessmen exerted pressure upon government through organisations such as the Association of Amazon Entrepreneurs (AEA) to change policies and divert funds to serve their interests, politicians and technocrats attempted to 'blame the victims' for the poor performance of small farmers in Amazonia (Wood and Schmink, 1979). Planning minister Reis Veloso, for example, condemned their 'predatory occupation' and called upon large companies to 'assume the task of developing the region' (Branford and Glock, 1985, pp. 70–1). Subsequently, directed colonisation was rejected by the Geisel government in its development plan (PND II, 1975–9) and public lands hitherto reserved for small farmers by INCRA were, from as early as 1973, sold off in large tracts.

Such a heavily biased policy could not be sustained for very long in the context of the massive influx of migrants into Amazonia which had been facilitated during the 1970s by the highway-building programme. While the vast majority of migrants came under their own steam in search of land, from the mid-1970s the government initiated directed resettlement programmes to encourage a more systematic occupation process in the western states of Rondônia and Acre. By 1983 up to 600,000 had been attracted to Rondônia, a large proportion by INCRA promises of land and support, but only 5,000 were absorbed by colonisation schemes. As the contributions to this volume by Martine (Chapter 2) and Bakx (Chapter 3) clearly show, however, mistakes witnessed a decade earlier on the *Transamazônica* were repeated, generating high colonist turnover. In a by now familiar pattern, a combination of poor planning and property speculation, along with widespread fraud in the allocation of titles, has resulted in the reconcentration of landownership, making the prospect of absorbing large numbers of landless farmers almost inconceivable without land reform. As early as 1975 over 80 per cent of Rondônia's farmland was in the hands of southern business enterprises benefiting from SUDAM incentives (Branford and Glock, 1985) a process which, as Bakx demonstrates, is now being repeated with a vengeance in Acre.

Thus, although development strategies for Amazonia have been motivated in part by the need to accommodate large numbers of dispossessed peasant farmers, as well as by macro-economic, political and geopolitical objectives, enormous contradictions become immediately apparent. For while there is undoubtedly a growing urgency and, indeed, considerable potential for the Amazon Basin to perform such a 'social' role, the wider forces which condition and constrain this occupation process seem, within the existing framework of national and regional development, destined to frustrate such ambitions. Both the policy bias in favour of large landed interests, as well as four-digit annual inflation rates, have encouraged land concentration in every successive Amazon frontier zone. Large landowners are the biggest culprits in this process, but a growing population of migrant farmers is under increasing pressure to engage in more destructive cultivation and occupation techniques. In 'mining' the soil farmers have engaged in a cycle of deforestation and ecological decay which can serve only to make extractive and agricultural activities unsustainable in the long term. This process of environmental degradation, while still in its infancy in Amazonia as a whole, has devastated localised areas and threatens to engulf a substantial part of the region within the next 10 years.

Another characteristic which has become automatically associated with frontier settlement in the Amazon Basin is that of rural violence, due to the incessant struggle over access to land among individual and corporate property owners, the mass of desperate rural migrants and indigenous Amerindian groups. The growing scale and intensity of land conflicts is closely linked with environmental destruction and land concentration, and is reflected in a large increase in associated deaths in Amazonia, from a few in the 1960s to about 60 in 1984 and almost 200 by 1986, or two-thirds of the total for Brazil (MIRAD, 1986; 1987). Several murders of well-known human rights activists have drawn attention to the severity of the problem, notably the assassination in May 1986 of Father Jósimo Tavares in Marabá, and that of the rural union lawyer Paulo Fontelles, in Belém, in June 1987. This series of atrocities continued with the murder by a local landowner of Francisco ('Chico') Mendes, leader of the rubber-tappers' movement, in December 1988, having survived two previous assassination attempts. As Treece points out in Chapter 11, indigenous groups have come under widespread attack, subjected to invasions of their territory by cattle ranchers, mining companies, loggers, government projects, encroaching small farmers, and massive waves of independent gold prospectors in areas such as Roraima, home of the Yanomami. So great has been the escalation in violence and intimidation that a special report by Amnesty International was devoted to the issue which, significantly, stresses its concern over 'evidence of inadequate state response to or even state acquiescence in these crimes' (1988, p. 1).

For a variety of ecological, social, economic and human rights reasons, therefore, the need for action in Amazonia could not be more pressing. While it is fashionable to express resignation about the region's fate, however, such pessimism is premature. The chapters in this volume by Fearnside (Chapter 8), Furley (Chapter 13) and Barrow (Chapter 14) all point towards the technical feasibility of exploiting Amazonia's natural wealth more productively to serve the interests of the majority small farmer population. This could be done through carefully-focussed programmes of research and development to diffuse sustainable farming and extractive techniques in specific, suitable areas. Of course, the wider structural problems of land access and stability of tenure would have to be addressed sooner or later if such measures were to be widely adopted. The collapse of the National Agrarian Reform Plan (PNRA) under pressure from the landowners' lobby, the inappropriately-named 'Rural Democratic Union' (UDR), and the plan's watering down under the 1988 Constitution (see Hall, 1989a), were major disappointments in this regard. The PNRA's fate was sealed in January 1989 when, as part of an austerity package, the Ministry of Agrarian Reform and Rural Development (MIRAD) was finally closed down after seeing its few powers whittled away during the three or so years of its precarious existence. Even bodies with special powers to regulate land issues in Amazonia, such as GETAT have, as Wagner de Almeida shows in Chapter 9, merely consolidated an unequal structure of property ownership, taking the political steam out of popular movements based on increasingly widespread peasant land invasions (see below) and contributed further towards the gradual monopolisation of Amazonian territory by military, landed and commercial interests.

Be that as it may, however, pressures for progressive change are mounting on a variety of fronts, sufficiently strongly to disturb a traditionally complacent government machine. A major source of protest at the inequities inherent in Amazonian development policy has been the peasantry itself. This has been manifested not through institutionalised political channels, from which the rural poor have, by and large, been excluded in Amazonia, as elsewhere in Brazil. Even rural trades unions (STRs), despite having grown in strength in some states such as Pará, have seen their effectiveness undermined by lack of funding, the difficulties of communicating with dispersed populations and, not least, by ideological conflicts arising from the CUT/CGT split within the whole syndicate movement (Grzybowski, 1987). Support from other bodies such as domestic and foreign NGOs as well as the Catholic Church through its 'pastoral land commissions' (CPT) and community religious groups (*comunidades eclesiais de base*, CEBs) has undoubtedly helped to offset this handicap. Yet some observers, such as Souza Martins (Chapter 10 in this volume), rightly or wrongly, view this support as a form of political patronage which does not assist the peasantry to challenge the status quo.

Farmers have themselves resorted to various forms of more spontaneous, unofficial action to press home their claim to a livelihood in the region. This is demonstrated most commonly as increasingly active resistance to land-grabbing, itself reflected in the escalating level of rural violence and numbers of hired gunmen killed by peasants no longer content to be unceremoniously evicted from their plots. The 'parrot's beak' area of Eastern Amazonia, where the states of Pará, Tocantins and Maranhão meet, is the most vivid example of such small farmer opposition. While such initiatives are usually fragmented and confined to individual estates, broader organisations may emerge from such conflicts; witness the formation in 1985 of the National Council of Rubber-Tappers (CNS) in Acre, which sprang from collective resistance to encroaching cattle ranchers. Another area of mounting popular resistance surrounds hydropower projects such as Tucuruí and the Xingú complex, which will undoubtedly generate serious opposition unless, as Mougeot maintains in Chapter 5, appropriate resettlement policies are devised and implemented adequately to compensate the thousands of displaced people involved.

One of the most effective yet least publicised types of peasant action in Amazonia takes the form of active estate occupations, or 'land invasions'. The relatively well-organised movement of landless rural workers (*Movimento dos Trabalhadores Rurais Sem Terra*, MST) has achieved a remarkable degree of success in southern Brazil, where 9,000 invasions have led to resettlement of 13,000 farming families. In the 'Brazil-nut polygon' of southern Pará, 43 out of 202 estates have been partially invaded by 2,000 families of small farmers. Overall within eastern Amazonia, possibly 6,000 families are involved in such occupations. As Hébette and Colares show in Chapter 12, hesitation and fear are gradually giving way to more aggressive and coordinated action by peasant groups in their on-going battles with the joint forces of landed interests and police. The success of such tactics is illustrated not just by the *de facto* land reform thus brought about, but also by the spur it sometimes gives to official action. For example, the rubber-tappers' movement managed to have the first 'extractivist reserve' of 100,000 hectares declared by the Acre state government in February 1988, while there is mounting pressure through the 'Amazon Alliance of Forest Peoples' for a consistent policy of setting up such areas to halt deforestation and protect the livelihoods of rubber-tappers. Likewise, the occupation of 200,000 hectares in the 'Brazil-nut polygon', referred to above, prompted their expropriation under the PNRA in 1988, one of the agrarian reform's few concrete measures. In the case of indian groups also, as Treece points out in Chapter 11, spontaneous forms of protest have often been the most effective in securing benefits in the face of outside threats.

Mounting pressure for change has also come from other sources, as a whole range of Brazilian and overseas non-government organisations (NGOs) have prevailed upon major multilateral aid bodies such as the World Bank to review their lending policies in the light of pressing environmental problems. At local, regional and national levels, domestic NGOs have come to form an influential force during the late 1970s and 1980s in Brazil (Hall, 1989b). Their use of 'popular education' (*educação popular*) methods, inspired originally by Paulo Freire, in working with communities in support of productive as well as more politically-based collective movements in areas such as land claims and human rights, has accorded them a clear role as agents of change. Despite some factionalism arising from ideological and political differences Brazilian NGOs, in collaboration with rural trades unions and Church-based community groups (CEBs) and pastoral commissions (such as the pastoral land commission or CPT), have been at the forefront of social protest in Amazonia. In particular parts of the region, such as Acre and Pará, the combined impact of these organisations in helping to articulate popular demands for change has been especially strong, reflected in the broader movements which have subsequently emerged and are referred to above.

A striking phenomenon amongst Brazilian NGOs in the late 1980s, as their role has expanded to encompass wider political and policy-formulation tasks, is the growth of the environmental lobby, embracing not just ecological but also related social and indigenous questions. Traditionally stronger in the South of the country, groups concerned with ecological issues have focussed their attention on Amazonia. Many conferences organised around this theme have taken place in Brazil during the past few years which have highlighted the process of Amazonian deforestation and its wider consequences. Whereas previously such initiatives were the almost exclusive preoccupation of foreign organisations, conference titles in Brazil such as 'Ecological Disorder in Amazonia' and 'Amazonia: the Agricultural Frontier 20 Years Later', are becoming increasingly common. One such meeting held in late 1988 as the new Brazilian Constitution was being published, brought together Church, rural unions and environmentalists and launched a campaign calling for a three-year 'Ecological Truce in Amazonia' during which all potentially destructive projects such as dams, cattle ranching and mining would be temporarily suspended.

Yet although internal pressures have fuelled the growing environmental debate within Brazil and contributed towards a significant if belated official response it is, arguably, international bodies which have been most influential in this respect. Foreign NGOs in Europe and particularly in the USA have constituted perhaps the single most important force, both directly and indirectly, for bringing about the re-examination of major environmental questions relating to Amazonia. This has been achieved partially through general public awareness-raising campaigns, but principally by means of direct challenges issued to aid organisations involved in funding major development programmes in the region. Three pertinent examples of this are offered by the Carajás iron-ore project, the North-West development programme for Rondônia and Northern Mato Grosso (POLONOROESTE) and the Second Power Sector Loan, which would support Brazil's expansion of its hydroelectric generating capacity in the Amazon Basin.

The Carajás iron-ore and railway project was granted $600 million in soft loans from the EEC Development Fund, as well as $405 million from the World Bank. Since the EEC loan was approved in 1982 non-government organisations have been concerned about the failure to build into the European agreements any significant independent provisions for dealing with Amerindian, social and ecological problems arising from the project's design and implementation. At successive meetings from 1982–7 the General Assembly of European NGOs has called upon the EEC to take corrective measures but these have been extremely tardy, taking the form of a special study commissioned in January 1988 of the project's impact upon indigenous groups, when most of the loans had already been disbursed. But although not immediately effective, European NGO pressure and the publicity surrounding projects such as Carajás have contributed significantly to increasing voters' awareness of environmental affairs. It is significant that ecological and social issues have been incorporated into the policies of a number of European parties and in some countries, such as West Germany, environmentalists or 'greens' have become politically strong enough to mount a serious parliamentary challenge.

The American environmental lobby has been in an even stronger position to exert pressure on decision-makers involved in funding environmentally sensitive development programmes. In the case of Carajás the World Bank is its second largest donor, and the US Congress is responsive, on both the right and centre of the political spectrum, to criticisms of Bank policies. Led by the Environmental Defence Fund, 29 international NGOs wrote an open letter to Bank president Barber Conable denouncing plans by the Brazilian government to set up pig-iron smelters along the 890-kilometre railway, with potentially disastrous consequences for the rainforest and its inhabitants. While it was recognised that the Bank was not directly responsible for this decision, nonetheless urgent demands were made that clauses in the original loan agreement pertaining to environmental protection should be respected (EDF, 1987). This was followed up by publication of further studies from Survival International and Friends of the Earth highlighting the project's harmful effects on the ecology and upon Amerindian groups (Treece, 1987). Due in part to such campaigns the Bank, in 1982, instituted a Special Amerindian Component in the form of an Accord between the State mining enterprise, CVRD, and Brazil's indian agency, FUNAI, for native groups to be compensated for the impacts of the scheme upon their land rights and livelihoods. Originally including some 4,500 indians, the project has been widened to encompass 13,000 from fifteen different groups in 24 reserve areas and is budgeted at $13.6 million, provided by the CVRD for reserve demarcation, project support and institutional strengthening of FUNAI (World Bank, 1988a). While a major step in the right direction, however, the Accord has come under intense criticism from anthropologists for failing to benefit the groups affected and channelling too much money into organisational support for FUNAI, deemed by many to be at best unrepresentative of and, at worst, directly opposed to, indian interests (Treece, 1987: Pacheco de Oliveira Filho, Chapter 7 and Treece, Chapter 11 in this volume).

The US environmental lobby showed its potential leverage more decisively in the case of POLONOROESTE when, in 1985, it temporarily halted disbursement of $256 million of a $434 million World Bank loan, protesting vehemently about the scheme's adverse impact upon the ecology, Amerindian land rights and rural conflicts. The decision was historic in being the first time that the Bank had halted a loan for environmental reasons. Funding granted by the Inter-American Development Bank (IDB) for a related project, the paving of the BR 364 highway which opened up the North-West region, has attracted similar criticism. A longer-term consequence of this pressure has been the re-examination of project design and the formulation of a remodelled 'North-West Consolidation Project' aimed at correcting past mistakes (World Bank, 1988b).

In the light of these experiences, as well as the treatment meted out to people displaced by the Tucuruí hydroelectric project, described by Mougeot in Chapter 5, domestic and foreign NGOs have learned to plan ahead and initiate their campaigns earlier rather than when it is too late for them to have any influence on project design or implementation. A case in point concerns the proposed $500 million Second Power Sector Loan from the World Bank, a major part of which would be used to finance an ambitious hydropower expansion programme in Amazonia involving 63 separate sites for reservoirs along the Araguaia–Tocantins and major Amazon tributaries, flooding some 100,000 square kilometres and displacing up to 156,000 people. A strong NGO protest movement has already begun against the Xingú river complex, whose five dams will flood 18,000 square kilometres and displace some 70,000, including ten indian communities. The first two dams, Kararaô and Babaquara, near the town of Altamira along the Transamazon highway, which will be partially flooded, are due to come on stream in 1998 and 2001 respectively (CPI, 1987). Mounting pressure from local groups and NGOs led to the organisation of a five-day protest meeting of 3,500 indians at Altamira in February 1989 to draw attention to their plight. World-wide publicity resulted from the Brazilian government's decision to prosecute two Kayapó chiefs along with an accompanying American anthropologist, Darrell Posey, for travelling to Washington to express their grievances to the State Department and World Bank about the dam project (The *Guardian*, 31 October 1988; *New Scientist*, 19 November1988). A series of such incidents prompted the Bank to issue an explanatory document giving reassurances of its concern for social and ecological issues in relation to power sector loans for Brazil (World Bank, n.d.).

Within this climate of growing popular opposition and sense of public outrage over the murder in Acre the previous month of Chico Mendes, the rubber-tappers' union leader, there was a significant build-up of pressure on the World Bank from US Congressmen and NGOs. The strength of international objections to Brazil's development priorities in the Amazon Basin was also clearly visible in the massive demonstrations at the World Bank's annual general meeting in Berlin during September 1988. Shortly afterwards, in January 1989, the Bank announced that a decision on the Power Sector Loan would be indefinitely postponed. While this move was due in part to American concern that the loan might be used indirectly to finance Brazil's nuclear energy programme, it was in large measure prompted by misgivings over the environmental and social consequences of the hydropower development programme for Amazonia. An open letter from one US Senator to the Bank's president, Barber Conable, insisted that the schemes would lead to 'the unnecessary destruction of thousands of acres of tropical rainforests and displace a commensurate number of people' (*Financial Times*, 12 January 1989).

Yet while a combination of better-informed decision-making along with external political pressures has induced short-run measures to enable re-examination of environmental aspects which would previously have been accorded marginal weight, more substantive changes have also been evident. The World Bank – without doubt the single most important aid organisation involved in the funding of development projects in Amazonia and, indeed, in many other parts of the globe – has taken significant strides towards acknowledging its tremendous responsibility in this field. Under its internal restructuring process a new Environmental Department has been formed as 'a direct response to criticism of the Bank's lack of ecological sensitivity' (*Financial Times*, 28 September 1988) which will incorporate environmental impact analysis more systematically and comprehensively into the project cycle. Another task will be to improve borrowers' environmental consciousness and encourage them to incorporate such concerns into their own planning procedures. The first systematic attempt at such a process in Brazil is manifested in the 'National Environment Project' which, with Bank funding will attempt, amongst other things, to strengthen state and federal environmental agencies as well as supporting the implementation of appropriate policies (World Bank, 1988b). A comparable preoccupation with the consequences of involuntary resettlement, a closely associated issue, has also led to the modification of Bank policy and implementation procedures (Cernea, 1988). In this case, however, internal pressures within the organisation were far more influential in bringing about these changes than in the case of ecological issues.

There are some indications that combined pressure from these Brazilian and foreign sources is beginning to have an impact on national policy-making. As perhaps a first cautious step, article 225 of Brazil's new Constitution, promulgated in 1988, declares that 'Everyone has the right to an ecologically balanced environment . . . [and that the Amazon and other forests] . . . are national patrimony, to be used within the law to ensure preservation of the environment' (Brazil, 1988, p. 51). This was possible only through the sustained efforts of environmentalists, represented in the Constituent Assembly by congressman Fábio Feldmann. Significant gains, on paper at least, were also won by indigenous groups and their NGO support lobby, which succeeded in strengthening indian rights within Brazil's new Constitution. As Treece notes in Chapter 11, however, although this is a promising step, there remains a huge discrepancy between theory and practice.

Within an international context of much greater sensitivity to environmental affairs, pressure from the World Bank and the International Monetary Fund were allegedly responsible for President Sarney's announcement in October 1988 of a package of measures entitled 'Our Nature' (*Nossa Natureza*). Most immediately, restrictions were placed on fiscal incentives and credits for Amazonian projects, eliminating those which might damage the region's ecosystem, and a ban on the export of logs was declared. In addition, measures were announced for protecting the region's ecology, including the setting up of six interministerial working groups, whose duty it would be to define appropriate policies for rationally exploiting Amazonia's resources (The *Guardian*, 13 October 1988; LARR, 24 November 1988). Another option not yet taken up in Brazil is that of debt-for-nature swaps linked to concrete environmental measures such as reforestation and conservation. Projects of this kind have been successfully initiated by aid agencies in collaboration with national governments in several countries such as Costa Rica, Bolivia and the Philippines.

However, Brazil's severe debt-related economic crisis provides a unique opportunity to link debt relief with international negotiations for a more systematic programme of environmental conservation and sustainable development for Amazonia. Although the recent steps taken by President Sarney are an encouraging start towards a less destructive harnessing of Brazilian Amazonia's huge resource potential, the road to sustainable development is long and tortuous. Even with continued pressure from environmentalists upon planners and policy-makers, there are strong constraints upon its effectiveness, from whichever source it emanates. The first clear danger is that of tokenism on the part of the Brazilian government in its handling of these issues; the desire to placate both public and donor opinion by constructing a false consensus around the problem of environmental preservation without effecting real change in planning priorities or procedures. This stems in large part, of course, from the continuing power of vested commercial interests which would undoubtedly suffer should such measures be implemented, and is reflected for example in the outcry from Brazilian businessmen with investments in agribusiness and land in Amazonia, following the *Nossa Natureza* package (LARR, 24 November 1988). Notable in this respect, too, is the current tendency of the Brazilian media, flying in the face of all the evidence, once again to place the entire blame for Amazonian deforestation upon small farmers, reminiscent of propaganda campaigns in the 1970s. Right-wing opposition to environmental controls in the Amazon Basin is epitomised by the popular slogan, 'Amazonia is ours' (*A Amazônia é Nossa*) which, of course, embodies the race for profit as much as any nationalist ideals. However, undoubtedly reinforcing the current 'anti-green' movement in Brazil is a clear resentment at such 'meddling' by foreigners in Brazil's internal affairs.

Commercial pressures for the perpetuation of past policies in Amazonia, with all their inbuilt subsidies to 'entrepreneurs' are matched by unrelenting military ambitions to strengthen its foothold on the region to avoid any possibility that national sovereignty might be forfeited. It is, perhaps, significant in this respect that the old National Security Council (CSN), renamed as the 'Advisory Secretariat on National Defence of the Presidency of the Republic' (*Secretaria de Assessoramento de Defesa Nacional da Presidência da República*), is actually in charge of the *Nossa Natureza* programme. Neither is there any reservation about publicly voicing anti-foreign sentiments as a weapon against the environmental movement. Army Minister General Leonidas Pires Gonçalves, in a statement strongly reminiscent of the 1960s, stated in January 1989, 'Repeated news about the devastation of the rainforest leads to the suspicion that there are unconfessed interests behind this tiring insistence ... however, those who intend to exercise undue influence over the Amazon are mistaken' (The *Guardian*, 20 January 1989).

There is also a danger, despite the adoption of revised policies and more sophisticated planning methods, that the integration of environmental and social concerns into the project cycle of international aid institutions will be less rigorous than is desirable, in order that the flow of funds from donor to recipient be interrupted as little as possible. This situation might arise due to a number of factors. The aid agency in question might wish to be seen to be meeting its allocation targets and thus automatically performing its 'developmental' role. Alternatively, diplomatic and political pressures to maintain aid transfers might stem from the financial problems associated with debt rescheduling and the pressure to strengthen foreign exchange reserves. Thus, despite current official attempts belatedly to introduce some form of control over the way in which Brazilian Amazonia is exploited, military, geopolitical and certain economic interests not necessarily compatible with environmental stability or sustainability, may continue to take precedence. We hope in this book to highlight some major issues relevant to this debate and, above all, to stress the urgent need for government measures to ensure that Amazonia's potential is developed in as balanced and equitable a way as possible, so that the region's wealth may serve the needs of its entire population.

References

Abel, C. and Lewis, C. (eds) (1989), *Welfare, Equity and Development in Latin America*, London: Macmillan.

Amnesty International (1988), *Brazil: Authorized Violence in Rural Areas*, London.

Baer, W. *et al.* (eds) (1978), *Dimensões do Desenvolvimento Brasileiro*, Rio de Janeiro: Campus.

Banck, G. and Kooning, K. (eds) (1988), *Social Change in Contemporary Brazil*, Amsterdam. CEDLA.

Bourne, R. (1978), *Assault on the Amazon*, London: Gollancz.
Branford, S. and Glock, O. (1985), *The Last Frontier: Fighting Over Land in the Amazon*, London: Zed Books.
Brazil (1988), *Constituição da República Federativa do Brasil*, n.p.: Editora Tecnoprint.
Brundtland Report (1987), *Our Common Future*, Oxford: World Commission on Environment and Development/Oxford University Press.
Bunker, S. (1985), *Underdeveloping the Amazon: Extraction, Unequal Exchange and the Failure of the Modern State*, Urbana: University of Illinois Press.
Cernea, M. (1988), *Involuntary Resettlement in Development Projects: Policy Guidelines in World Bank Financed projects*, Washington, D.C.: World Bank.
Cota, R. (1984), *Carajás: A Invasão Desarmada*, Petrópolis: Vozes.
CPI (1987), 'Hydroelectrics of the Xingú and Indigenous People', Saõ Paulo: Comissão Pro-Índio (mimeo).
Davis, S. (1977), *Victims of the Miracle*, London: Cambridge University Press.
Doria, P. *et al.* (1978), *A Guerilla do Araguaia*, São Paulo: Alfa–Omega.
EDF (1987), 'News Release: EDF, World Environmental Community Denounce World Bank Involvement in Amazon Deforestation Plan', Environmental Defence Fund, Washington, D.C. (7 August) (mimeo).
Financial Times, various reports.
Foweraker, J. (1981), *The Struggle for Land*, Cambridge: Cambridge University Press.
Gasques, J. and Yokomizo, C. (1985), 'Avaliação dos Incentivos Fiscais na Amazonia', Brasília: IPEA (mimeo).
Goodman, D. (1978), 'Expansão de Fronteira e Colonização Rural: Recente Política de Desenvolvimento no Centro–Oeste do Brasil', in Baer *et al.* (eds) (1978) pp. 301–37.
Goodman, D. (1988), 'Agricultural Modernisation, Market Segmentation and Rural Social Structures in Brazil', in Banck and Kooning (eds) (1988).
Grindle, M. (1986), *State and Countryside: Development Policy and Agrarian Politics in Latin America*, Baltimore and London: Johns Hopkins University Press.
Grzybowski, C. (1987), *Caminhos e Descaminhos dos Movimentos Sociais no Campo*, Petrópolis: Vozes.
Hall, A. (1987), 'Agrarian Crisis in Brazilian Amazonia: the Grande Carajás Programme', *Journal of Development Studies*, XXIII (4 July) pp. 522–52.
Hall, A. (1989a), *Developing Amazonia: Deforestation and Social Conflict in Brazil's Carajás Programme*, Manchester: Manchester University Press.
Hall, A. (1989b), 'Non-Government Organisations and Development in Brazil Under Dictatorship and Democracy', in Abel and Lewis (eds) (1989).
Hecht, S. *et al.* (1988), 'The Economics of Cattle Ranching in Eastern Amazonia', California: Graduate School of Planning, UCLA (mimeo).
IBASE (1983), *Carajás: O Brasil Hipotêca Seu Futuro*, Rio de Janeiro: Achiamé.
Isto É (1987), report 15 July.
LARR (1988), 'Sarney acts on the environment', *Latin American Regional Report*, Brazil (24 November).
Ledec, G. (1985), 'The Political Economy of Tropical Deforestation', in Leonard (ed) (1985) pp. 179–226.
Leonard, H. (ed.) (1985), *Divesting Nature's Capital*, London and New York: Holmes and Meier.
Mahar, D. (1988), *Government Policies and Deforestation in Brazil's Amazon Basin*, Washington, D.C.: World Bank (Environment Department) (June).
MIRAD (1986), *Conflitos de Terra*, Coordenadoria de Conflitos Agrários, Brasília: MIRAD.

MIRAD (1987), *Conflitos de Terra*, Coordenadoria de Conflitos Agrários, Brasília: MIRAD.
Moran, E. (1981), *Developing the Amazon*, Bloomington: Indiana University Press.
Myers, N. (1985), *The Primary Source*, London: Norton.
Pompermeyer, M. (1984), 'Strategies of Private Capital in the Brazilian Amazon', in Schmink and Wood (eds) (1985) pp. 419–38.
Reis, A. (1982), *A Amazônia e a Cobiça Internacional*, Civilização Brasileira, Rio de Janeiro: (5th edition).
Rich, B. (1985), 'The Multilateral Development Banks, Environmental Policy, and the United States', *Ecology Law Quarterly*, XII, pp. 681–745.
Salati, E. and Vose, P. (1984), 'Amazon Basin: A System in Equilibrium', *Science*, 225, pp. 129–37.
Schmink, M. and Wood, C. (eds) (1985), *Frontier Expansion in Amazonia*, Gainesville: University of Florida Press.
Síoli, H.(1985), *Amazônia: Fundamentos da Ecologia da Maior Região de Florestas Tropicais*, Petrópolis: Vozes.
Smith, N. (1982), *Rainforest Corridors: The Transamazon Colonization Scheme*, Berkeley: University of California Press.
Souza Martins, J. de (1984), *A Militarização da Questão Agráia no Brasil*, Petrópolis: Vozes.
New Scientist, report 19 November.
The *Guardian*, various reports.
Treece, D.(1987), *Bound in Misery and Iron: the impact of the Grande Carajás Programme on the Indians of Brazil* London: Survival International.
Wetterberg, G. (1981), 'Conservation progress in Amazonia: a structural review', *Parks*, VI (2) pp. 5–10.
Wilson, E. O. (ed.) (1989), *Biodiversity*, London: John Wiley.
Wood, C. and Schmink, M. (eds) (1978), 'Blaming the Victim: Small Farmer Production in an Amazon Colonization Project', *Studies in Third World Societies*, VII (February) pp. 77–93.
World Bank(1988a), 'The Carajás Iron-Ore Project: Amerindian and Environmental Protection Measures', Washington D.C. (September) (mimeo).
World Bank (1988b), 'World Bank Support to Environmental Programs in Brazil', Washington D.C. (September) (mimeo).
World Bank (n.d.), 'Brazil: Proposed Second Power Sector Loan, Social and Environmental Aspects, Background Note', Washington D.C. (mimeo).
WRI (1986), *World Resources 1986*, World Resources Institute and the International Institute for Environment and Development, New York: Basic Books.

Part I
Current Development Strategies and Frontier Integration

2 Rondônia and the Fate of Small Producers

George Martine

INTRODUCTION

The once-prevalent notion that the Amazon region possessed vast riches and that it would provide unlimited opportunities for agricultural expansion has been rather thoroughly discredited by recent events in Brazil. Successive grandiose settlement schemes, whether through government-oriented or privately-owned colonisation projects, or through large-scale capitalistic enterprises, have produced rather dismal results. The one exception to the general disenchantment with the chances for wide-scale development in the Amazon would appear to be Rondônia. This recently-created state seems to have somehow managed to survive the onslaught of migrant hordes and, apparently, emerged as a bustling if not prosperous frontier region.

The present chapter addresses itself to the issue of Rondônian growth. What factors account for its expansion? How successful has settlement really been? How have small agricultural producers fared in the overall scheme of things? What are the perspectives for the future maintenance and/or expansion of small farms in Rondônia? How does Rondônia itself fit within the more general pattern of regional development and population distribution?

THE SETTING

The state of Rondônia is located in the south-west portion of the Brazilian Amazon, covering about 5 per cent of that region (see Figure 2.1). Its land area is roughly equivalent in size to that of West Germany. The region's physical characteristics are highly variable. The dominant type of vegetation is upland wet forest although *cerrados* cover some 9 per cent of the total and, wet *várzea* forest, another 7 per cent. The climate is, for the most part, hot and humid, with a rainy season lasting from September to May and an annual rainfall between 2,000 and 2,200 mm.

The topography is varied. A wide diversity of soil groups can also be found, several of which are considered to be appropriate for both perennial and annual crops, as well as for pasture and forestry. Contrary to the situation found in most of the Amazon, areas of *várzea* in Rondônia are held to be of relatively little use for agriculture. The most promising soils are

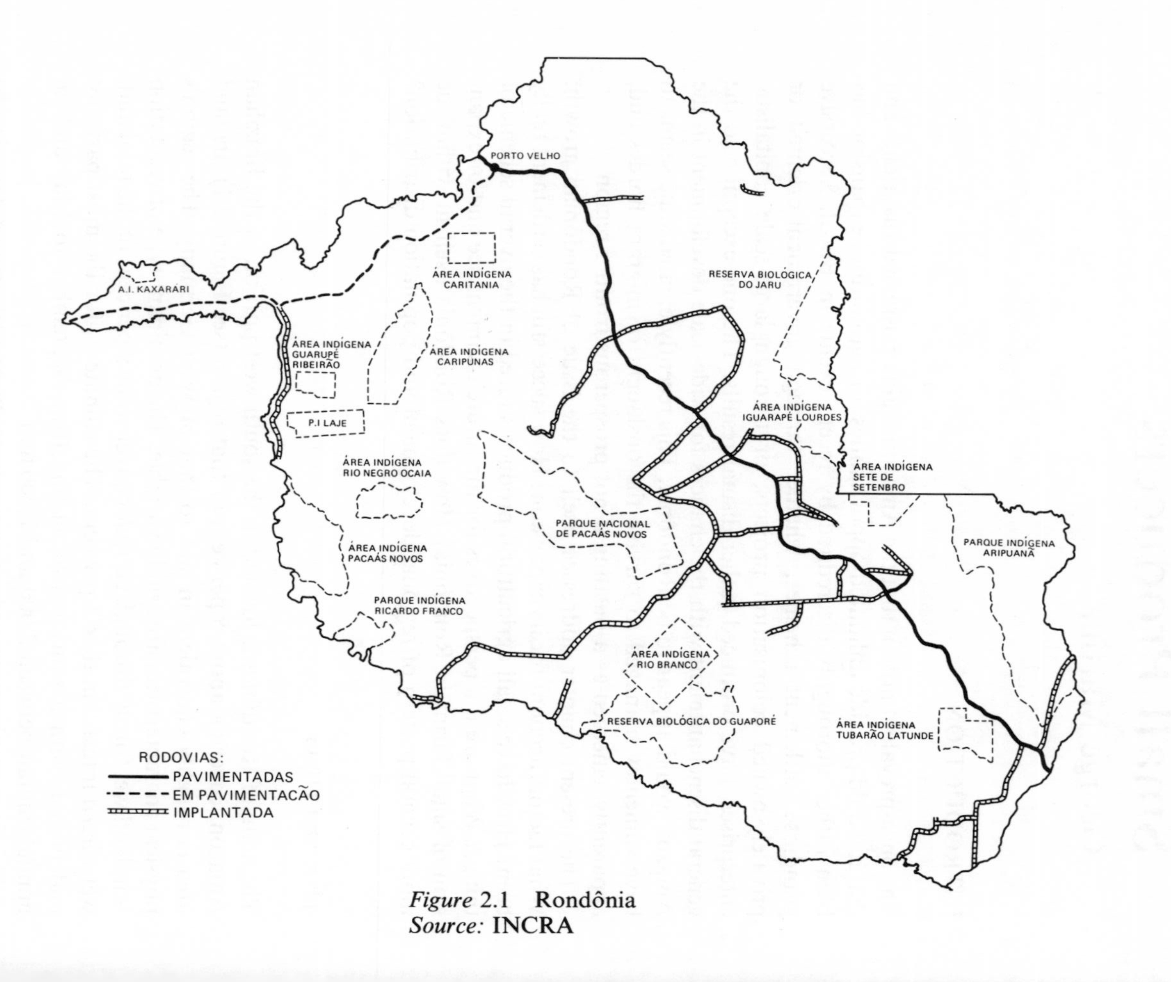

Figure 2.1 Rondônia
Source: INCRA

distributed from east to west across the central portion of the state, where many of the official colonisation projects have been located. Rondônia also has extensive areas covered with intermediate soil groups, appropriate for a more restricted range of uses, as well as a large group of soils whose utilization would require major engineering works.[1]

The first settlements in Rondônia occurred during the rubber boom of the nineteenth century but their dynamism was short-lived. The Second World War produced another brief rubber boom without lasting consequences, as attested by the fact that, in 1950, the then Territory of Rondônia had a population of only 37,000 people. Colonisation projects were initiated in Rondônia during the 1950s when a precarious dirt road, the BR-364, linking its capital city (Porto Velho) to Cuiabá was finally concluded. Tin oxide deposits were discovered in the late 1950s and Rondônia soon became Brazil's main producer of this mineral which, during the peak mining period, employed some 45,000 workers (Borges Ferreira, 1986). The main road linkage to the Centre–South was improved in the late 1960s, effectively integrating Rondônia with the rest of the country.

As a consequence of these several developments, Rondônia's population had grown to 111,000 by 1970. It was at this point that the events which concern us in this chapter began taking place. Land-hungry migrants began invading Rondônia in the early 1970s, accelerating the rhythm of population growth to an explosive pitch. Thus, between 1970 and 1980, Rondônia's population increased at a yearly rate of 16 per cent, elevating its total to 491,000 inhabitants in 1980. Since then, growth has continued at a rapid pace and it is estimated that, by mid-1987, Rondônia's population had reached 1,150,000.

A BRIEF HISTORY OF THE RECENT SETTLEMENT PROCESS

As has already been amply described in the literature, the settlement of Rondônia is part of the third phase of recent frontier expansion in the Amazon region (cf. Sawyer, 1984; Martine, 1981, 1982). The abandonment of colonisation schemes along the Transamazon highway, together with the rapidly-growing demand for land provoked by the massive expulsion of small producers in other regions, served to channel government-oriented colonisation projects to Rondônia and Mato Grosso in the early 1970s. Conditions in Rondônia seemed particularly propitious. In addition to the fact that Rondônian soils were reputed to be of much better average quality than in much of the Amazon region, most of the land area still belonged to the State. Another favourable factor was the existence of a passable highway between Porto Velho and Cuiabá; although not yet an all-weather road, the BR-364 actually made southern markets and manpower more accessible to Rondônia than had been the case of Transamazon settlements.

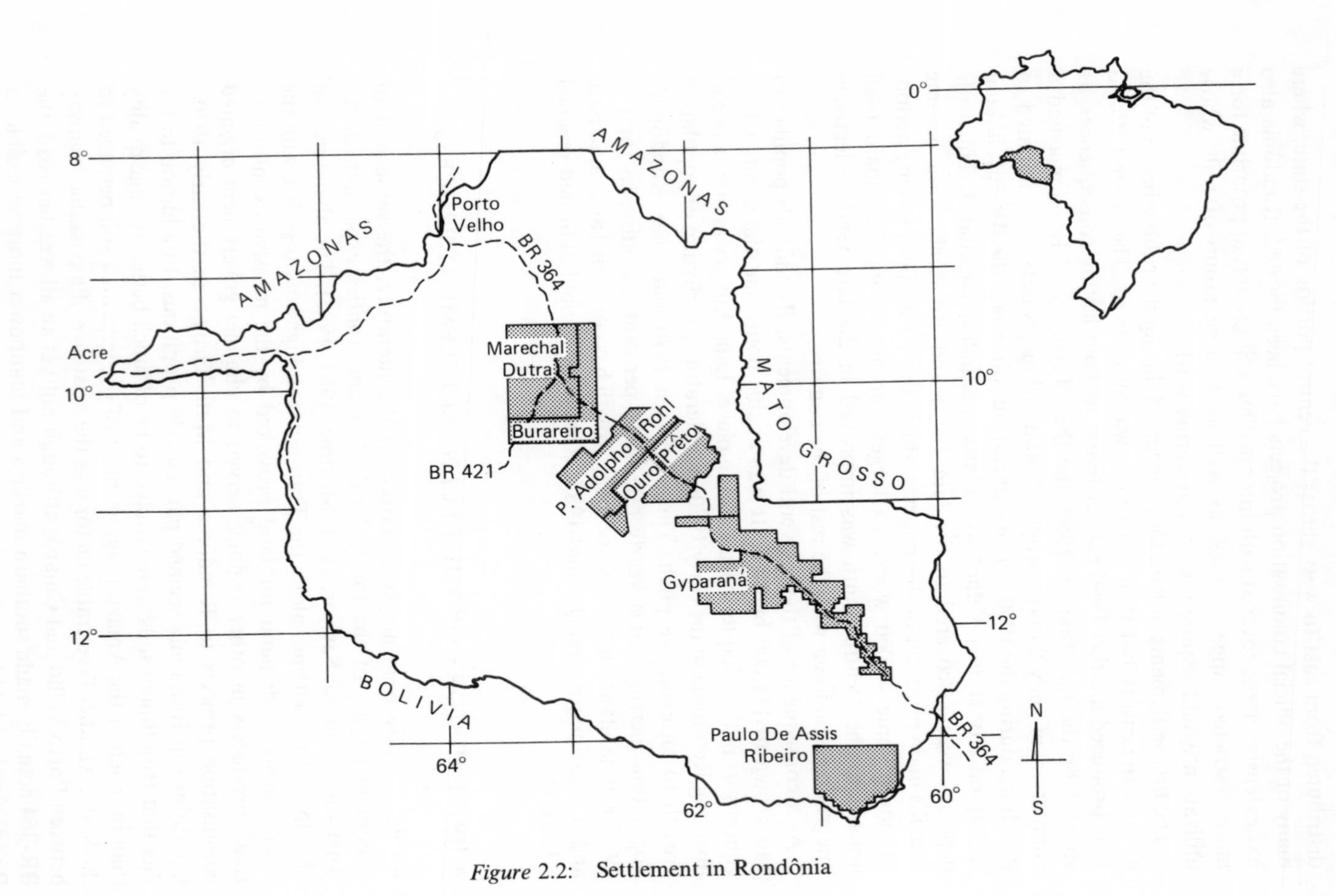

Figure 2.2: Settlement in Rondônia

The magnitude of migration flows to Rondônia in the 1970s went beyond the wildest expectations. The first project sponsored by INCRA – the federal government agency in charge of colonisation – was the PIC–Ouro Prêto settlement[2] (see Figure 2.2). It was set up in response to the demands made by 300 families from the state of Paraná who had been enticed into buying plots of Rondônian land by a private colonisation company and then been abandoned. Meanwhile, in other regions of the country, large masses of small producers – squatters, sharecroppers and owners – were being pushed off the land by the modernisation and speculation forces which were sweeping Brazil. Stimulated by ample doses of subsidised credit, new forms of agricultural production, centred on the adoption of the 'Green Revolution' technological package and on the strengthening of the agribusiness complex, spread throughout the Centre–South region. These had an explosive effect on out-migration. But even in those areas where agricultural production was not significantly affected, land speculation had the same impact on the expulsion of small rural producers (cf. Martine and Garcia, 1987, especially Chapters 1 and 2).

Some of the areas worse hit by the modernisation and speculation wave were in the north and north-west regions of Paraná and in the adjacent areas of São Paulo and Mato Grosso. These rich lands had been settled only a generation earlier, mostly by small-scale farmers who became easy prey to larger landowners and speculators. Thus, when news that the government was handing out free good land in Rondônia reached these farmers, the rush was on.

The trickle of migrants to Rondônia, initiated in the early 1970s, soon reached gusher proportions, due to the combination of its attractiveness to rural in-migrants and 'push' factors in more traditional agricultural lands. In practically all stages of Rondônia's recent occupation, the State has been swamped with many more migrants than could realistically be absorbed in an organised fashion.[3]

Four phases of Rondônian occupation are framed by changes in the form of government intervention in the settlement process. Prior to 1975, INCRA attempted to provide comprehensive orientation and control of the settlement process, from demarcation of the land area and selection of lots, down to the provision of roads, health and educational facilities, as well as technical assistance, credit, marketing and storage. But this system was obviously too cumbersome to cope with the rapid influx of migrants. The original PIC–Ouro Preto project, designed for 500 families, was obsolete before it was implemented. It was expanded in order to absorb 5,000 families but, in an attempt to keep up with the demand, a rush of six more projects, most of them catering to small-scale producers, were implemented by 1975. The total number of families eventually settled in these original projects amounted to 22,700 (see Table 2.1).

The year 1975 marks a clear departure from the original strategy aimed at

Table 2.1 Government-oriented settlement projects in Rondônia

Name or type of project	*Year initiated*	*No. of families settled or legitimated*
PIC–Ouro Prêto	1971	5,161
PIC-Jiparaná	1973	4,730
PIC-Assis Ribeiro	1974	3,076
PIC-Padre Adolfo Rohl	1973	3,462
PIC-Sidney Girão	1972	638
PAD-Marechal Dutra	1975	4,603
PAD-Burareiro	1975	1,540
Land tenure program	1975	13,146
Rapid settlement program	1979	23,098
Settlement program	1982	15,000
Total		74,454

Note:
1. Projected figure.
Sources: Galvão (1982); FIPE (1986b).

the paternalistic accommodation of small farmers. The high costs and low manageability of INCRA's colonisation model helped swing the balance to a more 'capitalistic' approach. But other factors were obviously at work. The reputedly good quality of Rondônian soil had attracted the attention of entrepreneurs as well as *grileiros*. At a time when speculation in land was at its height, considerable pressure was put on government authorities to adopt a more 'realistic' approach to Rondônian settlement. The associations of cocoa and coffee growers (CEPLAC and IBC, respectively) appear to have gained an inroad at this time. Thus, the last PIC implemented by INCRA, the Marechal Dutra project in Ariquemes, was planned as part of a more ambitious scheme involving three different strata of settlers in three adjacent tracts: Marechal Dutra was destined for subsistence farmers; the 'Licitação' project was sold to agricultural entrepreneurs who were expected to initiate varied forms of capital-intensive production in rubber, pepper, coffee and other export-oriented crops; and, the 'PAD-Burareiro' area was reserved for supposedly less-capitalised cocoa producers, originally expected to come from Bahia, Brazil's main cocoa producing centre (Hébette and Marin, 1982).

In retrospect, it is easy to see that the demarcation of three contiguous tracts destined for three different classes of producers, was aimed at making more 'rational' economic use of available factors than had previous colonisation projects. Producers of cocoa and other export-oriented crops need abundant manpower; this was to be provided by the subsistence farmers who, almost by definition, usually have a surplus of available manpower during a large part of the year. Hébette and Marin (1982) use the image of the demiurge to refer to INCRA's attempt to bring order to the chaos in Ariquemes, dividing up the land by social class as well as type of crop and projecting expected patterns of social interaction into the future.

Be that as it may, the three Ariquemes projects appear to be part of a concerted shift to 'economic realism' by government authorities in Rondônia around 1975. Thus, it would not seem merely coincidental that INCRA promoted its 'Land Tenure' project at this time. The object of this initiative was to separate out 'legitimate' title claims and weed out unwanted squatters (cf Galvão, 1982). Not surprisingly, the 13,000 families which eventually benefitted from this programme obtained title to a land area much greater than all other types of projects; some 7.6 million hectares were involved in the five Land Tenure projects. At the same time, attempts were made to curtail migration into Rondônia through disincentive campaigns in the principal sending areas of Paraná, and through the placement of roadblocks on the main artery between these states and Rondônia.

Despite these efforts, thousands of migrants continued trekking to Rondônia. But it was only in 1979, after the roadblocks and other disincentive measures had been slackened – in answer to the new government's ambitions of achieving statehood through a dramatic increase in population – that INCRA took up the task of delimiting new projects. Recognising that the flow of migrants had outrun the possibilities for organised settlement along the lines of the previous model, the government adopted the 'Rapid Settlement' programme. This amounted, in effect, to the legitimation of spontaneous settlement. Settlers would invade lands adjacent to whatever roads or trails existed (or were being programmed), or wherever the 'grapevine' indicated that public lands might be subject to distribution. They would select plots of a size more or less equivalent to that which INCRA was expected to hand out and begin cutting down the forests and producing subsistence crops. The government simply tried to put some sort of order into this process, controlling the amount of land area to be claimed by any one settler and then, hopefully, providing some minimal infrastructure in terms of access roads, storage facilities and the like.

If the unusual degree of consistency between INCRA data and that from the 1980 Agricultural Census is to be believed, INCRA had considerable success in controlling the settlement process of Rondônia during the late 1970s. According to INCRA, a total of 48,717 families were settled, with some form of assistance or control, during the 1970–80 period. This is actually a slightly higher figure than the total number of agricultural establishments uncovered by the 1980 Agricultural Census (48,371). It is legitimate to speculate that a large proportion of the farms which INCRA claims to have disciplined through 'Rapid Settlement' or 'Land Tenure' projects are actually the product of spontaneous occupation.

Be that as it may, the 'Rapid Settlement' programme continued until 1982, when a fourth phase was initiated. Main developments at this point were linked to the creation of a growth pole development programme (the POLONOROESTE), with World Bank assistance. The paving of BR-364 was completed in 1984, providing all-weather transport.[4] Meanwhile, a new form of colonisation programme, simply called 'Settlement Projects', but based on the IBRD's Integrated Rural Development concept, was initiated in Rondônia and Mato Grosso. The idea was to build on previous experiences and mistakes, both on the Transamazon Highway and in the PICs, in order to settle a large number of settlers as quickly, as cheaply and as productively as possible. Some 15,000 families were to have been settled in three sub-projects but, four years after the programme's inception, only 5,000 lots had actually been distributed (FIPE, 1986b, p. 3).

Meanwhile, the number of migrants arriving in Rondônia continued to grow at a rapid pace. Whatever the degree of the government's previous success in disciplining the settlement process, the 1985 data show an increase in the number of agricultural establishments out of all proportion to the number of new official settlements. Thus, according to the Agricultural Census, the number of establishments grew from 48,371 to 81,582 between 1980 and 1985. By contrast, available data indicate that, at best, some 15,000 families gained official access to land between 1980 and 1985, either through 'Rapid Settlements', 'Land Tenure' or 'Settlement' projects. In sum, at least 18,000 agricultural establishments sprouted up between 1980 and 1985 in areas and forms not under official control.

This rhythm of spontaneous occupation is still probably grossly underestimated in the light of the available figures on in-migration and demographic growth in Rondônia. Actually, such figures are not totally transparent, as is to be expected in a rapidly-expanding frontier area. Migration figures for 1980–5, collected in Vilhena – point of entry for migrants coming to Rondônia from all regions, except the trickle from the North itself – are presented in Table 2.2. These show a constant increase in the total yearly inflow of migrants; the total influx for the 1980–5 period amounts to 413,000

Table 2.2 Number of migrants arriving in Vilhena, Rondônia, by year of arrival and principal states of origin, 1980–7

Year	*No. of migrants*	*Principal states of origin*
1980	28,320	Paraná, Mato Grosso, Mato Grosso do Sul
1981	30,072	Paraná, Mato Grosso, Amazonas
1982	48,851	Paraná, Mato Grosso, Minas Gerais
1983	92,723	Paraná, Mato Grosso, São Paulo
1984	153,327	Paraná, Mato Grosso, São Paulo
1985	151,621	Paraná, Mato Grosso, Minas Gerais
1986	165,899	Paraná, Mato Grosso, Minas Gerais
1987	103,654	No information

Source: Governo do Estado de Rondônia, SEPLAN (1986).

in-migrants. Our best estimate of net migration for this period is around 340,000.[5] In any case, the important point is that there can be little doubt that Rondônia continued to attract a volume of in-migration out of all proportion to the actions and absorptive capacity of official colonisation agencies. The demographic growth rate of Rondônia was 16 per cent per annum between 1970 and 1980 and 13 per cent between 1980 and 1985. The main questions which have to be answered in this context are: (1) Where are the strong migration flows to Rondônia ending up? (2) What are the chances for subsistence and/or growth of the majority of small-scale settlers who continue to move into Rondônia?

IMPACTS OF THE SETTLEMENT PROCESS

Observers are usually most impressed by the accelerated pace of city growth in Rondônia. Indeed, urban growth in this state, as elsewhere on the frontier, has been extremely rapid in recent years. Yet, there can be no question that the present vigour of urban growth is dependent on rural settlement, either through colonisation programmes or spontaneous occupation. In this sense, towns and cities increase at approximately the same speed as the arrival of settlers and their needs for support.[6] In addition, however, the urban areas end up accommodating, for better or for worse, those colonists who have difficulty in obtaining a plot, or who are eventually pushed off the land. Thus, whether urban growth rates are excessive or not is ultimately determined by the correspondence between the rhythm of continued in-migration and the possibilities for stable absorption of colonists. The following pages address this issue directly.

What exactly has happened to the structure of agricultural production in Rondônia during the last few years? The information presented in Table 2.3 reveals enormous changes. From 7,100 agricultural establishments and 1.6 million hectares which it had in 1970, Rondônia progressed to 81,600 establishments and 6.1 million hectares in 1985. The land area under cultivation has increased from 45,000 hectares in 1970, to 539,000 in 1985; more than 40 per cent of this cultivated land area is covered with permanent crops. The number of persons occupied, part or full time, in agricultural activities went from 21,000 in 1970 to 325,000 in 1985; the number of tractors increased from 52 to 1,007 in the interim and the herds of cattle increased from 23 to 768 thousand head. Agricultural production has proceeded apace, at least in some crops; thus, Rondônia currently contributes some 7 per cent of the national production of cocoa and bananas, as well as 2–3 per cent of the national production of rice, beans and manioc (CEPAGRO, various years).

In short, an initial overview of changes in agricultural activity reveals considerable dynamism. The evolution of the land tenure system also

Table 2.3 Changes in the structure of agricultural production in Rondônia, 1970–85; selected indicators.

Year	*No. of establishments*	*Total area (in 000 ha)*	*Average area (in ha)*	*Area under permanent crops*	*Area under temporary crops*	*No. of persons in agricultural activities*	*No. of tractors*	*Cattle*
1970	7,082	1,632	230.4	12,273	32,363	20,563	52	23,175
1975	25,483	3,082	120.9	45,763	147,700	103,992	68	55,392
1980	48,371	5,224	108.0	170,178	203,253	176,934	570	251,419
1985	81,582	6,091	74.6	223,800	315,326	325,086	1,007	768,411

Source: IBGE, *Agricultural Census* (various years).

Table 2.4 Evolution of land tenure in Rondônia, 1970–85, percentages.

Size of establishments (ha)	Establishments				Area			
	1970	1975	1980	1985	1970	1975	1980	1985
Less than 10	8.0	19.1	25.1	27.8	0.2	0.6	1.0	1.6
10–50	29.1	17.5	15.0	24.9	2.7	3.2	3.5	9.0
50–100	10.0	10.6	25.9	27.6	2.9	6.7	18.5	24.1
100–200	13.0	47.3	29.1	16.4	7.1	40.2	28.2	23.7
200–1,000	38.4	4.5	3.7	2.7	48.7	16.1	10.9	11.9
1,000 and above	1.6	1.1	1.1	0.5	38.4	33.2	37.9	29.6
Total ($N = 100\%$)	7,082	25,483	48,371	81,582	1,631,640	3,082,852	5,223,631	6,090,647

Source: IBGE, Agricultural Census (various years).

presents an interesting study (Table 2.4). Obviously, the most drastic changes occurred between 1970 and 1975, when the large-scale distribution of 100-hectare plots in previously-unoccupied land had a dramatic deconcentration effect. Between 1975 and 1985, one can observe the growing importance of smaller establishments, particularly those within the 50–100 hectare category, and a gradual reduction in the importance of the 100–200 and 200–1,000 hectare categories. This is again in consonance with aforementioned changes in colonisation policy.

The sweeping trend to deconcentration shown in Table 2.4, produced by the large-scale distribution of small and medium-sized plots, masks two other patterns which are also of considerable significance: the spread of the *minifúndios* and the growing importance of the *latifúndios*.

The multiplication of *minifúndios* (i.e., farms of 10 hectares or less) is clearly part of the survival strategy of migrants who have been unable to obtain or retain a plot of their own in settlement projects; of the 23,000 *minifundistas*, 69 per cent had less than 5 hectares of total area. Some 43 per cent of the total were squatters, 39 per cent sharecroppers or tenants and 17 per cent declared that they owned the land.

At the other extreme, the larger plantations (1,000 or more hectares) have tripled their land area between 1970 and 1985; as intimated earlier, the greatest boost was received during the 1975–80 period, when the 'Land Tenure' programme was put into effect. Despite a large relative decline between 1980 and 1985, this category, which accounts for only 0.5 per cent of all establishments, still controls 29.6 per cent of all land area in Rondônia. By contrast, in Brazil as a whole, the 1,000 or more hectare category accounted for 0.9 per cent of all establishments and 16 per cent of the total land area in 1980.

The fact that the continued distribution of small and medium-sized plots in Rondônia disguises a parallel movement towards land concentration is confirmed by the FIPE survey of land tenure in colonisation projects along the BR-364. (FIPE, 1986a) The 150 lots sampled in this study had an initial average size, when distributed by INCRA, of 135 hectares; when visited in 1986, these lots averaged 198 hectares apiece. A very small proportion of this total land area was under cultivation; the process of concentration is thus explained as a result of land valorisation, caused by government investments in infrastructure and speculation.

Some of the recent land concentration and speculation may well be linked to expected profits, not only in the agricultural sector, but in mining. Rondônia is Brazil's main producer of tin oxide and third most important producer of gold. The largest tin oxide deposits are located in the northern half of Rondônia while the main gold deposits are found along the Madeira basin between the Bolivian border and Porto Velho. But minor gold fields have also been located in southern Rondônia and prospectors range far and wide, lending greater substance to the overall speculative fever. Theoretically, since mineral rights belong to the State and are not acquired along with land rights, mining prospects would not, in a more closely-controlled system, provoke abrupt land valorisation. These are not the concrete conditions prevailing in Rondônia (cf. Erse, 1984).

In short, land concentration in Rondônia seems to stem from a variety of somewhat ill-defined expectations of future profit. Land rights are often sold by small farmers in colonisation projects because they acquired the land for next-to-nothing and because the lump sums they are offered appear attractive (cf. Vieira, 1987). The amount of public funds continually being invested in Rondônia, whether in colonisation projects or infrastructure, leads potential investors to believe that land will inevitably continue to increase in value; land transactions based on this hypothesis inherently produce further valorisation. Meanwhile, numerous examples of rapid fortunes being amassed in *grilagem* (in urban real estate), in mining or in other productive urban activities prompted by the current boom, all provoke further speculation and valorisation. In short, the apparent prosperity generated by the rapid influx of migrants and the quick changeover of property feeds a constantly-growing balloon of land valorisation out of all kilter with increases in production and in conflict with attempts at small-farmer settlement.

However this may be, the important point is that, despite the largest government efforts at small-scale colonisation ever carried out in Brazil, Rondônia tends to reproduce the deformed landholding structure which has historically prevailed in the rest of the country.

What do these broad trends, depicted by Census data, mean in terms of the satisfaction of migrant aspirations? The answer to this is facilitated by the fact that, during the last 10 years, Rondônia's settlers have been repeatedly studied in a succession of sample surveys. Most of the field research in Rondônia has been carried out in official colonisation projects. Since spontaneous occupation accounts for a significant proportion of all settlements, this may cause a certain bias. Nevertheless, the convergence of the pictures painted by these independent sources lends them considerable strength (cf. Martine, 1987).

Space limitations prevent the reproduction of comparative survey data. Nevertheless, a review of this information would highlight the following main aspects. First, the various surveys corroborate the fact that the land area being distributed to colonists has been progressively reduced over time. This does not seem to be a major factor for the survival of the colonists since all of the surveys note that the forest-clearing capacity of small farmers in Rondônia is quite limited. In fact, even in older projects, the area which has effectively been cleared is small and that under cultivation, even smaller. The 1985 Agricultural Census data confirm this point, showing that 96 per cent of the farms with declared land area in Rondônia had less than 20 hectares of cultivated land and 81 per cent had less than 10 hectares.

Second, in all the field surveys carried out in Rondônian colonisation projects, the difficulties of establishing a permanent, profit-making enterprise, or even just a bare subsistence venture, are highlighted. In a stylised summary of the information presented in the various studies, the following trajectory would seem to be the most common. After the colonist arrives with his family and belongings on a bus or back of a truck, he is subjected to a more or less extended period of searching and waiting: searching for an ongoing colonisation project or an area of relatively open land which he can invade, and/or waiting for INCRA to provide him with a plot of land. Depending on luck, ability, timing and resources, the waiting period may vary from a few days to a few years. During this period, in order to survive, the would-be colonist is usually forced to find gainful employment. The two most common forms of sustenance for waiting families involve working as sharecroppers on somebody else's lot or as unskilled labourers in the construction, services, transport, sawmill or commercial sectors of the growing cities. Often the sharecropping arrangement, devised as a temporary survival strategy, will become a more or less permanent setup – a fact which helps explain the growing importance of *minifúndios*.

If and when the migrants do find a piece of land, either on their own, through INCRA, or through sharecropping arrangements, the difficulties of coping with a strange jungle environment have to be faced. In addition to the aforementioned difficulties of clearing the land, the farmer faces the problem of what and how to plant. Clearing 2 to 3 hectares in the first year, the settlers tend to plant subsistence crops, mainly rice, beans, manioc and corn. Permanent crops such as coffee and cocoa are highly recommended but they require a larger initial investment and the capacity to wait 3–5 years for the first returns; few colonists have access to such resources nor to the subsidised credit which would permit diversification.

Consequently, after 3 or 4 years, if the settler has managed to stay on his lot despite the lack of infrastructure, the isolation and the physical hardships, he will have cleared some 10 to 15 hectares. By this time, however, the area which had first been cleared off will have lost its natural fertility and gone back to scrub or, under the best of circumstances, become 'pasture'. The total area under pasture tends to grow rapidly in Rondônia since it is the only manner in which colonists can postpone the total loss of previously-cleared land. On the other hand, very few of the colonists have cattle; therefore, the distinction between pasture and scrub land is largely theoretical. Overall, the land covered by forests, scrub and pasture accounts for close to 90 per cent of the total area of the plots which have been distributed in colonisation projects in Rondônia (FIPE, 1986a, p. 19).

In short, despite the apparent dynamism of agricultural production and urban growth at the macro-level in Rondônia, the images generated by the various surveys carried out on individual farms in colonisation projects in Rondônia do not permit an optimistic perspective for the permanent establishment of a landed middle class, nor even for the large-scale absorption of subsistence farmers. Since conditions in areas occupied spontaneously (i.e., without the benefit of government tutelage or support) are likely to be even worse than in official projects, the probability that a large number of small farmers who arrived in Rondônia during the past few years will have to look elsewhere for survival is great. Indeed, the different data on turnover of colonists, although not entirely consistent, all suggest that a rather large proportion of settlers choose to sell, trade or abandon their lots fairly soon after arrival. The FIPE sample survey of colonisation projects along the BR-364 highway showed that only 41 per cent of the lots were occupied by the original settlers (FIPE, 1986a, p. 4). This figure on frequency of turnover is not incompatible with other sources. For instance, a longitudinal survey in the recently-created Machadinho project showed that 24 per cent of the settlers interviewed in 1985 were no longer on their plots in the 1986 survey (Gama Torres n.d.).

PROBLEMS OF RONDÔNIAN SETTLEMENT

In order to understand the rather dismal results of colonisation efforts so far and to judge the possibilities of future projects more accurately, it is worthwhile attempting to break down the difficulties faced by colonists into their principal components.

Migration Intensity, Land Speculation and Violence

The first years of Rondônian settlement were particularly marked by violence in the struggle for land. The causes of migration intensity to Rondônia have already been discussed. The relative paucity of official resources allocated to the business of settling migrants facilitated the speculative activities of *grileiros*, who formed armed bands and pressure groups to guarantee their claims to public land, which was then resold to unwary settlers. The chaotic nature of this process, in addition to generating considerable violence and death, made it impossible for government agencies to inject some semblance of order into the process (Martine, 1982).

In more recent years, the level of violence has become less obtrusive, but the intensity of both land speculation and in-migration has been greater than ever. The number of families awaiting relocation on a piece of land has probably also reached new heights. Given the difficult conditions for small farmers already on the land, selling out may often be the most rational alternative. On the other hand, this may result in a considerable loss to public coffers. INCRA estimates that it costs at least US$5,000 to set up each colonist in a project such as Machadinho; in some instances, these lots are being sold for less than one-tenth of their cost to INCRA (Borges Ferreira, 1986, p. 29).

Present Technological Limitations[7]

Agricultural models which have been tried in the Amazon region cover a wide range of alternatives – few of which have been proven feasible or durable. Large-scale agroindustrial projects have attempted to transplant the technological package which prevails in the Centre–South, based on the intensive use of mechanical and petrochemical inputs. Not only is this expensive in terms of transport costs but the results have been disappointing in terms of production.

Much the same difficulties are being faced by the technified settlers in Rondônia who, having proceeded largely from Paraná and São Paulo, attempt to recreate the practices of the prevalent technological package in the Amazon. In recent years, because of their greater resources, these technified farmers have tended to supplant the original itinerant and subsistence farmers in Rondônia. So far, however, their efforts have provided low productivity yields, low returns and contributed to more or less serious environmental problems. The main exception has been in cocoa production where heavy support by CEPLAC in terms of credit, technical assistance, warehousing and transport has yielded significant growth in Rondônia. But even here, the quality of the product is considered to be inferior. In any case, the question which arises in this context is: How long will this be economically and ecologically viable? Declines in international prices for cocoa underlined the inherent vulnerability of Rondônian production in competition with the state of Bahia – Brazil's number one producer of cocoa – which is thousands of miles closer to seaports.

At the other extreme, the most prevalent type of agriculture in the Amazon and in Rondônia is the itinerant slash-and-burn model. This practice tends to deplete the soil's natural fertility in two or three years' time, after which the land becomes useless as scrub, or is sown with pasture grass. The 'caboclo' type of agriculture is practised by natives of the Amazon who make more selective use of the land, utilising local seeds and long-term rotation practices. Soil conservation and environmental balance are achieved, but at the cost of very low intensity of total land-use.

In short, the various technological patterns utilised in the Amazon region have, for different reasons, attained only low levels of productivity. Both large and small-scale agricultural holdings have high social and environmental costs. As a result, forest devastation and erosion are occurring at an alarming pace. It is estimated that at present rates of deforestation, Rondônia will lose close to half of its forest area by the end of the century. One of the major culprits is the use of heavy machinery by *grileiros* to tear down the forest for the sole purpose of demonstrating that they are 'working' on the land and thus legitimise their claim to it (Alencar, 1987). The failure of colonisation projects to produce the expected bumper crops which would permit the Amazon region to become the 'granary of Brazil' derives in large part from the inedequacies of existing models for Amazonian conditions. In turn, this inadequacy makes large-scale settlement in the region a difficult proposition.

The Habitat and Malaria

There can be little doubt that the climate, the vegetation, the isolation, the topography and the fauna constitute formidable natural obstacles to occupation in the Amazon region. But perhaps the most important natural threat to Amazonian settlement at the present time is the high incidence of malaria transmission.

According to information from SUCAM – the malaria control agency in Brazil – Rondônia had, in 1985, 42.1 per cent of Brazil's total proven malaria cases (as compared to 0.66 per cent of its total population). This disease typically plagues new settlements which 'are optimal for transmission and difficult for control because of high vector density, high exposure to vectors, shelters inappropriate for residual spraying, logistic problems and deficient treatment and health care' (Sawyer and Sawyer, 1987, p. 9). Malaria incidence tends to recede with age of a given settlement but this fact is of little help to new settlers whose very ability to survive in colonisation projects is often jeopardised by its effects. 'High malaria prevalence may contribute to "negative" selectivity with regard to both attraction and fixation of settlers who have more resources and skills . . . Malaria is a particularly serious problem for family farmers because they cannot easily substitute for disabled labor and because they must bear the direct and indirect costs of malaria on the entire family. Speculators and commercial farmers, on the other hand, need not reside on the farm and can hire labor as needed. Turnover of settlers, to which malaria contributes, leads to reconcentration of property, defeating the social purposes of colonization' (Sawyer and Sawyer, 1987, p. 66).

Researchers and even casual visitors to Rondônia are inevitably struck by the virulence and pervasiveness of malaria's impacts. In the earlier projects, it was common for colonists to give up their lot in exchange for 'a treatment'. It is reputed that several of the pioneer medical doctors in Rondônia became large landowners as a result of these practices.

The prospects for the immediate future are not promising, as perceived in the CEDEPLAR study. 'Malaria control in Rondônia has been particularly difficult. Shortages of personnel, vehicles, insecticides and drugs have plagued the program . . . Logistics are extremely difficult in remote rural areas, where roads are often impassable in the rainy season, if they exist . . . The strategy based on use of parasiticides and insecticides faces new difficulties in Rondônia. There are serious problems of drug resistance . . . Although there is no evidence of insecticide resistance, there may be insecticide avoidance' (Sawyer and Sawyer, 1987, p. 13).

It is interesting to note that the recent 'Settlement' projects sponsored by the POLONOROESTE went to special pains to reduce malaria prevalence, as part of the gamut of measures adopted to increase the rationality of official projects. Land was partitioned in such a way as to guarantee access to running water to as many settlers as possible; building materials were to be provided in the hope that board houses built at some distance from water would help reduce the incidence of malaria. A special malaria control project, involving an interdisciplinary mobile team, was instituted (Sawyer and Sawyer, 1987, p. 14). Despite all these precautions, in 1986, at a time when Machadinho was still an incipient colony, it already was responsible for much of the increase of malaria in Ariquemes, the municipality with the largest number of malaria cases in Brazil (p. 9). In short, the perspectives for malaria-free colonisation in Rondônia would seem rather dim. This may unwittingly provide some perverse social benefits; one poor settler was quoted as saying – 'Thank God there is so much malaria here. Otherwise, this land wouldn't be here for people like us' (p. 46).

The Distance Factor

A review of frontier expansion in Brazil during the last half century shows clear comparative advantages of earlier settlement regions over more recently-occupied lands (Martine, 1987). A major difficulty of the settlement process in the Amazon and which serves to magnify most of its other problems is the distance factor. For instance, inferior soil conditions in most of this region would require more intensive use of chemical correctives. However, given the rising costs of fuel and the great distances over which inputs have to be transported, this system would greatly increase production costs. By the same token, in order to reach the markets of the Centre–South, agricultural produce has to travel thousands of kilometres, thereby greatly reducing the producer's profit margin.

Thus, inferior soil conditions and greater distance to markets would have to be compensated by higher productivity levels in places such as Rondônia. But that, in turn, is made impossible by the higher costs of the inputs which are necessary to increase productivity. As explained by Osório de Almeida, this leaves only two alternatives for agricultural expansion on a large scale in the Amazon: subsidise production and/or find a way of reducing transport costs (Osório de Almeida, 1987, pp. 4–6). Subsidies to credit, to the purchase of land and agricultural machinery, to warehousing, to transport and to prices have been commonplace, although patchy in the last 15–20 years. Nevertheless, it is unlikely that the rest of the society would agree to permanent large transfers of resources for the sole purpose of keeping Amazonian agriculture alive. The reduction of transport costs would involve considerable investments in the implementation of an integrated policy involving highway, river and railroad transport. Again, the costs of such a system would have to be weighed against other alternatives for agricultural expansion in the Centre–South.

The frequently-voiced expectation that agricultural produce could be industrialised on a large scale in the Amazon, thereby creating solid jobs, boosting profits and integrating the region's economy, appears to be another pipe dream. Aubertin, for instance, has convincingly argued that the conditions of agricultural production in places such as Rondônia simply cannot guarantee the quality, quantity and stability which are necessary for progressive industrial growth. The substantial incentives necessary for industrial takeoff in the region would create even greater social and regional imbalances (Aubertin, 1986, pp. 2–4).

In short, the economic rationalisation of agricultural production in places such as Rondônia does not appear to be easy to carry out on a large scale, nor is it likely to be a viable alternative within the immediate future.

FUTURE SETTLEMENT PERSPECTIVES

Rondônia currently arouses the observer's curiosity with several paradoxes. Thus, the influx of migrants in search of land has increased yearly but their aspirations are being met less and less due to the reduction in the availability of open and/or government-distributed land. The number of would-be settlers grows rapidly despite the repeatedly-demonstrated failure of colonists and the high turnover rates in existing colonisation projects. Yet, at the same time, the latest 'Settlement' projects seem to have trouble filling their quotas. Production has increased significantly in some crops but, often, the fertility of the land is quickly depleted. Land speculation, presumably founded on expectations of probable future use, continues unabated yet, lasting successful models or farming experiences are hard to find. Colonisation tends to deconcentrate the agrarian structure, yet some of the fastest growth is occurring at both extremes of the scale – in *minifúndios* and *latifúndios*.

In short, Rondônia is at once the new Eldorado to which land-hungry migrants are still flocking in droves, as well as the proverbial inhospitable Amazonian jungle habitat, subject to isolation, infertility, erosion, malaria and assorted other ills. Perhaps the key to this paradox is the time lag between positive and negative events. Migrants are still coming to Rondônia on the basis of favourable information which began to spread more than a decade ago; obviously, a large number of the early settlers have already moved out but knowledge of widespread failure is not yet available or it is unconvincing to land-hungry potential migrants. Agricultural production has apparently been increasing at a satisfactory rate on the basis of the steady incorporation of new lands which – within the space of a few years – are likely to become less fertile. Landholding is being deconcentrated as a result of the distribution of a large number of small plots and the recreation of *minifúndio* relations within them; but, for the state as a whole, the tendency is towards concentration of land. Meanwhile, the *minifúndios* become more prevalent as a survival strategy among the 'leftovers' and as a source of manpower for larger farms. Towns are booming, not only because of the services they provide for primary sector activities but also because they are constantly besieged by the new arrivals who provide both a cheap labour force and a captive market.

If this perspective is correct, then the Rondônian situation is potentially explosive. The different time lags – between rapid in-migration and wholesale occupation, between intensive settlement and ecological disaster, between government handout of small plots and the re-concentration by 'market' forces – permit a certain temporary accommodation. However, it may be only a question of time before the various lags combine to promote rapid out-migration – as has happened in other recent frontier regions.

How could this prospect be avoided? Several alternatives to this dilemma have been suggested or implemented. Perhaps most important among these is the attempt to resume the wholesale distribution of lots to small farmers. Thus, in 1986, the Ministry of Agrarian Reform proposed to settle 21,300 families on individual plots by 1990 (MIRAD, 1986). However, results of such an undertaking are not promising. Available data indicate that, by mid-1987, few new families had actually been settled. Second, the demand continues to outstrip supply; according to the Ministry's own estimate, 40,000 families had been awaiting a plot of land as of January, 1986 (MIRAD, 1986); an additional 166,000 new migrants were counted in 1986 (SEPLAN/RO, 1987). Thirdly, even if, by some miracle, land redistribution were to keep up with the demand, past experience suggests that most of the settlers will soon be hampered by serious problems and, consequently, many of them will not remain for long on their lots.

Several observers have ascribed the difficulties of settlers in colonisation projects to defects in the design, organisation and/or implementation of colonisation models. These analysts argue, for instance, that decisions taken with respect to such questions as the size, shape and location of individual lots, administration and personnel issues, provision of basic services and infrastructure, etc. are the major determinants of success or failure in colonisation projects (World Bank, 1980). These would appear to be the main concerns which guided settlement of the Machadinho and Urupá projects financed in recent years by the World Bank; despite special administrative precautions, however, these two projects are faltering seriously, as shown earlier. Indeed, a recent report prepared by the Bank has suggested the evacuation of Machadinho (População e Desenvolvimento, 1987, p. 47).

Another leading current deposits considerable expectations in the potential of technological upgrading (Alvim, 1972, 1980). In this view, the goal to be pursued is increased production and productivity – which will eventually improve the general prosperity of the region and its population – rather than the distribution of lots *per se*. Nevertheless, just how technological progress is to be achieved remains unclear. The adaptation of the prevalent technological package to Amazonian conditions is, as shown above, difficult to implement on a wholesale level. Even when such progress is feasible, the problem of distance to markets is virtually insurmountable, except for crops of high unitary value. There now seems to be a consensus among experts that perennial crops such as cocoa, coffee, rubber, etc. are more appropriate to the Amazonian agroecosystem than the short-cycle crops. But, with few exceptions, such crops can be grown more cheaply in areas closer to markets. There is always the hope that technological breakthroughs of a yet-unforeseen nature will capitalise on some of the region's natural advantages. However, as Fearnside puts it – 'The faith that research results will someday overcome any given agronomic and environmental limitations is pernicious: it can and does lead planners to dismiss concern for future consequences of present development decisions' (Fearnside, 1983, p. 66).

Another set of proposals centres on the need gradually to develop the skills and know-how to deal with the Amazonian ecosystems on a piecemeal basis, thereby avoiding the disastrous consequences of massive deforestation, as well as the transplant of alien and inappropriate agricultural practices. One view argues for a more incremental approach to colonisation planning, given the 'experimental nature of the Amazonian adventure in development' (Moran, 1984, p. 297). This is compatible with the notion that 'no single development option should be promoted but rather a carefully planned mosaic of natural ecosystems and agroecosystems with different intensities and types of management' (Fearnside, 1983, p. 118).

In this connection, the possibility of adapting indigenous or *caboclo* technology has received increased attention. Based on careful management of the existing ecosystem and rotation of crops, this approach is land-extensive and low in productivity. Nevertheless, its obvious ecological advantages and the fact that it is centred on small farmers have made it attractive to some analysts. In Peru, for instance, a hybrid form of settlement, adapting the native indigenous model to small-scale farming, is being attempted (Durand, 1987). This model evidently eschews individual and commercially-oriented enterprises for communitary or associative forms of production which are just slightly above the subsistence level.

Despite their relative appeal to different sectors, none of the above 'solutions' appears to pave the way for the absorption of large masses of small farmers and landless rural workers. The cautious 'ecological' approaches to settlement at least have obvious environmental advantages, but it is difficult to see how they could involve great numbers of migrants. In addition to the cultural obstacles (i.e., individual migrant families come in search of their own piece of land, to work as best they know how), such 'romantic' alternatives would hardly drum up political and bureaucratic support for anything but marginal programmes in Brazil's present stage of capitalist expansion.

In short, the possibilities for continued, large-scale and permanent absorption of excess rural population in Rondônia appear to be rather bleak. It is unlikely that broad, encompassing, once-and-for-all solutions (such as massive land redistribution schemes or the application of a specific technology) will be encountered. Rather, painstaking and piecemeal approaches suited to the ecological, economic and social conditions of specific sub-areas will have to be implemented. In other words, the alternatives which will have to be worked out belong not to the level of grandiose 'Brasil Grande' schemes, but to more modest and realistic handicraft-type management.

For instance, the smaller-scale exploitation of agricultural products with a high unitary value or low transport costs is a viable alternative for certain regions and groups. Elsewhere, large-scale cattle ranching has also proved to be a feasible alternative. Even large-scale soybean production has been profitably carried out, under given circumstances. Mining and rubber extraction will, for some time to come, continue to offer real possibilities in some areas. Moreover, to the extent that frontier expansion in the last 15 years has brought considerable urban growth and that much of this growth is irreversible, then there exists an expanding market for locally-produced vegetables, fruits, poultry, meat, dairy products and the like. Much of the food consumed in Amazonian cities is brought in (sometimes by plane) from the Centre–South; perversely, some of this produce originates in the region, is transported to São Paulo and then carted back. Clearly, there is room for rationalisation of the local market structure which could make small-farm production of subsistence and cash crops feasible and sustainable.

On the national market, however, it is difficult to conceive how agricultural production in places such as Rondônia could ever compete on a large scale with produce from richer lands in more accessible areas. If it is impossible, due to the distance factor, to overcome the natural disadvantages of the Amazon region in terms of soil, climate, topography, etc. then it seems inevitable that the profit margin of farmers in this region will be permanently squeezed. Altogether, the possibility of competing on an equal basis with agricultural production from other regions in the foreseeable future is slim. Without continuous subsidies, in other words, large-scale agricultural production in the Amazon region is unlikely to prosper. This should help reverse the upward spiral in land prices caused by speculation and re-open the possibilities for small-scale, low-income farming. In this picture, places like

Rondônia could eventually become a more or less privileged habitat for subsistence farmers. Given the patchy nature of fertile soils in the Amazon and the region's vulnerability to the effects of large-scale monoculture, occupation by small farmers may make more economic and ecological sense anyway (Fearnside, 1983, 1984).

The fact that Rondônia's rural development and absorption possibilities are inherently limited is apparently beginning to seep through to both policy-makers and potential migrants – finally! Thus, the Rondônian government has re-activated the disincentive campaigns of the mid-1970s in selected municipalities of Paraná and Mato Grosso. On the demand side, migration data show a reduction from 166,000 to 104,000 in-migrants between 1986 and 1987. Meanwhile, evidence of out-migration from Rondônia to other less-densely populated regions of the Amazon indicates that the same problems are probably being recreated elsewhere. Since they are not likely to be resolved in those regions either, the end-destination of most migration streams will be the towns and cities, both within the region and outside it.

Observation of interlinkages between the rural exodus in older agricultural areas, the relatively reduced absorption of colonists in frontier regions, and the accelerated growth of cities – particularly of larger metropolitan regions – provokes sobering thoughts. First, it helps bring home the realisation that the problems of absorbing excess rural manpower freed from the land in traditional agricultural areas will eventually have to be faced seriously within these same regions – and not shunted off to the Amazon frontier. Secondly, since the various panaceas for productive incorporation of Amazonian lands and massive absorption of migrants have not yielded promising results, a scaling-down of expectations and a more painstaking patchwork approach to Amazonian development will have to be worked out.

Brazilian policy-makers and development plans in recent decades have, by and large, favoured grand designs and sweeping solutions. In this light, the massive urban concentration registered in recent decades has tended to be shrugged off as an inevitable sequel of development. But the rhythm of urban concentration and its attendant problems now urge consideration of alternative styles of development which will provide breathing space for smaller-scale farmers. Promotion of associative forms of production, preferential credit treatment and technological R & D directed to small-scale enterprises are among the obvious components of such an approach. Small-farmer settlement of frontier regions – even in the piecemeal manner suggested above – should help stem the tide of the uprooted population which eventually treks towards the cities, if combined with measures emanating from a similar philosophy, aimed at increasing the viability of small farmers in traditional areas. Whether or not such time-gaining alternatives will be pursued is ultimately dependent on changes in the correlation of political forces which, at the time of writing, are extremely difficult to predict.

Notes

1. Background information presented here on Rondônia's physical setting was obtained from Menezes (1981) Furley (1980) and Mueller (1980).
2. The acronyms 'PIC' stand for Integrated Colonisation Project and 'PAD' for Directed Settlement Project. The former involves rather comprehensive tutelage by government agencies whereas the second type supposedly involves more entrepreneurial-type settlers and less government interference.
3. The list of authors who have generally contributed to our understanding of Rondônian settlement include at least the following: Mueller (1980, 1982); Galvão (1982); Turchi (1979); Hébette and Marin (1982); Borges Ferreira (1986); PERSAGRI II (1982); Moran (1984); FIPE (1986a); Léna (1986); Coy (1986); Wesche (1978).
4. The term 'all-weather road' has to be used broadly in Rondônia. The BR-364 has been repeatedly washed out and rerouted as the result of erosion provoked by deforestation in the strip between Pimenta Bueno and Vilhena, cf. Alencar (1987).
5. This estimate is based on the mini-census carried out in 1985 by the IBGE (Brazilian Census Bureau) in Rondônia which showed a total population of 909,000, indicating an increase of 418,000 people between 1980 and 1985. Considering that natural increase would have contributed at least 70,000 individuals to Rondônia's population in this period, then net migration should not be much beyond 340,000. The discrepancy between the two estimates (i.e., 340,000 versus 413,000 from SIMI) could be attributed to the fact that ours is an estimate of net migration while SIMI's is a count of total in-migration at Vilhena, the main gateway to Rondônia.
6. Assuming that the same ratio of persons per establishment, and establishment per rural population, existed in 1980 as in 1985, then it can be estimated that Rondônia's rural population grew from 262,530 to 442,800 between 1980 and 1985. The urban population would thus have experienced a faster increase (from 228,500 to 466,200) in this period.
7. This section is largely based on Osório de Almeida 1984, 1987).

References

Alencar, J. R. (1987), 'Rondônia, uma fronteira sem futuro', *Guia Rural*, 1(1) pp. 147–50, São Paulo.

Alvim, P. (1972), 'Potencial agrícola da Amazônia', *Ciência e Cultura*, 24, pp. 437–43, São Paulo.

Alvim, P. (1980), 'Agricultural production potential of the Amazon region', in Barbira-Scazzocchio (ed.) (1980) pp. 27–3600.

Aubertin, C. (1986), 'Industrialiser les frontières?', *Cahiers des Sciences Humaines*, 22 (3–4) pp. 419–28, Paris: Orstom.

Barbira-Scazzocchio, F. (ed.) (1980), *Land, People and Planning in Contemporary Amazonia*, Cambridge: Cambridge University Press.

Borges Ferreira, A. H. (1986), 'Settlement in Rondônia', Belo Horizonte: CEDEPLAR/UFMG, (mimeo).

Coy, M. (1986), 'Realidade atropela os planos do Polonoroeste', in *Pau Brasil*, 11(II, March–April) São Paulo.

CEPAGRO/IBGE/SEPLAN (various years), 'Levantamento sistemático da produção agrícola', Rio de Janeiro: IBGE/CEPAGRO.

Durand, C. (1987), 'Tecnologia para la intensificación del uso de las areas colonizadas de la selva alta del Perú', *Seminario sobre Tecnologias para los Assentamientos Humanos en el Trópico Húmedo*, Manaus: CEPAL/IPEA.
Erse, F. (1984), *Posição Mineral*, Brasília: Câmara dos Deputados, Coordenação de Publicações.
Fearnside, P. M. (1983), 'Development alternatives in the Brazilian Amazon: an ecological evaluation', *Interciencia* 8(2), pp. 65–78, São Paulo.
Fearnside, P. (1984), Brazil's Amazon settlement schemes: conflicting objectives and human carrying capacity', *Habitat International*, 8(1), pp. 45–61.
FIPE, Universidade de São Paulo (1986a), 'Tendências da Estrutura Fundiária em Rondônia', Relatório de Avaliação, São Paulo: FIPE/USP/POLONOROESTE (mimeo).
FIPE, Universidade de São Paulo (1986b), 'Avaliação dos modelos de colonização de Urupá e Machadinho', São Paulo: FIPE/USP, POLONOROESTE (mimeo).
Furley, P. A. (1980), 'Development planning in Rondônia based on natural renewable resource surveys', in Barbira-Scazzocchio (ed.) (1980), pp. 37–45.
Galvão, M. (1982), 'A contribuição do INCRA no processo de ocupação do território de Rondônia', in *Doenças e Migração Humana*, Brasília: Ministry of Health (SUCAM).
Gama Torres, H. C. (n.d.) 'Desistência e substituição de colonos em projetos, de colonização na Amazônia: o caso de Machadinho' (mimeo).
Governo do Estado de Rondônia, SEPLAN/NURE (1987), *Boletim de Migração, 1986*, Porto Velho.
Hébette, J. and Rosa E. Azevedo Marin (1982), 'O Estado e a reprodução da estrutura social na fronteira: Ariquemes, Rondônia', Série Seminários e Debates Belém: NAEA/Universidade Federal do Pará.
IBGE (various years), *Censos Agropecuários*, Rio De Janeiro: Instituto Brasileiro de Geografia e Estatística.
Léna, P. (1986), 'Aspects de la Frontière amazonienne', *Cahiers des Sciences Humaines*, 22 (3–4) pp. 319–344, Paris.
Martine, G. (1981), 'Recent colonization experiences in Brazil: expectations versus reality', in Jorge Balán (ed.), *Why People Move*, Paris: UNESCO Press, pp. 270–92.
Martine, G. (1982), 'Colonization in Rondônia: continuities and perspectives', in Peter Peek and Guy Standing (eds), *State Policies and Migration*, London: Croom Helm, pp. 147–72.
Martine, G. (1987), 'Migração e absorção populacional no Trópico Úmido', *Seminário sobre Tecnologia para os Assentamentos Humanos no Trópico Úmido*, Manaus: IPEA/ECLA (mimeo).
Martine, G. and Ronaldo Garcia (1987), *Impactos Sociais da Modernização Agrícola*, São Paulo: Editora Hucitec (especially Chapters 1 and 2).
Menezes, M. A. (1981), 'O atual estágio de conhecimento sobre os recursos naturais da Amazônia: o pressuposto para a definição de uma política de ocupação', *Anais do Segundo Encontro Nacional*, São Paulo: Associação Brasileira de Estudos Populacionais, pp. 11–82.
MIRAD (Ministério da Reforma e do Desenvolvimento Agrário) (1986), 'Plano Nacional de Reforma Agrária (PNRA) do Estado de Rondônia', MIRAD/INCRA (January) (mimeo).
Moran, E. F. (1984), 'Colonization in the Transamazon and Rondônia', in Schmink and Wood (eds) (1984) pp. 285–306.
Mueller, C. C. (1980), 'Recent frontier expansion in Brazil: the case of Rondônia', in Barbira-Scazzocchio (ed.) (1980) pp. 143–7.
Mueller, C. C. (1982), 'O estado e a expansão da fronteira agrícola no Brasil', in *Anais*

do Seminário Expansão da Fronteira Agropecuária e Meio-Ambiente na América Latina, UnB, 1, Brasília.
Osório de Almeida, A. L. (1984), 'Selectividade perversa na ocupação da Amazônia', *Pesquisa e Planejamento Econômico*, 14 (21) pp. 353–98, Rio de Janeiro.
Osório de Almeida, A. L. (1987), 'Tecnologia agrícola moderna para o pequeno produtor na Amazônia', *Seminário sobre Tecnologia para o Assentamento Humano no Trópico Úmido*, Manaus: IPEA/ECLA.
PERSAGRI II (1982), 'Relatório Final: município de Ariquemes, Rondônia', Rio de Janeiro: Convênio Ministério da Agricultura/Fundação Getúlio Vargas (mimeo).
População e Desenvolvimento (1987), 'Colonização: um projeto que pode fracassar', *População e Desenvolvimento*, 21 (144) pp. 47–8, Rio de Janeiro.
Sawyer, Donald (1984), 'Frontier expansion and retraction in Brazil', in Schmink and Wood (eds) (1984) p. 180–203.
Sawyer, Donald and Diana Sawyer (1987), *Malaria on the Amazon Frontier: Economic and Social Aspects of Transmission and Control*, Belo Horizonte: CEDEPLAR/UFMG (mimeo).
SEPLAN – Governo do Estado de Rondônia, (1986), 'Estudo Sócio-Econômico de Populações Assentadas nos Projetos de Colonização do Estado de Rondônia: Gy-Paraná', Porto Velho: SEPLAN/RO (mimeo).
Schmink, M. and Wood, C. H. (eds.), *Frontier Expansion in Amazonia*, Gainesville: University of Florida Press.
Turchi, L. (1979), *Colonização Dirigida: Estratégia de Acumulação e Legitimização de um Estado Autoritário*, Master's thesis, University of Brasília.
Vieira, M. A. da Costa (1987), 'A venda de terras do ponto de vista dos lavradores: a venda como estratégia' (mimeo).
Wesche, R. (1978), 'Modern agricultural settlement in Rondônia, Brazil', *Notes de Recherches* 17, University of Ottawa.
World Bank (1980), *The Integrated Development of Brazil's Northwest Frontier*, Report no. 3042a-BR, Washington, D.C.

3 The Shanty Town, Final Stage of Rural Development? The Case of Acre

Keith Bakx

INTRODUCTION

The corporate invasion of Amazonia, characterised by the implantation of agro-ranching and mineral enterprises in areas that had formerly been occupied by peasant farmers, rubber-tappers and indigenous groups, has severe social and economic repercussions for the region as a whole. In Acre, which had no known mineral resources, the principal activity of the newly arrived entrepreneurs was ranching, although in many instances this was merely a façade for speculation in land. Such ranching activity, whether real or phantom, involved the clearance of large areas of rainforest which, firstly, destroyed the economic basis of traditional extractivist and subsistence activity and, secondly, disrupted whole communities as the rural population was forced to abandon its land to swell the ranks of the urban poor. In opposition to this process, the principal role of rural communities has been to struggle to impede its development, to maintain access to land. Thus, whereas ranchers seek to claim exclusive use of and definitive rights to land, to deny others access to it, the struggle of the rural population and allied groups represents a move towards the redemocratisation of land, the promotion of its social use.

This chapter will examine State measures which have sought to defuse violent confrontation in the Acrean countryside and so stem the rural–urban exodus that it engendered. An assessment will be given of the federal use of colonisation schemes and producer cooperatives as methods of permanently settling peasants on the land. This will demonstrate that such projects functioned as mere stepping stones on the road to landlessness and poverty in the shanty towns of the state capital. An analysis will also be made of the differential response of the residual rural population to State intervention as it sought to avoid this fate. The bulk of the primary source data presented here was collected by interview during three extended field trips to Acre between October 1982 and September 1987 (see Bakx, 1986, 1987, 1988). Where a secondary source has been used, this will be noted in the text.

THE DEACTIVATION OF ACRE'S RUBBER ESTATES

Unlike its other Amazonian counterparts, the Territory of Acre began the post-war era with its economy firmly based on the extraction of natural rubber (see Figure 3.1). Social relations of production were still characterised by traditional forms of debt peonage which tied the direct producer, the rubber-tapper, to the rubber estate owner. Less than thirty years later, the situation had changed markedly. State intervention, both direct and indirect, had eroded the power base of the traditional landowning class and the now autonomous rubber-tappers were engaged in an increasingly violent struggle as they attempted to halt the advance of the ranching front which had been facilitated by that very intervention.

The demographic expansion of Rio Branco, the state capital, played an important role in this transformation. Following the cessation of hostilities in 1945, those who had opted to produce rubber for the Allies in Western Amazonia rather than participate in the Italian Campaign were released from their obligations and many of these *soldados da borracha* began to drift into the urban areas. Then, in 1962, the elevation of Acre to statehood led to an expansion in the capital's bureaucratic apparatus with a corresponding increase in the number of bureaucratic personnel. Towards the end of the decade, the BR-364 highway, which links Rio Branco to Brasilia, was constructed. This not only permitted the development of direct commercial links with the industrialised South, it also facilitated in-migration. As a result, Rio Branco's population rose from 9,371 in 1950 to 34,988 in 1970, an increase of 273 per cent in just twenty years (Guerra, 1955, pp. 118–76; Brasil, 1980a).

Prior to the Second World War, the Acrean rubber-tapper was tied to the estate through debt peonage and coercion with the estate owner controlling both access to the means of production and the labour process itself. Of particular importance was the prohibition of subsistence cultivation which obliged the tapper to exchange rubber for basic subsistence goods, the price of both being set by the estate owner. After 1960, many rubber estate owners began to engage in the more lucrative commercial activities in the rapidly expanding urban centres. This loosened the ties that bound the tapper to the estate in that the rubber-tapper was now allowed to engage in subsistence activity, thus relieving the estate owner of the responsibility of providing such subsistence goods. It also signified that the tapper was no longer dependent on the exchange of his rubber product for subsistence items which were sold at a high price in the estate owner's store. This had the effect of breaking the cycle of debt which bound the tapper to the estate which in turn monetised the economy of the interior and also allowed the tapper to move from estate to estate in search of a better price for his labour (Bakx, 1986, pp. 45–83).

At the national level, federal policies and programmes also functioned to

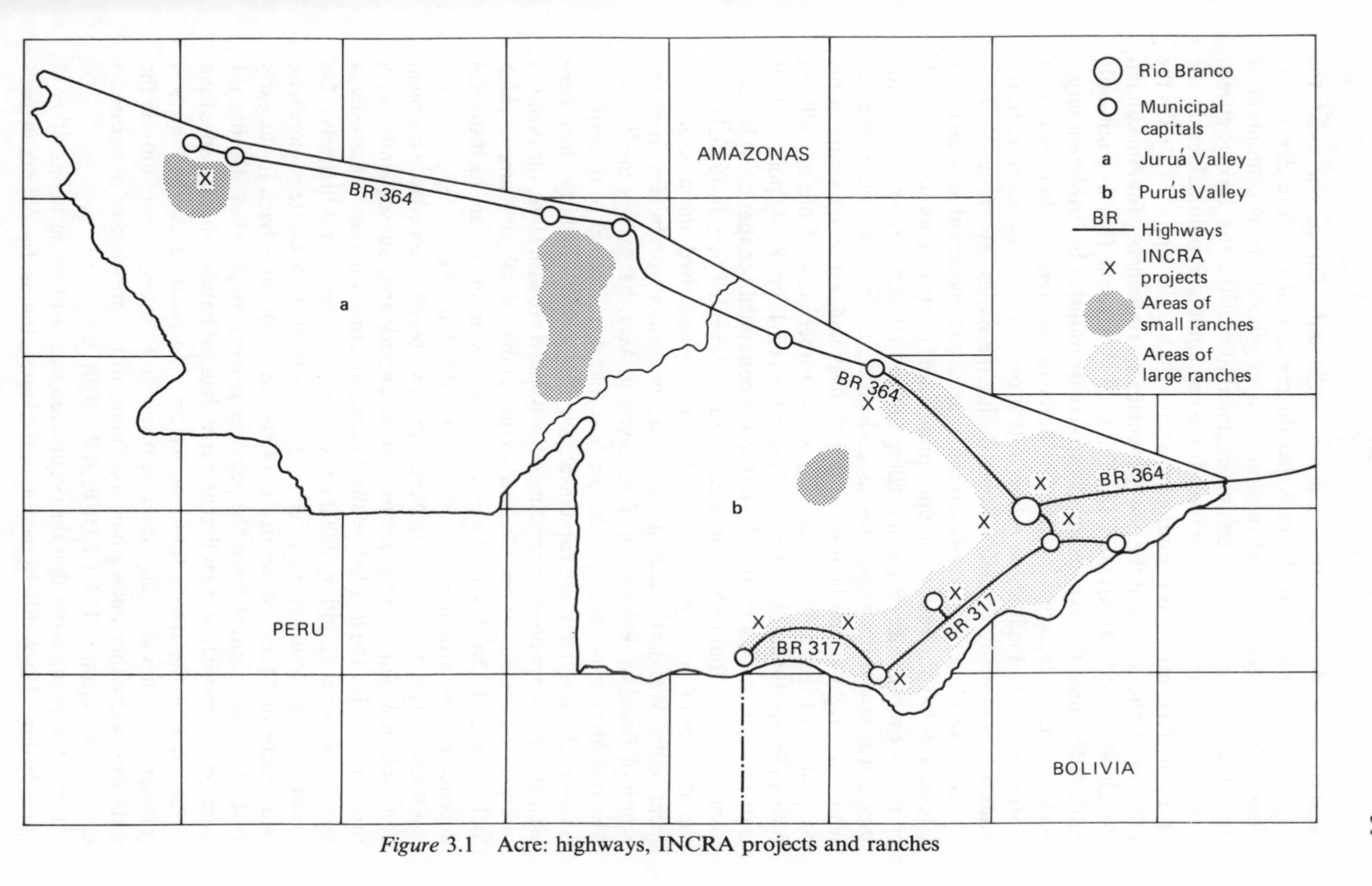

Figure 3.1 Acre: highways, INCRA projects and ranches

weaken the traditional power structure in the state. On the one hand, the national development model provided the contextual structure that underpinned the struggle for land in Amazonia in general: the subsumption of agriculture to industry; widening regional disparity; the links between national and international capitals at the economic and political levels; the creation of a centralised planning apparatus (Bakx, 1987, pp. 533–7). On the other hand, the actions of the newly created Superintendency for Amazonian Development (SUDAM) and the Bank of Amazonia (BASA) not only militated against the expansion of extractivist industry by attracting large-scale 'modern' capitalist production units to the region; their policy of restricting credit facilities to rubber estate owners was instrumental in the bankruptcy of many of their number during the sixties (Mesquita, 1977).

The net result of this process was that many estate owners abandoned their estates and left the resident tapper population to survive as best it could, cultivating subsistence crops and selling rubber to itinerant river traders. The tapper was now transformed into an authentic *posseiro* – i.e., a homesteader who has a right to the land on which he lives and works by virtue of his occupation of that land for one year without interference from any alleged owner (Brasil, 1964). However, as Foweraker (1981, pp. 83–4) points out, it is one thing to have a right to land through continued occupation, it is quite another to have that right translated into a documented title. In 1970, before the situation of Acre's *posseiro* population was regularised, the state government under Wanderley Dantas issued an invitation to ranchers from the South of Brasil to 'Produce in Acre, Invest in Acre, Export via the Pacific'. The invitation was based on the premise that Acre needed to diversify its economic base and develop capitalistic activities such as ranching which were assumed to be capable of generating self-sustained growth. Here, the position of the state government was identical to that of the federal government which had charged SUDAM with the task of freeing Amazonia from its historical dependency on extractive activities (Mahar, 1979, p. 14).

Following this invitation, the Acrean *posseiro* became locked into a violent confrontation with, firstly, former estate owners and, subsequently, with ranchers. During the first half of the decade, land prices in the state rose by at least 1,000 per cent and by 2,000 per cent near to the federal highways. On hearing of the rapidly rising price of land, many of those estate owners who had deactivated their estates in the 1950s and 1960s came back in the early 1970s to wrest control from the resident *posseiro* population. During the same period, more than 5 million hectares changed hands – that is, more than 30 per cent of the state's total surface area. In most cases, the land was transferred from a generally absentee owner to an investor from outside the state who was either unaware or indifferent to the presence of *posseiros* on the land (Mesquita, 1977; CEDEPLAR, 1979).

It must be pointed out that the occupation was not homogeneous either in space or time. The southern section of the state – that is, the old municipalities of Rio Branco, Xapuri and Brasiléia – was occupied in the early 1970s. This area accounted for 44.5 per cent of the state's rural population and 67.5 per cent of its urban population in 1970. For the most part, the new arrivals were small to medium ranchers with their own capital who bought areas of up to 1,000 hectares. They were attracted by the 'availability' of land, its low price relative to that in the area from which they originated, and the existence of a functioning road network which guaranteed access to the capital and facilitated commercialisation of any product. The second area to be 'occupied', in the mid-1970s, consisted of the central part of the state – that is, the valleys of the Rivers Iaco, Envira and Tarauacá. This contained 33.7 per cent of the rural and 16.3 per cent of the urban population of the state in 1970. It was here that large groups from the South bought extensive areas of up to 10,000 hectares. These groups were attracted by the fiscal and financial incentives offered by SUDAM and were more interested in land speculation than its productive use. The third zone, that of the Rivers Juruá, Moa, Liberdade and Gregório, accounted for 21.8 per cent of the state's rural population and 16.2 per cent of its urban population in 1970. This has been virtually untouched by the 'ranching front'. It was the last area to be crossed by the BR-364 highway and this section is trafficable only during the months of July and August. Its relative isolation has signified that rubber tapping on the basis of traditional debt peonage remains the predominant labour relation in this area (Brasil, 1980a; Aquino, 1982, p. 105; Bakx, 1986, pp. 202ff).

THE STRUGGLE FOR LAND

Once the estates had been transferred, the new 'owners' began to use increasingly violent methods to evict the resident population notwithstanding the *posseiros*' legitimate claim to the land on which they lived and worked. The federal agency responsible for the issuing of land titles, the Institute for Colonisation and Agrarian Reform (INCRA), was conspicuous by its absence throughout this initial period of confrontation. By 1974, INCRA had regularised the titles on just 81 properties covering a mere 7,737 hectares, less than 1 per cent of the area that changed hands during the first half of the decade (Brasil, 1981a). It quickly became obvious to *posseiro* families that only two courses of action remained open to them: to migrate away from the affected areas or to organise themselves to combat the violence perpetrated against them.

The mid-1970s were marked by a significant population movement out of those areas affected by the violence, principally the south-eastern corner of the state. The rural sectors of the municipalities of Rio Branco, Xapuri and Brasiléia each show negative population growth rates for the decade as a whole. The migration took two principal forms: rural–rural, to the rubber estates of Bolivia's Pando province; and rural–urban, into the shanty towns of the state and municipal capitals. According to the Church's Land Commission in Acre, over 10,000 *posseiros* moved south crossing the international boundary to tap rubber in Bolivia. Generally, these were unaccompanied males whose families stayed behind in the shanty towns to survive as best they could on remittances. Acre's urban population rose by 38.2 per cent between 1970 and 1975, with Rio Branco receiving the lion's share of the migratory flow. The state capital's population rocketed from 36,095 in 1970 to 92,304 in 1980 (Acre, 1980; Brasil, 1980a).

The urban sector was totally unprepared for the influx of such large contingents of migrants. On the one hand, the urban labour market was incapable of expanding in the short term and thus unable to absorb unskilled labour to any significant extent. As late as 1982, Acre's entire industrial force numbered just 4,900 persons, the majority being employed in the traditional industries of food and drink (30.9 per cent), sawmills and furniture-making (29.3 per cent) and brickworks (24 per cent) (Brasil, 1980b). In addition, the capital's administration could not meet the social needs of the migratory contingent. To the west of the city centre, on the low ground near to the river, are the migrant reception areas, densely crowded wooden shacks with no sanitation, water or power supply. There are no schools or health posts. The narrow tracks that meander in between the shacks are covered in a choking dust during the summer months and ankle deep in mud in the rainy season. There are no areas where kitchen gardens can be planted, and no space for leisure activities.

Rather than accept this fate, large numbers of *posseiro* families decided to remain on their land and make a stand against the actions of the ranchers and their labourers. The summer months in Acre, May until September, are dry and it is at this time of year that the rural population burns down small areas of forest to plant subsistence crops. During the 1970s, these months were the months of rural confrontation. On the one hand, the ranchers attempted to prohibit the clearing of subsistence plots. On the other, as deforestation for pasture became more generalised, so the *posseiros* themselves began to organise to prevent it.

Prior to the summer of 1974, there was an almost total lack of Acrean press coverage of the conflict that was occurring in the countryside. Where it was reported, it was couched in terms of the criminal acts of *posseiros* against the property and persons of the new 'owners'. Given the high incidence of violence at this time, coupled with the close proximity of social, political and economic power in small towns like Rio Branco, it is probable that the local press houses and the Dantas administration colluded so as to avoid the state being classed as an area of social tension at the very time the state government was trying to attract capital investment from the South.

In 1974, the situation changed dramatically. During that summer, the grouping together of *posseiros* to prevent deforestation became generalised throughout the southern half of the state. In July, at the height of the violence, a *posseiro* shot and killed the leader of a group of labourers hired by the rancher to evict him from his land. It was obvious that this was not an isolated case, but part of a more widespread phenomenon and it became increasingly impossible for the press and state government to maintain their manifest indifference to the process by which *posseiros* were being evicted. The situation clearly worried the state authorities. This was not only an election year, but also the first summer that *posseiro* families had arrived in significant numbers in the state capital.

It was at this time that INCRA's policy underwent an apparent about-turn. INCRA opened up a new office to coordinate its activities in southwestern Amazonia – that is, Rondônia and Acre. It must be pointed out that Rondônia then was receiving approximately 10,000 migrant families per annum, fleeing from the capitalisation of agriculture in the South, and that this current was likely to overflow into Acre in a relatively short space of time (Mahar, 1981, p. 16). Subsequently, given the federal government's stated policy of promoting capital investment in the region and the increase in the incidence of rural violence, INCRA inaugurated a programme in Acre based on the following three priorities;

1. The regularisation of the situation of those who were legitimate *posseiros* on lands owned by the federal government – that is, *terras devolutas*.
2. The regularisation of the situation of those large property owners with legitimate title so as to facilitate their access to credit and federal incentives.
3. An enquiry to ascertain which areas of the state were suitable for expropriation in the 'social interest' and which would be distributed at a later date to those rural families who had already migrated to the urban centres.

To implement this programme, INCRA created a special land commission which was to deal with the crucial question of ascertaining which lands were *terras devolutas* and which were privately owned. According to the Brazilian Land Statute of 1850, *posse* is possible only on *terras devolutas* and the latter lose their status as public lands only when they come into private ownership – that is, through the issue of a legitimate title by a recognised authority. Given that Acre was annexed by Brazil in 1903, the competent authorities in this case were: the Province of Amazonas (prior to 1898); the Independent State of Acre (1898–1904); The Republic of Bolivia (prior to 1903); and the Republic of Peru (prior to 1903). Land titles not descended from or based upon titles issued by the above were deemed invalid and the land itself regarded as *terra devoluta* and so subject to the possibility of *posse*.

However, INCRA's acceptance of the right of *posse* did not automatically ensure that titles would be forthcoming for the *posseiro*. The Commission concentrated upon the issuing of titles for either those large properties which were proven to have been legitimately titled, or for smaller properties on the agricultural nucleii close to the urban centres. Titles, then, were issued to *latifúndios* or to *minifúndios*, both of which INCRA had promised to eradicate. By concentrating on these two numerically small groups, the vast majority of Acre's *posseiro* population were still open to expropriation through violence. In addition, the enquiry into land suitable for expropriation in the social interest was specifically aimed at the settlement of those already dispossessed and not the permanent settlement of those who were still fighting to maintain access to land. Thus, despite its rhetoric to the contrary, INCRA's land discrimination programme was specifically aimed at resolving the crisis in the urban areas rather than that in the countryside.

While INCRA–Acre openly stated that de facto *posseiros* should have their claims to land legitimised, the agency issued a total of only 8,514 titles in the state between 1972 and 1981. This demonstrated to *posseiros* that their actions to halt deforestation and the implantation of pasture were justified and in addition that this struggle had to continue because of bureaucratic delays. During this period, the *posseiros* of Acre developed a significant level of organisation. This took the form of grouping together for what is known locally as an *empate*, a joint show of force to prevent the rancher or his hired hands from carrying out further deforestation before the area had been the subject of an INCRA enquiry. At first, *empates* were localised, but by 1976 they had taken on statewide dimensions with *posseiros* from all Acre's municipalities uniting in displays of solidarity.

Faced with this impasse, INCRA began to intervene directly in land disputes as and when they arose. It saw its role as an 'independent' arbiter between the *posseiro* and those who sought to dispossess him and focussed on persuading the *posseiro* to accept an alternative parcel of land in exchange for the area on which he had previously worked. In theory, this could have benefitted both the parties concerned, but in practice it undermined the already tenuous hold that the *posseiro* had on his land. On the one hand, the *posseiro*'s acceptance of another plot of land confirmed the other's right to that land in that he was effectively relinquishing his right of *posse* to it. In addition, these alternative plots were generally of much smaller size, in many cases less than 10 hectares, and often distant from access roads and river banks. This signified that in a very short space of time the *posseiro* was likely to abandon his new plot and migrate to the already congested urban areas.

In sum, although INCRA's policy in Acre seemed to have moved through 180 degrees in the mid-1970s and appeared to support *posseiro* claims to land, it has been shown that the two principal modes of action undertaken by INCRA during this period, discrimination and arbitration, both functioned to subvert INCRA's stated policy commitment of permanently settling the *posseiro* on the land.

MANAGING THE CRISIS

In 1976, INCRA's policy regarding its treatment of *posseiro* claims to land underwent another significant change. This was in part a response to the failure of its previous two initiatives, discrimination and arbitration, but was also directly related to the arrival in 1975 of a representative of the Confederation of Brazilian Agricultural Workers (CONTAG), who began to organise *posseiros* into municipally-based unions (STRs). This period also witnessed the increased socio-economic engagement of the Church following the death of the Bishop of Acre–Purús and his replacement by a Bishop who followed the Liberation Theology line of Vatican 2 and Medellin.

Both of these facilitated the consolidation of *posseiro* organisation in the state and legitimated their claims to land. INCRA countered this development by using the other weapon in its armoury and expropriated 700,000 hectares in Acre in December of that year. These areas were destined to become the sites of two Directed Settlement Projects (PADs) on which a total of 8,000 families were to be settled. The basis of settlement on these projects as outlined by INCRA was:

1. That the occupation should be by families and not by private enterprises, the majority coming from the dispossessed living in the shanty towns of Rio Branco.
2. That each family should be allocated a plot of up to 100 hectares.
3. That, while a variety of crops should be cultivated, the long-term aim was not subsistence, but the creation of small rubber plantations of between 3 and 10 hectares on each plot.

INCRA began the process of land expropriation in the summer of 1977, but by the end of the following year only 50,000 hectares were actually in INCRA's possession and only fifteen families settled on them. The year of 1978 had been one of heated debates over the two PADs, particularly by the landowners who were facing expropriation. The latter called in 'expert' witnesses to the INCRA tribunals who testified that the areas under expropriation order were infertile, that they were rife with malaria and other infectious diseases, that the *posseiros* were ignorant of modern cultivation techniques and would starve to death. These delaying tactics brought about a resurgence of rural conflict during the summer of 1979 and INCRA was forced to accelerate its programme and by mid-1980 1,436 families had been settled, approximately 18 per cent of its original target. However, the summer of 1980 was also marked by a high incidence of rural violence which was brought to a head with the assassination of the Rural Workers' Union president in the municipality of Brasiléia and the subsequent lynching of the rancher who was alleged to have ordered his death. The result was the expropriation of a further 187,072 hectares also destined to become PADs.

The expropriated areas had a dual purpose: the absorption of labour from the state capital's shanty towns and the diffusion of tension in the countryside. However, the form that the projects took and INCRA's mode of operation directly contributed to the continuation of the rural exodus, rather than helping to stem it. First, as has been noted, there was a long delay between the initial expropriation of the land and the actual settlement of it. When interviewed, INCRA staff blamed the delay on the high staff turnover and lack of cooperation from the other state agencies involved and the low settlement numbers on the non-take up of lots by *posseiros* following a campaign waged by the Rural Workers' Union which publicised the failure of the state consortium to keep its infrastructural promises. This campaign was seemingly quite justified in that by January 1984, some six years after the initial expropriations, the projected access roads were far from completion, health posts were few in number and schools had yet to be constructed.

Secondly, INCRA based its selection criteria for prospective colonists on financial or economic capacity and knowledge or experience of modern agricultural techniques. It justified the inclusion of these criteria on the grounds that those who met the required standard would be more likely to succeed and, in addition, fewer state resources would be needed over the long term. However, the tradition of rudimentary slash and burn agriculture and primitive extractivist techniques in Acre, coupled with the history of debt peonage which denied the possibility of capital accumulation to all but a few rubber-tappers, signified that the former *posseiro* living in the shanty town was extremely unlikely to meet either of INCRA's principal selection requirements. Yet it was precisely for the settlement of this group that the PADs were created. By December 1983, approximately 25 per cent of all PAD colonists in Acre were from outside of the state and on some of the newer-settled PADs out-of-state migrants accounted for over 50 per cent of the colonists (Bakx, 1987, pp. 545ff).

Thirdly, the colonist had great difficulties obtaining credit. The State Bank (BANACRE) did not extend credit to the majority of PAD applicants because they did not have a definitive title and therefore were not the legal owners of the land on which the credit was to be utilised. INCRA officials in their turn stated that titles would not be granted to plot holders until sufficient improvements had been made. Two of the three INCRA requirements were the construction of a house and the clearing of a variable percentage of the area for cultivation (the percentage depended upon a number of criteria – e.g., number and age of family members). The third was the planting of permanent crops such as rubber and cocoa as opposed to short-term subsistence crops. This left the colonist in the impossible situation in which he would receive credit for permanent cultures only if he was in possession of a definitive land title, yet the latter was denied to him until he had already planted those crops.

The above has shown that, in their present form, INCRA's projects in

Acre cannot provide a long-term solution to the problem for which they were originally created. Throughout this period, a second federal agency, the Superintendency for Rubber Production (SUDHEVEA) was also operating in the state. Unlike INCRA, which had a global brief, SUDHEVEA had but one concern – the increased production of rubber. From 1972 onwards, especially after the oil crisis of 1974, SUDHEVEA initiated a series of programmes to encourage natural rubber production. The first two programmes (PROBORs 1 and 2) were not aimed at the small producer, but rather sought to increase home rubber production through the creation of plantations on larger properties. SUDHEVEA was faced with three major problems in terms of programme implementation in Acre. First, property owners diverted financial incentives from plantation creation to other uses, especially commercial activities. Secondly, the struggle on the part of the *posseiro* population against deforestation did not and could not distinguish between clearing land for pasture and clearing land for plantation rubber as both inevitably meant their forcible expulsion from the land. Thirdly, the deactivation of the estates during the late 1960s and early 1970s led to a dramatic fall in the production of rubber in some municipalities in the state. This was compounded in the mid- and late 1970s by deforestation. The municipality of Brasiléia, for example, produced 1,250 tonnes of rubber in 1971 and 403 tonnes in 1981, a fall of 67.8 per cent (Brasil, 1981b).

Unlike its predecessors, PROBOR 3 was aimed at the small landowner. It was felt that the latter, given adequate technical and financial assistance, was more likely to meet the production targets set by the programme. This too ran into severe difficulties as the majority of recipients of SUDHEVEA's technical assistance were PAD colonists who were unable to obtain credit facilities because they were not in possession of a definitive title to their land. In direct response to this dilemma, the state government in conjunction with SUDHEVEA launched a programme known locally as PROBORZINHO whose aim was to construct a series of *mini-usinas* throughout the state. A *mini-usina* is a small rubber processing plant, capable of being run by a labour force of three, that converts latex into high quality smoked rubber sheets. It was envisaged that each *mini-usina* would serve between twenty and forty small producers. Under the provisions of PROBORZINHO, the state bank, BANACRE, guaranteed credit facilities to any small producer, regardless of land title status, provided that SUDHEVEA guaranteed technical assistance to that person. A prime motivating factor for this action was the realisation on the part of the state authorities that the creation of plantations in other states such as Bahia represented a direct threat to Acre's position as the leading producer of natural rubber.

SUDHEVEA launched PROBORZINHO in early 1982 and within two years nineteen *mini-usinas* had been constructed, fourteen in Brasiléia, three in Xapuri and two in Tarauacá. The concentration of *mini-usinas* here was not accidental. First, this area had the highest numbers of *posseiros* still producing rubber, an independent labour force experienced in rubber production that was to be transformed into producer cooperatives based around the *mini-usina*. Secondly, this portion of Acre had been the scene of major confrontations for over a decade with the result that production levels were extremely low and an obvious target for SUDHEVEA intervention. Thirdly, although this area had witnessed much deforestation, the resistance put up by the *posseiros* meant that there was still a significant number of tapping posts still in production, albeit on a much reduced level given that the *posseiros* need to engage in subsistence activity. Again, if sufficient incentive were offered, the deactivated trails on these tapping posts could be brought back into production in a relatively short space of time.

According to SUDHEVEA's coordinator in Brasiléia at this time, the only problem that really concerned the agency was the lack of trust that *posseiros* had in state agencies, an inheritance from their experience with INCRA. The position of SUDHEVEA improved following a policy change by the local Rural Workers' Union after the assassination of its president in 1981. The new leadership, whilst still insisting on the right of *posse*, moved away from its earlier intractable position in which only the recognition of existing *posses* was acceptable. The union now recognised that the position of its members could not be resolved without access to state resources, and that if its long-term aim of land titling was to be achieved, then it must be prepared to cooperate with the state. Any residual resistance on the part of the union membership was overcome by a massive publicity campaign in the municipality by SUDHEVEA which offered a comprehensive package founded upon the following five points:

1. INCRA agreed temporarily to suspend its settlement of PAD Santa Quitéria and PAD Quixadá in the municipality which would give *posseiros* breathing space during which time their cases could be considered.
2. The Bank of Amazonia (BASA) agreed to provide credit to *posseiros* at preferential rates. Loans were given on the understanding that SUDHEVEA underwrote them and supervised the construction of the *mini-usinas* and their management in the early stages.
3. SUDHEVEA agreed to train *posseiros* in the operation of the *mini-usinas* and guaranteed to purchase their rubber for a higher price than that produced by the traditional method.
4. SUDHEVEA entered into an agreement with the national Food Supply Company (COBAL) to provide subsistence and other essential items at prices lower than that of local river traders and city merchants.
5. *Posseiros* could participate in the programme only if they guaranteed to produce a minimum of 200 kilos of rubber per season and formed producer cooperatives with their neighbours.

Thus, in contrast to INCRA's experience on the PADs, SUDHEVEA seemingly managed, in Brasiléia at least, to overcome four of the principal stumbling blocks to the accumulation of capital on the part of the small rural producer: the commercialisation of the product; the provision of essential supplies at a low price; the provision of cheap credit without the need of a definitive land title; and readily available technical assistance.

EXTRACTIVE RESERVES, AN ALTERNATIVE MODEL FOR AMAZONIA?

The municipality of Xapuri was also targeted by SUDHEVEA as a possible site for the location of tappers' cooperatives organised around the *mini-usina*. Like Brasiléia, Xapuri's rubber production declined dramatically because of rural conflict and deforestation, falling from 1,132 tonnes in 1971 to 679 tonnes in 1981. However, SUDHEVEA faced two further problems which led to the failure of its programme in that municipality. On the one hand, the area in which the bulk of Xapuri's *posseiros* lived was located in the interior and not served by any access roads. The land on either side of the principal highway, the BR-317, had already been almost totally deforested and most of the rural population expelled from it. This created difficulties for the commercialisation of rubber production by SUDHEVEA and the provision of essential items by COBAL. In addition, the local branch of the Rural Workers' Union continued to maintain its stance of non-cooperation with state agencies until the problem of land titling had been resolved. The positions of the two STRs essentially reflect those of the two main opposition parties prior to the demise of the military regime. The PT argued that the tapper had an inalienable right to his *posse* and that he should hold his ground at whatever cost. The PMDB, on the other hand, argued that there had to be a dialogue between the parties if the dispute was ever to be resolved.

SUDHEVEA succeeded in persuading only three groups of *posseiros* to form cooperatives. *Mini-usinas* were constructed at these sites in January and February 1983, but all had ceased to operate by December of that year. The first was located at the edge of an INCRA colonisation project and was expected to serve colonists with small rubber plantations. These had not begun producing and, as there were few traditional tapping posts left in the vicinity, the *mini-usina* was shut down for lack of latex. The second closed because of the exceptionally bad summer that year. *Posseiros* were unable to burn down an area sufficient to meet their subsistence requirements. Many were forced to exchange their rubber for subsistence items with local merchants as and when required, rather than wait for it to be processed by the *mini-usina*. Again, it closed for lack of latex. Several members of the third cooperative were not entirely convinced of the viability of the enterprise and sent only a portion of their product to the *mini-usina* and sold the rest to local traders. The cooperative's revenue was insufficient to pay off credit payments and after three months over half of the members withdrew from the cooperative and it ceased to operate as a unit.

It is important to reiterate that SUDHEVEA's prime directive was the increased production of rubber and that the creation of cooperatives was merely a means to achieve this end. The *mini-usina*, then, was not of itself an agent of social differentiation. This is illustrated by SUDHEVEA's actions in Tarauacá. Although there were several deactivated rubber estates close to the town itself which were occupied by *posseiros*, rubber production in the rest of the municipality was still based on the traditional debt peonage system and it was here that SUDHEVEA concentrated its effort. Of the two *mini-usinas* already constructed, the first is owned by an estate owner and the second was sold by SUDHEVEA to a local merchant, himself an ex-estate owner. Both use the *mini-usina* as a means of continuing the traditional relation of debt peonage at a time when the system has fallen into disuse in other areas. As such, the *mini-usina* in this municipality is an instrument of class oppression, rather than the means by which that oppression can be overcome.

The non-cooperation policy of the Rural Workers' Union in Xapuri grew partly out of the union's own experience with INCRA during the previous decade, but was also a product of an educational programme organised by the Centre for Documentation and Research in Amazonia (CEDOP). In late 1981, CEDOP approached and received finance from OXFAM (England) to teach literacy and numeracy to a group of *posseiros* who lived in the interior of Xapuri. After three months in Rio Branco, this group returned to their tapping posts and began to teach their family and neighbours. Lessons were loosely based on the Paulo Freire Method and centred on issues relating to the historical and contemporary experiences of Acrean rubber-tappers. Within twelve months of completing the construction of the first school, five more had been built. During 1983, as a direct result of the high political content of these lessons, producer cooperatives were set up at two of these rural schools. The cooperatives which received loans from the Church's Land Commission (CPT) are loosely held together units in which approximately fifteen individual producer families combine at certain times of the year communally to sell their rubber product and buy essential supplies. This action enables them to by-pass local middlemen and deal directly with the wholesalers of the state capital and so obtain a better price for their rubber and supplies at a lower cost.

The principal advantages of this form of joint action are threefold. Each family benefits in direct proportion to its level of productivity, thus making it more advantageous to have as many rubber trails open as family labour will allow. It also opens up the possibility of employing non-family labour and the sharecropper, *meieiro*, has recently made an appearance in this area. Secondly, each family benefits from the joint sale of rubber and purchase of supplies. Interviewees stated that goods were purchased in Rio Branco at approximately 60 per cent of the price that was previously paid to merchants and river traders in Xapuri, while rubber sold for 38 per cent more. Thirdly, the cooperatives were able to create reserve funds by putting a levy of 10 per cent on the price of goods bought. This was seen as vital for the continuation of the movement in terms of coming to a member's aid at times of crisis, but also in terms of its expansion with the purchase of pack animals and motorised river transport.

Another, though less tangible, product is the experience that this form of organisation has given its members, an experience that has been incorporated into the fund of knowledge of the whole community. In practical terms, it motivated the local STR to formulate new policies based upon it. First, in terms of its own organisation, in early 1984 the STR leadership began the task of persuading its members to take a more active role in union affairs and discussions are under way concerning the allocation of official positions on a rota basis. Secondly, the STR is promoting the formation of similar cooperatives as the principal means of loosening the ties between *posseiros* and the river traders who now occupy a similar position to the former estate owners in relation to the monopoly control over the purchase of rubber and sale of merchandise in the interior of the municipality.

Towards the end of 1983, just at the time the Xapuri cooperatives were being formed, a serious problem arose for members of several cooperatives in the Brasiléia area. Without forewarning SUDHEVEA, the local union branch, or the cooperatives themselves, INCRA reneged on its agreement and went ahead with the settlement of PADs Santa Quitéria and Quixadá. Much of the 70,000 hectares involved was already occupied by *posseiros* who had already constructed *mini-usinas* with BASA loans and had begun production. When confronted by a delegation of *posseiros*, INCRA stated that it was forced to settle the land because of the large numbers of out-of-state migrants that had arrived in Acre during the summer of that year. In addition, even though the *posses* of tappers varied between 200 and 500 hectares because of the extensive nature of extractivist production, INCRA stated that, because they were resident within the boundaries of a PAD, they would be reduced to the INCRA norm of 100 hectares and the remainder re-allocated to new colonists.

Here, INCRA was behaving in both an arbitrary and illegal manner by ignoring the rights of *posse* of the resident population. INCRA's action posed a serious threat to the *posseiros* and for the future of rubber production in the municipality. First, if the landholding of the *posseiro* was limited to 100 hectares, then this would considerably reduce his ability to earn a living by tapping rubber. Secondly, it would also prejudice his ability to put his children to work once they reached adulthood. As has already been noted, one of the factors which motivated SUDHEVEA to operate in this municipality was the large number of deactivated rubber trails which could be reactivated with the use of family labour. Thirdly, settlement by colonists necessarily signified further deforestation in an area that had already borne the brunt of this process. Not only would colonists' subsistence plots cut right through existing rubber trails and disrupt production, they would also reduce the rubber tree capital of the municipality and severely limit possible future production. This placed in jeopardy the whole concept of the *mini-usina* as a viable proposition.

Throughout the month of January 1984, tripartite discussions were held between the *posseiros*, SUDHEVEA and INCRA. The latter argued that the fault lay with the *posseiros* themselves in that they had not taken advantage of SUDHEVEA's PROBOR programmes and created small rubber plantations on their land. The *posseiros* countered this by pointing out that not only had credit not been forthcoming for small producers under these programmes, but also that once credit became a possibility under PROBORZINHO the agency charged with the responsibility of providing rubber tree saplings, COLONACRE, had been unable to meet even a small fraction of the demand. A compromise was reached whereby INCRA provisionally agreed to colonise 20,000 hectares of the PADs and freeze the settlement of the remaining 50,000 hectares for a period of between 8 and 12 years. During this time, the *posseiros* could create small 3-hectare rubber plantations and have them producing before the next plase of the settlement of the PADs. Finance for the plantations was to be provided by BASA at an annual rate of 35 per cent. It is interesting to note here that the finance obtained by large estate owners during the same period for the setting up of similar small rubber plantations under PROBORS 1 and 2 was just 12 per cent.

The next four years were again marked by problems related to the apparent contradictory functioning of state agencies. Firstly, COBAL ceased to function over a great portion of the state. By early 1986, only two COBAL agencies were still operational: the mobile unit in Rio Branco itself and the unit which supplies the rubber estate owners in Tarauacá. The *posseiro* without transport was now once again forced to exchange his product for supplies with the river trader at prices set by the latter. Secondly, Brasiléia's *posseiros* found that they could not meet the payments of the credit they had recieved from BASA. This was especially problematic after the announcement of the *Plano Cruzado* in 1986 when the market price for rubber was frozen without a corresponding freezing of the price of supplies bought from the river trader and small town merchant. Thirdly, SUDHEVEA announced that it was going to cease operating in the state and by early 1987 it had closed its head office in Rio Branco as well as its outreach units in the interior. This completed the dismantling of the accord reached by SUDHEVEA and Brasiléia's *posseiros* only five years earlier.

A further blow came in mid-1987 when the Acrean office of the Brasilian Forestry Commission (IBDF) announced a ban on all deforestation in the state. This specifically aimed to bring a halt to large-scale destruction of the natural environment although, given the IBDF's lack of manpower and resources, it is difficult to imagine how it would implement its policy. However, it contained no provision for the small-scale, 2–3-hectare, burning of subsistence plots by *posseiros*. This placed *posseiros* in the impossible position of having to break the law and face possible heavy fines or else starve on their tapping posts. It was at this point that the *posseiros* of Brasiléia changed their position and united with those of Xapuri and the rest of the state behind a single banner, that of the fight for the creation of extractive reserves.

The concept of an extractive reserve emerged in the struggle that took place at both the local and the national levels throughout this period. It represented a further step on the road to self-determination for the rural population of Acre which began with the *empates* of the early and mid-1970s and the cooperatives of the early 1980s. In May 1985, during the fourth National Congress of Brazilian Rural Workers, President Sarney announced his proposal for the elaboration of the first National Plan of Agrarian Reform (PNRA). The principal stated objective of the latter was the implementation of the Land Statute of 1964 – that is, the eradication of *latifúndios* and *minifúndios* and their replacement with *rural enterprises*: small agricultural units greater than the INCRA-defined regional rural module, but smaller than 600 times that module (Brasil, 1973). By announcing the PNRA at a rural workers' congress, something quite unheard of previously, the regime was seeking to diffuse tension in the rural areas where petty commodity production prevailed, while promoting the expansion of capitalist agriculture into those very areas. As the Ministry of Agrarian Reform's report (Brasil, 1986a) notes, rather than declining, the violence related to land disputes increased following the proposal of the PNRA. During the twelve months prior to its publication, 261 persons died violently in incidents related to land disputes. Of these, less than 11 per cent were ranchers or their gunmen.

Given the climate of political upheaval following the demise of the military regime, the generalised lack of confidence in state agencies charged with implementing agrarian reform and the high incidence of rural violence, the *posseiros* of Acre conceived of a national congress of rubber-tappers at which they could air their grievances and discuss possibilities for joint action. At the very moment that the PNRA was being enacted in Brasília (Brasil, 1985a) local meetings were held in Rondônia, Acre and Amazonas to select representatives and propose an agenda and the first National Congress of Rubber-Tappers met in Brasilia in mid-October. It was decided to form a National Council of Rubber-Tappers (CNS) which could act as a pressure group so as to influence the government to formulate a specific policy on extractivism in the region, rather than allowing things to happen by default as in the past. In particular, they wished to ensure that the regional agrarian reform plans would take into account the peculiar nature of extractivism in general and the specificity of Amazonian non-plantation natural rubber production. In addition, it would publicise the current conditions of existence of Amazonian rubber-tappers and *posseiros* and what the new reforms would mean in terms of the disintegration of this class of worker.

The first act of the CNS was to propose the creation of extractive reserves in all the Amazonian rubber producing states. These were to be based along the lines of indigenous reserves and essentially represented an attempt to preserve the status quo by delineating areas in which further road building and deforestation would be prohibited, as would further ranching by implication. The rationale here was founded upon a number of ecological, social and economic imperatives. First, current deforestation was taking place without sufficient controls being placed upon the timber companies and ranchers to ensure that forest loss was kept to a minimum, was restricted to areas not already occupied by extractive workers and generated revenue for the states concerned. Secondly, according to the 1980 census of the IBGE, 304,023 persons were engaged in extractivism in legal Amazonia. Given an average family size of five persons, this signifies that over 1,520,000 persons were dependent upon extractivism for their livelihood, approximately 32 per cent of the region's total population. If these communities were not to be broken up and forced into the urban shanty towns, then a halt had to be called to the depredation of the environment. Thirdly, in 1980, Amazonia's extractivist production was valued at US$75 million dollars (Brasil, 1983). A large proportion of this revenue would be lost if such deforestation continued. The federal government's apparent discouragement of extractivist production was seen as especially problematic given that Brazil has steadily consumed more natural rubber than it has produced. In 1980, for example, Brazil consumed 81,059 tonnes of natural rubber while it produced only 27,813, or 34.3 per cent (Brasil, 1981b).

In December 1985, the preliminary version of the Regional Plan for Agrarian Reform in Acre was drawn up and published by INCRA (Brasil, 1985b). The regional plan's urban bias is demonstrated by the fact that, rather than resolving the titling problem of existing *posseiros*, it continued to uphold the principle of settlement projects. During the first four years of the plan, a total of 212,196 hectares were forecast to be expropriated for the settlement of 7,300 families. Given the five million hectares that changed hands in the 1970s and the thousands of displaced persons already living in Rio Branco's shanty towns, as well as the large contingents of migrants spilling over into Acre from Rondônia, the regional plan did not represent a threat to the existing agrarian structure. In response to this in early 1986, the *posseiros* of Acre instituted their own *de facto* agrarian reform and marked out their own extractive reserve in the municipality of Xapuri. This action was soon to be repeated by *posseiros* in the municipality of Aripuanã, Amazonas, who also marked out their own reserve. Although limited to approximately fifty families in each case, this represented an important step in the struggle.

In reaction to this and a measure of its not merely symbolic nature, almost as if it were an attempt to nip the movement in the bud, INCRA–Acre issued its own internal document (Brasil, 1986b) setting out its proposals for establishing 'Extractive Settlement Projects', *Projetos de Assentamento Extrativistas*. These were to be based along the lines of the PADs, but specifically concerned with the production of natural rubber and Brazil nuts. It is interesting to note that at the time of writing INCRA–Acre had proposed the setting up of such a project in the Acrean municipality of Sena Madureira on land that had previously been earmarked as a PAD and which has one of the least active Rural Workers' Unions in the state.

CONCLUSION

The occupation of Amazonia by agro-ranching and mineral enterprises was not only facilitated by the infrastructural developments of the military regime, but actively promoted through its use of fiscal and financial incentives. The resolution of the regime's twin problems of the occupation of its territorial space and the generation of export earnings through the implantation of 'modern' capital-intensive enterprises in the Amazon, not only ensured that the countryside would again become the locus of a violent struggle, but also that agrarian reform would continue to be the principal banner around which the rural population would unite. However, as has been shown, State-directed agrarian reform has been used as a delaying tactic to defuse social tension in rural areas while allowing the continued expansion of capitalist units into those very areas.

In terms of Acre, direct and indirect State intervention at the national level, mirrored by the actions of the local state government, eroded the power base of the traditional landed class in the state to the extent that, when the 'ranching front' arrived, it found land available, cheap and occupied by a fragmented peasantry. The isolation of the area, the lack of publicity of the violence and the limited actions of the federal agency responsible for land questions, compelled a disunited peasantry to organise itself. Only then was it able to reverse the process under which it was being expelled from the land and forced to join the ranks of the urban unemployed and under-employed. The social and economic consequences of the rural exodus generated by this intervention were unforeseen by the Acrean administration, with tragic results. First, the shanty towns of the urban centres grew at an alarming rate without the provision of even basic services and with an urban labour market unable to absorb more than a fraction of this unskilled labour pool. Secondly, the migratory flow coupled with extensive deforestation caused a dramatic decrease in the quantity of rubber produced by the state and a consequent fall in its revenue, 80 per cent of which is now derived from federal sources.

The crisis that affected both the rural and urban areas became unmanageable locally, and two federal organisations, INCRA and SUDHEVEA, were empowered to resolve it. On the one hand, this demonstrated to *posseiros* that their struggle to maintain access to land was legitimate. On the other hand, however, bureaucratic failures and lack of political will on the part of INCRA, SUDHEVEA and other state agencies also proved to *posseiros* that, if their struggle was to succeed, then they must carry out their own agrarian reform, or at least defend the gains that they had already achieved. This was re-enforced by the State's rhetoric concerning agrarian reform under the *Nova República*. At the time of writing, almost two years after the regional plan for Acre was produced, no areas had been expropriated under its provisions. The state's land commission which had been charged with implementing its recommendations had neither met, nor had its members even been chosen.

Almost a quarter of a century has lapsed since the enactment of the Land Statute of 1964 whose express purpose was the eradication of inefficient forms of land occupation. Over 80 per cent of the productive land in Acre is still in the hands of *latifundiários*. This is not to suggest that reform of the agrarian structure has not taken place during this time. State intervention has produced two new antagonistic classes: capitalist *latifundiários* and an independent peasantry. Under traditional extractive relations, the class struggle between the estate owner and the rubber-tapper centred on the use of land, on control over the means of production. In the 1980s, following the dynamisation of the land market in Acre after the arrival of the 'ranching front', the new struggle became centred on the issue of whether the land was to be used or not – that is, on the role of land as either an object of labour or as an object of speculation. Thus, whereas traditional rural social relations in Acre were related to the appropriation of the labour power of one class by another, the new struggle concerns the actual exclusion of one group by the other from the land itself.

The notion of extractive reserves emerged as a vehicle for halting or at least delaying this process of exclusion from the land. However, whether or not it is a means of carrying the movement into the 1990s and onwards is a matter of conjecture. The battle lines are continuously being drawn and re-drawn. In economic terms, SUDHEVEA's strategy of locating the majority of its new rubber plantations outside of Amazonia in states much nearer to the industrial market suggests that the continued extraction of rubber in the Amazon will be short-lived. In social and political terms, the federal and state governments are still faced with the problem of large numbers of families without access to land and with no prospect of employment in the urban environment. The resolution of this problem depends as much, if not more, on political will than economic laws. Such political will was demonstrated by SUDHEVEA in Brasiléia although for only a short period of time. However, given the crisis management approach of previous regimes, it is likely that more weight will be given to overcoming the urban crisis than the rural and that landlessness will be treated as a function of urban unemployment and not a problem *per se*.

References

Acre (1980), *Anuário Estatístico Acreano*, Rio Branco: Estado do Acre.

Allegretti, M. H. (1987), *Reservas Extrativistas: Uma proposta de Desenvolvimento da Floresta Amazônica*, Curitiba, Paraná: Instituto de Estudos Amazônicos (mimeo).

Aquino, T. Vale de (1982), *Indios Kayinauaá: De Seringueiro Caboclo ao Peão Acreano*, Rio Branco: Empresa Gráfica Acreana.

Bakx, K. (1986), *Peasant Formation and Capitalist Development: The Case of Acre, South–West Amazonia*, unpublished Ph.D. thesis, University of Liverpool (January).

Bakx, K. (1987), 'Planning Agrarian Reform: Amazonian Settlement Projects, 1970–1986', *Development and Change*, 18, pp. 533–55.

Bakx, K. (1988), 'From Proletarian to Peasant: Rural Transformation in the State of Acre, 1870–1986', *Journal of Development Studies*, 24(2) pp. 141–60.

Brasil (1964), *Estatuto da Terra*, Campinas: Julex Edições.

Brasil (1973), *Urbanismo Rural*, Brasília: INCRA.

Brasil (1980a), *Censo Demográfico*, Rio de Janeiro: IBGE.

Brasil (1980b), *Censo Industrial*, Rio de Janeiro: IBGE.

Brasil (1981a), *Cadastro da Estrutura Agrária – Acre*, Rio Branco: INCRA.

Brasil (1981b), *Anuário Estatístico: Mercado Nacional*, Brasília: SUDHEVEA.

Brasil (1983), *Anuário Estatístico do Brasil*, Rio de Janeiro: IBGE.

Brasil (1985a), *Plano Nacional de Reforma Agrária*, Rio de Janeiro: Livraria Freitas Bastos.

Brasil (1985b), *Plano Regional de Reforma Agrária – Acre*, Rio Branco: MIRAD/INCRA.

Brasil (1986a), *Conflitos de Terra*, Brasília: MIRAD.

Brasil (1986b), *Reservas Extrativistas: Minuta de Proposta com Vista a Integração dos Órgões Participantes no Desenvolvimento do Projeto*, Rio Branco: INCRA (mimeo).

CEDEPLAR (1979), *Migrações Internas na Região Norte: O caso do Acre*, Belo Horizonte: Universidade Federal de Minas Gerais.

Foweraker, J. (1981), *A Luta pela Terra: A Economia Política da Fronteira Pioneira no Brasil de 1930 aos Dias Atuais*, Rio de Janeiro: Zahar.

Guerra, Antonio Teixeira (1955), *Estudo Geográfico do território do Acre*, Rio de Janeiro: IBGE.

Mahar, D. (1979), *Frontier Development Policy in Brazil: A Study of Amazonia*, New York: Praeger.

Mahar, D. (1981), *Brazil: Integrated Development of the Northwest Frontier*, Washington: World Bank.

Mesquita, G. (1977), 'Evidence Given to the Commission on Agriculture of the Chamber of Deputies', *Diário do Congresso Nacional*, Seção 1, (December) Brasília.

Schwartzman, S. (1987), *Extractivist Production in the Amazon and the Rubber Tappers' Movement*, paper presented to *Forests, Habitats and Resources: A Conference in World Environmental History*, 30 April 1987, Duke University, Durham, North Carolina.

4 Private and Public Colonisation Schemes in Amazonia

Emilio F. Moran

INTRODUCTION

Colonisation schemes in the Brazilian Amazon have been a major force behind social and environmental change in the region since the 1970s (Moran, 1981; Smith, 1982; Fearnside, 1985). Although small-scale colonisation had been taking place for several decades before then, except for the construction of the Belém–Brasilia highway and its associated spontaneous colonisation, their character had been small scale and their impact not felt at a regional scale (Tavares *et al.*, 1972).

Beginning with the Plan for National Integration (PIN) and its associated Transamazon Highway Colonisation Scheme, the Brazilian State began to take an active and commanding role in colonisation. Highways were built in record time and colonists were brought by jet plane, bus, and boat to the Amazon to occupy planned government communities built literally overnight (Kleinpenning, 1977).

The Geisel administration that came into power in 1974 announced, just as swiftly, that the 1971 Transamazon Scheme had been a failure. It announced a new policy that reorientated public sector efforts at regional development away from government-directed small-scale farming towards private colonisation and large-scale development.

This chapter provides an assessment of the nature of both kinds of settlement schemes, their relative performance, and suggests avenues that might be followed in the future based upon these past experiences. Emphasis will be placed upon the failure to carry out environmental assessments in advance of settlement, a lack of fine-grained detail about local habitats, the absence of institutional capacity and coordination in adjusting the programme to field realities, and on a misperception of the characteristics of migrants who come to forested frontiers in the early stages of settlement.

The discussion will focus on the Transamazon Colonisation Scheme in examining public colonisation efforts because this Scheme was touted as the 'litmus test' of government-directed settlement in South America (Nelson, 1973). As I have shown in *Developing the Amazon* (Moran, 1981), the Scheme was riddled with many of the same problems that have characterised public colonisation efforts elsewhere. For the discussion of private colonisation, I will focus on the Tucumã Colonisation Scheme, a project executed by a top Brazilian multinational corporation, and expected to deal more effectively with past administrative oversights and seeking to attract more middle-class farmers. Both the Transamazon and Tucumã can be taken to represent 'best cases' from the point of view of planning but, as we shall see, their performance falls well within the range of other projects, whether public or private.

THE TRANSAMAZON COLONISATION SCHEME

When Brazil undertook the construction of the Transamazon Highway and its associated settlement of small-scale farmers along its margins in 1971 it had a grand vision: an undertaking of comparable scale to placing a man on the moon. At the time Brazil was the world's darling – experiencing the 'economic miracle' of 11 per cent rates of real growth and low rates of inflation (c. 15 per cent). The plan was indeed grand: to construct 5,400 kilometres of highways through largely unknown territory, to settle 100,000 families within the first five years along a 20 km-wide strip of the highway, to produce a surplus of rice and beans to replace the production of Rio Grande do Sul which had been diverted to soybean production for export, and to reduce pressure for agrarian reform in areas already occupied (Ministério da Agricultura, 1972).

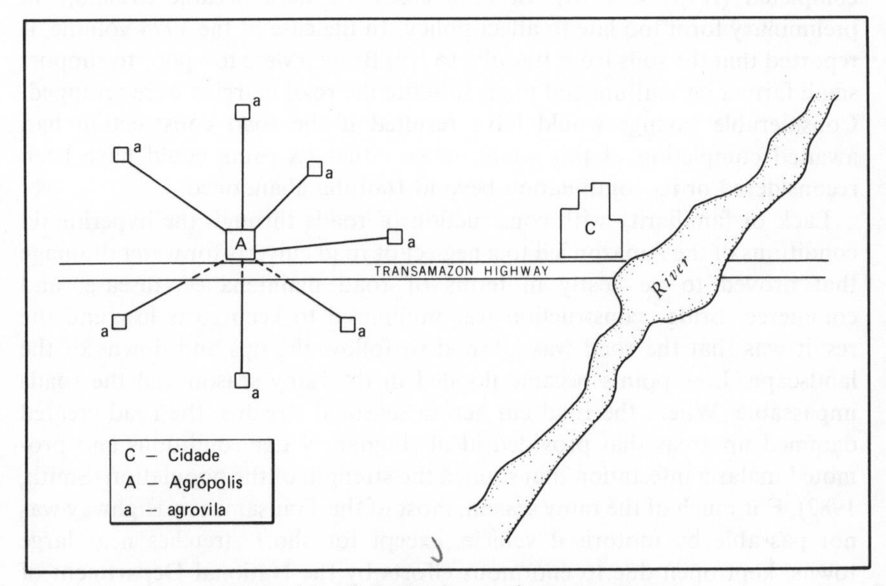

Figure 4.1 Colonisation settlement pattern

Such a grand task could not be accomplished by hesitation or partial commitment. Given the urgency and short-time frame envisioned, government set out to recruit potential settlers throughout the country, to build entire agricultural communities at regular intervals along the highway, to build a system of hierarchical communities with predefined missions and size (see Figure 4.1), to provide inputs at favourable rates to guarantee that the farmers would not be limited by access to necessary credit and technology, and to reduce risk to the small farmer by guaranteeing purchase of the output at least at the cost of production. With all these considerations in mind, could anything go wrong?

Knowledge of the Environment

One of the persistent problems that has plagued settlement of farmers in the humid tropics has been the absence of environmental surveys before project implementation (Crist and Nissly, 1973). This problem recurred in the Transamazon highway. While the government commissioned studies as soon as it announced the PIN, it did not allow sufficient time for the results of these studies to affect project design and implementation. For example, the first approximation of the soils along the Transamazon Highway up to Itaituba (see Figure 4.2) did not appear until 1972 by which time the road had already reached Itaituba (Falesi, 1972). The report on the soils from Itaituba to Rio Branco did not appear until 1974 when the road was nearly completed (IPEAN, 1974). In both cases the data became available in preliminary form too late to affect policy. In the case of the 1974 volume, it reported that the soils from Itaituba to Rio Branco were too poor to support small farmer agriculture and plans to settle the road margins were scrapped. Considerable savings would have resulted if the road construction had awaited completion of this study, when either its route could have been reconsidered or its continuation beyond Itaituba abandoned.

Lack of familiarity with construction of roads through the hyperhumid conditions of the Amazon led to a neglect of road culverts for water drainage that proved to be costly in terms of road maintenance, disease, and commerce. Bridge construction was minimised to keep costs low and the result was that the road was allowed to follow the ups and downs of the landscape. Low points became flooded in the rainy season and the roads impassable. Where the road cut across seasonal streams, the road created dammed up areas that provided ideal stagnant water conditions and promoted malaria infestation that sapped the strength of the population (Smith, 1982). For much of the rainy season, most of the Transamazon Highway was not passable by motorised vehicle, except for short stretches near large towns, kept open due to enormous efforts by the National Department of Roads (DNER).

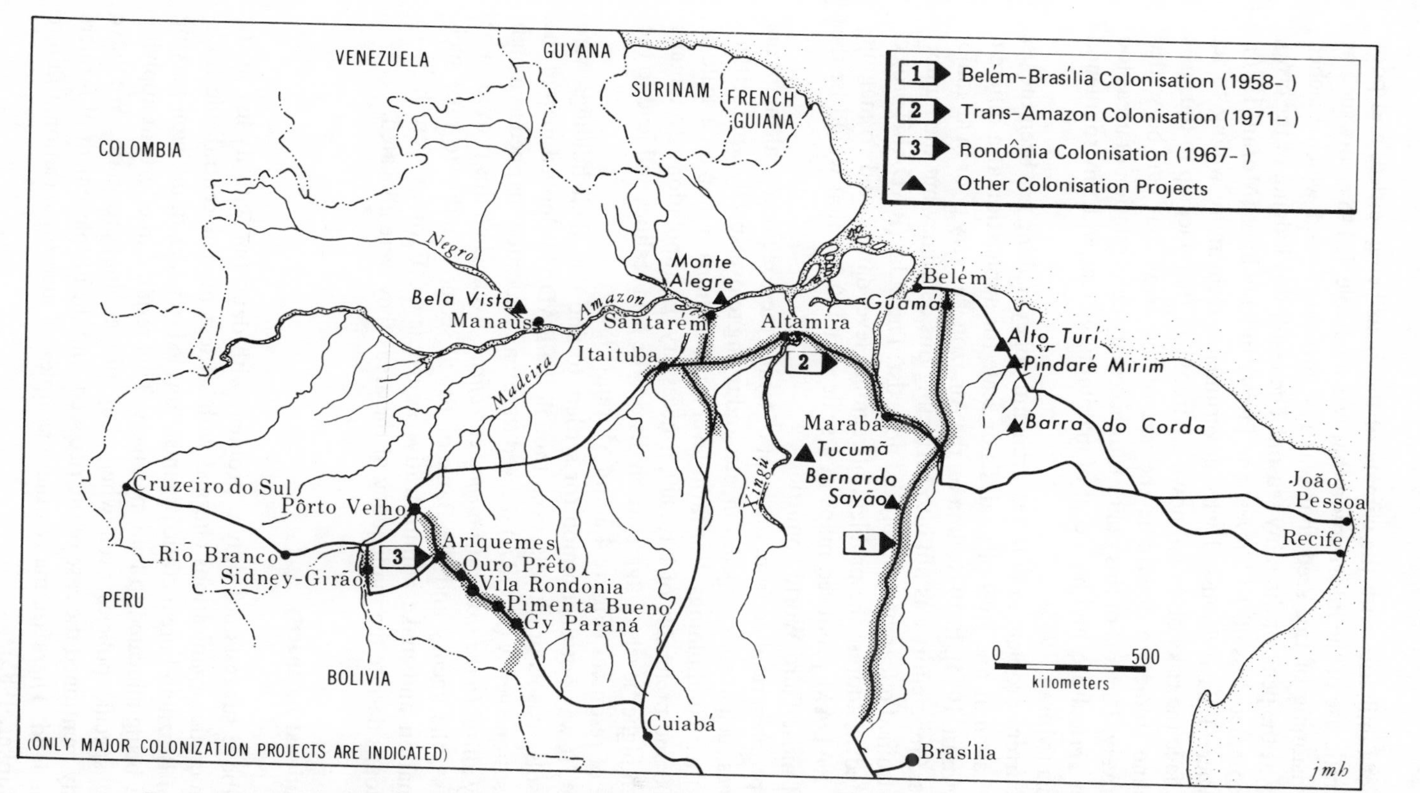

Figure 4.2 Roads and colonisation in the Brazilian Amazon, only major colonisation projects are indicated
Source: Moran (1984, p. 274)

Lack of familiarity with the climate of the areas being settled, and poor bureaucratic use of the information that was available, led to errors such as the provisioning of rice seed that was fast growing – and which reached maturity at the peak of the rainy season! The result was lodging of the crop, almost complete loss of the harvest and farmer indebtedness (Moran, 1976).

Elsewhere I have discussed why governments seem more willing to let farmers face the risks of an unknown environment than face up to the costs of baseline studies to determine the size of the areas that can be settled productively. The cost of this policy is unnecessary deforestation of areas too poor for agriculture, and high cost of maintaining areas with environmental problems (Moran, 1983).

A recurrent problem in both research and policy-making has been the use of data taken at one level of analysis or scale for application at a different level (Moran, 1984). In 1971, data on the Amazon was very poor in all major respects. Great effort was put into correcting this deficiency but the number of well-trained researchers simply did not exist. The radar (RADAM) project attempted to address the problem of regional level data. It did so magnificently. RADAM produced maps at a scale of 1:100,000 that would be the envy of most Third World countries.

The problem resulted from the error of taking these regional scale maps on resources such as soils, vegetation, etc. and trying to overspecify what settlers did on their landholdings. For farm-level investigations, a map on a scale of 1:20,000 and preferably of 1:10,000, is necessary. Extrapolations from maps at much higher levels of aggregation distort the field realities and lead to the misuse of resources (Figure 4.3 and Moran, 1984).

Thus, it was not uncommon for settlers to find that their holdings were spodosols, oxisols or ultisols, even though the RADAM-derived map at the colonisation agency (INCRA) showed the area in question mapped as a high quality alfisol (*terra roxa estruturada eutrófica*). Even when this proved to be the case, the agency and the Bank of Brazil seemed unable to take such information and make administrative adjustments. Farmers were held to their loans despite the inadequacy of the areas they were allocated.

Institutional Incapacity

The above suggests not only a problem with data, but with its use in the bureaucratic organisations charged with settlement. The centralising tradition of Brazilian bureaucratic organisations led to a lack of decision-making power being allocated to field managers. This resulted in efforts at applying highly specific policies even when they proved inappropriate. We have already mentioned the case of the rice seed, and the misplacement of farmers on the land. There are many other examples of such organisational failure (cf. Bunker, 1985).

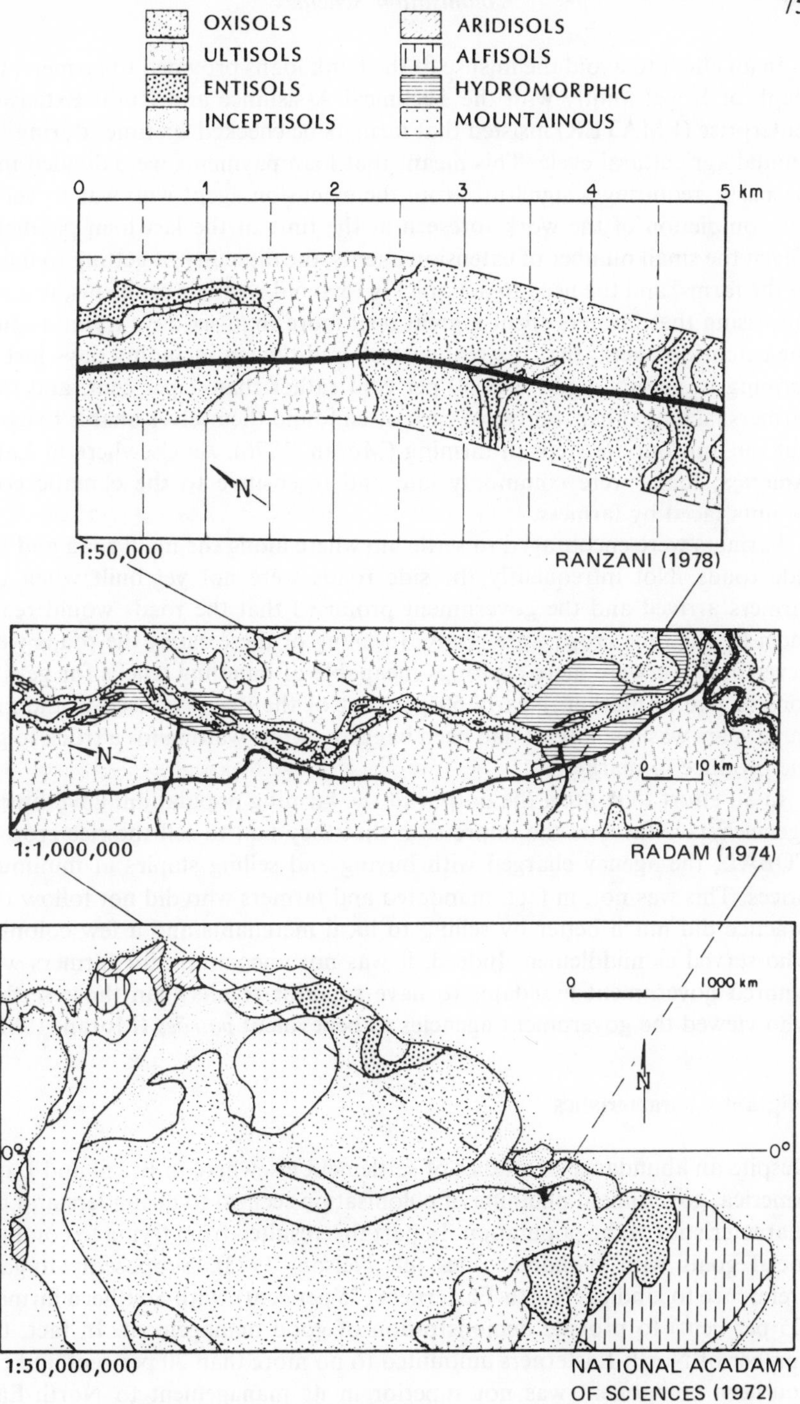

Figure 4.3 Changing soil data as a function of scale

In an effort to avoid the misuse of the bank loans provided to farmers, the Bank of Brazil jointly with the Technical Assistance and Rural Extension Enterprise (EMATER) insisted that farmers be checked six times during the annual agricultural cycle. This meant that loan payments were divided into six parts, requiring a signature from the extension agent who was to verify the completion of the work foreseen at the time of the last loan payment. Given the small number of extension agents, the difficulty of getting to many of the farms, and the need to act quickly to avoid weather problems, it is not surprising that farmers spent a great deal more than six days trying to find the extension agent, that the real cost of credit when including days lost to farming in getting credit was 50 per cent rather than 7 per cent, and that farmers lacking other sources of cash often found themselves unable to carry out the appropriate steps in farming (Moran, 1976). As elsewhere in Latin America, loans were commonly late and insensitive to the climatic constraints faced by farmers.

Farmers were encouraged to settle anywhere along the main road and the side roads. Not infrequently the side roads were not yet built when the farmers arrived and the government promised that the roads would reach them by the time crops were ready. In many of these cases, the roads were never built, or were mere clearings incapable of handling trucks for getting tons of rice to market. Nearly 50 per cent of the rice produced in the first three years was lost after it had been bagged – while waiting for a truck to get the produce to market.

In an effort to control the output of the farming population, government agencies gave farmers the impression that they had to sell the rice crop to COBAL, the agency charged with buying and selling staples at minimum prices. This was not, in fact, mandated and farmers who did not follow this practice did much better by selling to local merchants and a few colonists who served as middlemen. Indeed, it was not uncommon for farmers who ignored government mandates to have superior results than those settlers who viewed the government agencies as benevolent *patrões* (Moran, 1981).

Migrant Characteristics

Despite an abundance of studies of settlement elsewhere in Brazil and Latin America, government planning of colonisation seemed to misunderstand the characteristics of the population. In part this resulted from trying to combine social goals with economic ones (cf. Nelson, 1973). Supposedly it gave preference to landless North-Easterners (75 per cent), with Southern farmers (25 per cent) to provide demonstration of good management. In fact, the number of North-Easterners amounted to no more than 30 per cent and the Southern component was not superior in its management to North-Easterners, Amazonians, or Central Brazil immigrants (Moran, 1979).

As in frontiers elsewhere (Moran, 1988), the Transamazon farmers were a varied lot. Rather than the stereotyped landless North-Easterners from rural areas, the population that came was highly heterogeneous, and included nearly 30 per cent of non-rural people with limited experience in agriculture but with important skills to bring the frontier. However, their contribution proved to be often non-agricultural and after failed efforts at farming, they cashed in on the increased value of land a decade later and returned to their original craft occupations, but now providing their skills in frontier towns. Table 4.1 summarises the economic performance of farmers in one agricultural community. Note the importance of having owned or managed a farm before and the role of residential stability in the agricultural performance of the population.

A presumption underlying much of the government scheme was that the real constraint to small farmers was lack of land and that making land available would resolve most other problems. This naive view was not confirmed by the Transamazon experience. Land is an empty gift if farmers are not given sufficient time to adapt to their new situation. Human cultural adaptation is a slow and gradual process that requires not less than 10 years, more in some cases. Human behaviour needs to test past procedures to see if they work in new environmental and socio-economic circumstances. If they do not work, time is necessary to experiment with alternatives that do not put the survival and persistence of the households at risk.

In the Transamazon only 21 per cent of farmers had previous experience

Table 4.1 Economic variation among farmers in the Transamazon settlement scheme

	Previous owners		*Non-owners*	
	Urban experience	*Non-urban*	*Urban experience*	*Non-urban*
Initial capital[1]	571	96	242	20
Average debt	5,391	1,785	949	749
Non-farm income	6,857	800	2,609	642
Farm income	714	1,256	536	381
Gross income	7,571	2,056	3,145	1,023
Expenses	3,428	1,157	1,312	1,153
Net income	4,142	900	1,833	−129
Liquid assets	7,214	1,142	52	62

Note:
1. Figures are on an annual basis, except for initial capital and average debt owed the Banco do Brasil and the Colonisation Agency (INCRA), and are listed in 1974 dollars. The data are based on a 50 per cent sample of one community intensively studied. This community was not found to differ significantly from data collected in other communities in the scheme, using region of origin as the criteria for random sampling.
Source: Data adapted from Moran (1981).

with bank credit. Time is needed to learn to use such credit productively, and not see it as a source of easy access to consumer goods – or as a way of maximizing the forested area cleared in order to convert a largely free good into a marketable one. As is true in frontiers elsewhere (Moran, 1988) first-decade settlers tend to clear far more land than they can reasonably manage, and to become excessively indebted in an effort to maximise the land area they can claim.

FROM GOVERNMENT TO PRIVATE COLONISATION

When the Geisel administration reorientated its efforts at Amazonian development from government-directed to privately-sponsored colonisation it hoped not only to reduce the costs to government, but also to see a superior form of management less subject to the inefficiencies of centralised bureaucracies.

The first effort at this new policy was called POLAMAZONIA, which focused on large-scale agricultural, mining, hydroelectric, and forestry projects. Private colonisation took off, too. 25 such projects came into being in the second half of the 1970s (Schmink, 1981). Private colonisation companies have shown a preference for settlement in southern Pará and northern Mato Grosso.

This form of land occupation is not new. A considerable literature was produced in the 1950s on the success of the Northern Paraná Land Company in developing coffee lands in Paraná (cf. Margolis, 1973). Based upon this past research, the limited literature emerging on private colonisation in Amazonia seems intent on not seeing the similarities that it has with other forms of colonisation. Butler (1985) suggests that the farmers attracted to these projects are different from those going to government-directed projects, as are the communities that develop within them.

This difference is alleged to derive from the fact that private firms are in the business of selling land, and that they seek to attract settlers who will be able to pay for the costs of infrastructural development and the margin of profit projected by the company. In the rest of this discussion, I will focus on the Tucumã project in Pará, which was studied by Butler (1985) and Moran (1987). See Figure 4.2 for the location of the project in Amazonia.

Unlike government-directed colonisation, private firms' chief criterion for selling land is the ability of a farmer to pay for it (Butler, 1985, p. 17). The question arises: Why would anyone buy land in a privately-run colonisation project,when land was available free from the public domain? One important reason seems to be that some farmers in southern Brazil have read about land violence between immigrants and squatters, between immigrants and indians, and of the difficulty of obtaining clear title to land (Schmink, 1982). The farmers in Tucumã claimed that they preferred to pay for the land because in that way they avoided those potential problems. Private projects are thus likely to attract farmers committed to small farm operations, rather than take the risks involved in making larger land claims in public domain lands.

The Construtora Andrade Gutiérrez obtained 400,000 hectares of land from INCRA in public auction for US$0.87 per hectare but with the obligation of building infrastructure and liberating the project within 6 years (Butler, 1985, p. 41). As would be expected of any private undertaking, the Company attempted to restrict access to its territory with security personnel. Nevertheless, because of the extent of the area and the limited number of security guards, *grileiros* were still able to lay claims to land and force the Company to purchase their claims in order to liberate the contested land for sale to incoming colonists (Butler, 1985, p. 44).

The problem of security got out of hand with the discovery of gold within the project area in 1981. Since the subsoil belongs to the State, according to Brazilian law, miners claimed that the Company had no right to restrict access to these deposits. Earlier clashes at another private colonisation project, at Alta Floresta, served to remind the company of the potential for violence and the miners were allowed to return to the gold find. A boom town, Ourilândia, developed just outside the project and it grew even larger than Tucumã itself. Let us now turn to an assessment of the private project's performance.

Knowledge of the Environment

Land occupation generally proceeded outwards from the immediate area of the town. This may have been a result of the generally high quality of the soils near town, the advantage of proximity to town markets, and the affordability of smaller lots. The latter factor permitted these holders to pay for these lots in one instalment and avoid monetary correction faced by those who bought larger lots and tried to pay them over two years. An important factor for some farmers was the lower price per hectare of land further away from town, permitting the purchase of more land for the same amount of capital – thereby overcoming the *minifúndio* problem that brought them to the area.

Most farmers interviewed indicated that proximity of water and ease of access were the most important factors they considered in choosing their land's location. Few recognised the vegetation on their land. More value was placed on the colour of the soil, with red being preferred, rather than on other indicators of fertility. After two years hardly anyone seemed to have acquired any notable knowledge of local soils – not surprisingly given the absence of Amazonian *caboclos* in the project areas (Moran, 1977, 1981). Without a local population to serve as experts, the newcomers were totally dependent upon agronomists, which numbered *one* in the project area. No one had considered turning to the nearby Kayapó indigenous population as a source of knowledge about the area's resources and appropriate forms of agriculture (cf. Posey, 1982).

As in the Transamazon highway project, the Tucumã project worked with data that was at a scale of aggregation inappropriate for farmer decisions. The town of Tucumã was built on superb alfisols, while only 2 kilometres away were poorer soils that would have permitted the area on which the town was built to become farmland. The maps used were RADAM's at a scale of 1:100,000 which, as we have seen earlier in this chapter and elsewhere (Moran, 1984), are inappropriate for farm management. Climatologic data was sorely lacking and farmers experienced very different rainfall seasons in their first two years on the project. Despite the availability of many studies that suggest the inappropriateness of mechanical land clearing (for example, Seubert *et al.*, 1977), most farmers on arrival cleared the land with bulldozers at a cost ten times that of manual clearing and obtained lower yields than if they had cleared manually.

After the first year or two most recognised the error of having cleared mechanically in both economic and agronomic terms. Farmers pointed out to me the size of corn plants in areas cleared mechanically in contrast to the larger plants and ears of corn in areas cleared manually in the second year. Those intending to clear land for the second year or third were planning smaller areas, and most of the new clearings would involve removal of secondary growth from the first year, since few of the newcomers had been able to keep up with the vigour of secondary succession in the generally large areas cleared initially.

The emphasis of settlers on clearing forested land has two primary causes: on the one hand, settlers can convert a generally inexpensive good into a valuable and marketable commodity by clearing the land of forest, since this gives it greater market value to potential buyers; on the other, bank credit at the favourable rate of 35 per cent (in an economy with inflation in excess of 200 per cent) constituted a subsidy that encouraged land clearance beyond the ability to manage it. The highest correlation I found among the variables used in the Tucumã study was between the amount of bank financing and area cleared ($r = 0.70$ $p = 0.0001$). Thus, although the credit was to cover clearing, planting, weeding, and harvesting, a disproportionate amount of the total credit was used to maximise the area cleared.

While they came to recognise the error of mechanical clearing within a year or two, problems remained. 91 per cent of farmers interviewed did not recognise differences in agricultural management between the Amazon and southern Brazil. Most of the farmers associated the dense vegetation with high soil fertility, an error often noted in the literature (see Jordan, 1982). Nor did they recognise the need to use fertilisers in the soils of the region for at least 20 or 30 years – opening up the potential for soil mining and degradation in the years ahead.

Soils in the area were highly diverse as the ranges in soil samples taken suggest (Table 4.2). This diversity was not acknowledged, since the experimental work was sited on the town's superb alfisols. Only 4 of the 23 soil samples taken did not have some nutrient in sufficiently short supply to cause concern. Phosphorus and potassium were low in one-third to one-half of the samples (Table 4.2).

Institutional Incapacity

While the Company had a much greater capacity for responding to colonists than did government bureaucracies, it suffered from the economic downturn in Brazil after 1981. Staff was cut by one-half on the project to reduce costs, and experimental work initiated in the first year to develop adaptive cropping for the project was largely abandoned after one year. Only one agronomist remained as an extension officer, with one assistant to serve the farming population. Like the Transamazon project, the staff was willing to help, but it was presented with an implementation task beyond the normal means of any person or institutional capacity.

Unlike the Transamazon, the roads and bridges built by the Company were of excellent quality and the roads were passable most of the time, even during the rainy season. The bulk of the malaria associated with the Tucumã project was a result of the open pit gold mining activities of prospectors, a few of whom were colonists.

Table 4.2 Soil analyses from a Carajás colonisation area

Property	*Mean value*	*Range*	*N*
pH	5.76	4.20– 7.30	23
% clay	19.90	3.69–43.18	23
Available P (in ppm)	4.50	2.00–10.00	23
Available K (meq/100cc)	0.20	0.08– 0.46	23
Available Ca (meq/100cc	4.14	0.49–12.72	23
Available Mg (meq/100cc)	0.95	0.30– 2.60	23
Available Al (meq/100cc)	0.38	0.00– 4.43	23
Available ECEC (meq/100cc)	5.66	1.86–15.49	23

Notes:
pH: soil acidity
P: phosphorous
K: potassium
Ca: calcium
Mg: magnesium
Exch. Al: exchangeable aluminium
ECEC: effective cation exchange capacity
ppm: parts per million
meq:/100cc: milli-equivalents per hundred cubic centimetres.
Source: Moran (1984) soil sampling.

The Company tried to sell the land in an orderly manner by showing visiting farmers from southern Brazil only the next section away from those already sold. This was done on the principle that by keeping settlement dense the Company could more economically keep up the roads serving the farms. Farmers in some cases felt that this restricted their access to the best lands on the project. Given the lack of expertise about local soils noted above, it is unlikely that even open access to any land in the project would have resulted in significantly better soil identification by buyers.

Characteristics of the Migrants

As in settlements elsewhere, the colonists did not conform to the ideal profile envisioned by planners. While a majority did come from the states of Rio Grande do Sul and Santa Catarina, 43 per cent of those who bought land had never owned land before. This group included persons who came to the area to work for the company as engineers, draftsmen, and in services (e.g., pharmacists, restauranteurs, supermarket owners, accountants, and carpenters). Thus, like the Transamazon farmers, these settlers were far from homogenous and far from entirely competent to farm. Nevertheless, they provide important skills in a frontier society without which it probably could not develop over time.

Of the 602 lots sold by 1984, there were fewer than 60 settlers engaged in some farming (see Figure 4.4). The great majority of colonists came from Rio Grande do Sul, Santa Catarina, and less frequently from Paraná and Goiás. Most of them came from Southern counties characterised by *minifúndia* resulting from the traditional inheritance pattern of dividing land equally among all heirs. Over time, this had created holdings too small to farm effectively given the recent emphasis on soybean production for those areas of Brazil (IBRD, 1979). By contrast, those coming from Goiás and Paraná had responded to publicity and hearsay.

Real estate agents were the principal recruiters of the southern colonists, presenting lectures on the project in areas of origin and inviting them to sign up for a bus trip to look at the project. As soon as the rains eased at least one bus per week made the 5,000-kilometre trip. About 30 farmers came each week, of which the company said about two to five bought land during their visit.

From an income point of view, the settlers tend to have less favourable situations than those who choose not to come. Whereas those who went to the lectures averaged holdings of 29.7 hectares, those who came to look at the project owned 23.2 hectares, and those who actually bought land owned on average only 21.8 hectares. This decreasing scale of property suggests that only those in the more marginal positions in the area of origin are willing to risk a move at this time to a new area. Other indicators support this suggestion.

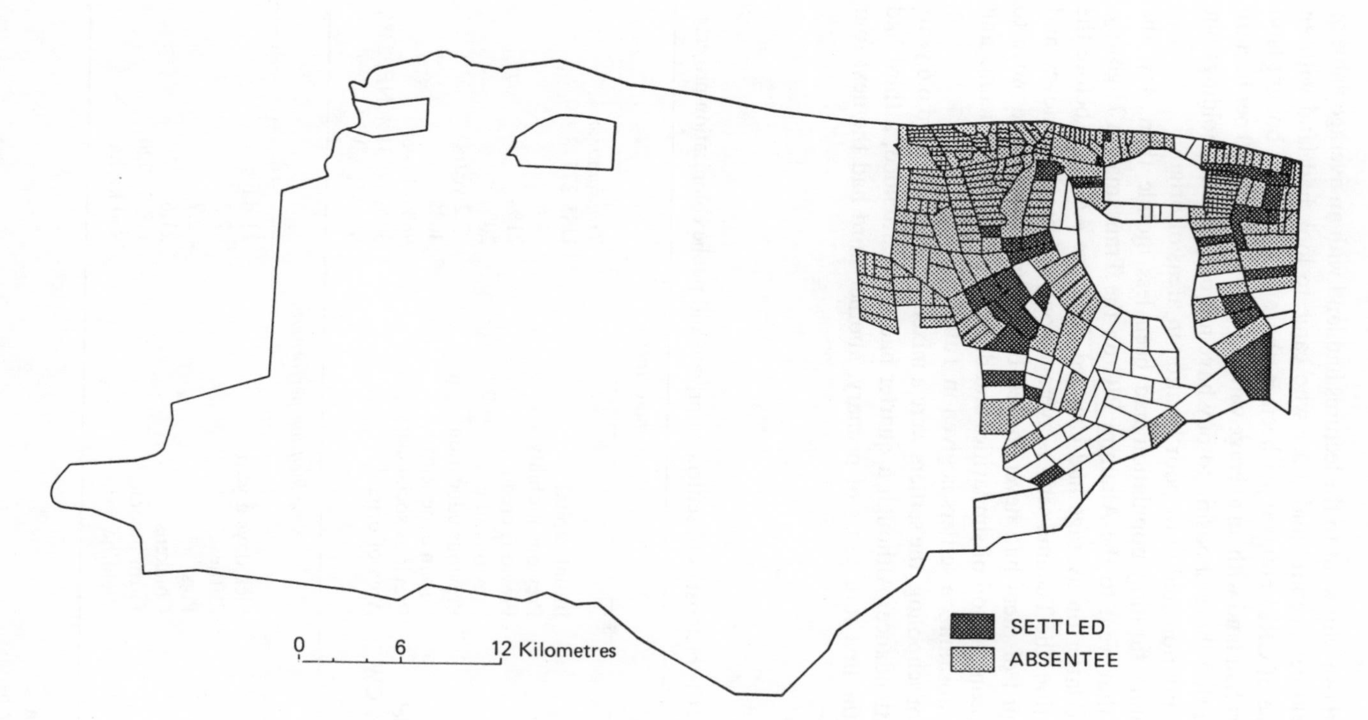

Figure 4.4 Distribution of plots: Tucumã Project

Whereas those who went to the lectures had land with an average value of CR$1,400,000 per hectare, and those who came to look had land with an average value of CR$1,600,000 per hectare, those who actually bought land in the project had land with an average value of only CR$750,000 per hectare – suggesting that those choosing to buy have not only smaller holdings but also land of inferior quality or poor location in areas of origin.

The Tucumã farming population had been less mobile than, say, the population that went to the Altamira PIC on the Transamazon Highway. Whereas the latter on average had changed residence 4.6 times before the latest migration, the Tucumã settlers averaged only 2.6 previous moves, and fully one-third of them had never moved before undertaking the move to Tucumã. A comparison of similarities and differences between Tucumã and Altamira Transamazon settlers is given in Table 4.3.

In terms of schooling, the settlers were a mixed lot who averaged 6.6 years of school attendance. Although a quarter had never attended, a third had completed the first four years of primary, another third had the next four

Table 4.3 Characteristics of settlers on private and public colonisation projects

	Basic indicators	
Tucumã		*Transamazon*
US $4,000	Initial capital	US$ 275
2.6	Previous mobility	4.6
73%	Previous credit	21%
57%	Previous title	20%
6 years	Previous education	3.2 years
75%	Urban experience	30%
97% 'middle'	Previous social class	96% 'lower'
80% S; 20% CW	Area of origin[1]	30% N; 31% NE; 23% S; 14% CW
	New location performance	
12 days	Sick days a year	11 days
5.4	Cattle	2.3
6.9	Pigs	2.7
42.5	Chickens	21.6
US $2,000	Farm income	US $ 720
900 kg/ha	Yield/ha	950 kg/ha

Note:
1. N: North
NE: North-East
S: South
CW: Central–West

years of school, and the remainder had 11 to 18 years of schooling. This latter group included the area's physicians, engineers, agronomists, and businessmen who bought land.

In terms of age and household composition, the settlers were relatively young, with an average age of 37 for the male head of household, and with relatively small families averaging 3 children each. This compares with the 5.7 national average and the even higher household size in the Transamazon Colonization Scheme due to the priority given in that project to large families in selection. Despite their relative youth, 75 per cent had previous experience with bank credit, as compared with only 21 per cent in the Transamazon. Because of the requirement that land had to be paid within the first two years, in 1984 67 per cent of settlers had already paid for their land. Those who bought their land outright, were in a much better position than those who scheduled their payments given the 230 per cent inflation to which their payments were adjusted.

The average initial capital of settlers was CR$ 5,237,000 with the range being as low as 0.28 million to as high as 38 million cruzeiros (US$175 to 23,750), with the modal capital being 3 million (US$1,875). Of the average capital of 5.2 million, 4.2 million on average was spent at the outset on the purchase of the land. The average landholding purchased was 86 hectares, with a range of 16 to 400 hectares. The company required a minimum down payment of 30 per cent and the rest had to be paid within either one or two years corrected for inflation. The average farmer spent 1.8 million cruzeiros in the first year clearing his land. Given the scarcity of labour, and unfamiliarity with clearing forest, most of the farmers opted for mechanical clearing – the cost of which was ten times the cost of manual clearing. The average size of the area cleared in the first year was 11.4 hectares – with the range being between 5 and 40 hectares.

Like the Transamazon farmers, they quickly became decapitalised and faced the problems associated with scarce capital. Like the Transamazon farmers, they cleared far more land than they could manage and asked for loans to cover most of the cost of land clearing – loans that they had trouble paying within a year of settlement. Like the Transamazon farmers, within three years they had reduced the total area being cultivated, and allowed forest regrowth to claim the other areas.

Yields obtained by farmers varied a great deal, reflecting both differences in soil quality and management skill. Most farmers on the very best soils are more experienced farmers, with over 20 years' experience as land owners, they have more labour available within the household, and combine this expertise with wage labour. By contrast, those on the medium to poor soils are young farmers with still undeveloped expertise and urbanites who have capital but little farm experience. Average yields for the population as a whole were 900 kg per hectare, mostly accounted for by rice production in both years. Beans have been disappointing due to disease, and corn yields have been low. While the average production in 1983 was 900 kg per hectare, some of the better managers obtained 1,500 to 1,800 kg per hectare, and one producer obtained 3,840 kg per hectare. This very high productivity by one farmer occurred on a small area of 0.5 hectares with intensive management and the use of inputs. His small farm of 20 hectares, is planted in a diversified manner: bananas, papayas, peanuts, pasture, cocoa, and most major subsistence crops. In addition he raises chickens, pigs and rents a small high quality pasture to neighbours with cattle but still undeveloped pastureland. The lack of familiarity with Amazonian farming of most farmers led to comparable results in Tucumã as in the Transamazon project (Table 4.3). Yields were about half of average regional yields, but there was a lot of between-farm variability. As in the case of the Transamazon highway, a small proportion of independent operators relying heavily upon family labour and choosing superior soils performed up to expectation, while the majority of the population divided its effort between subsistence production, wage labour, and other forms of gainful occupation other than farming.

In no small part, farmers coped with their sudden decapitalisation by means of off-farm work. This is not new to this frontier. In head-of-household salaries alone, Tucumã farmers earned 143 million cruzeiros or an average 2.9 million per head of household. In addition, the wives averaged 557,000 cruzeiros each and the sons 106,000 each. For most of the population, farming was insufficient to cover the relatively high expenses that they had incurred – average debt per household was 4.6 million cruzeiros.

FUTURE PROSPECTS OF AMAZONIAN COLONISATION

Forecasting is a perilous activity for any scientist, particularly in the social sciences. Human agency has proven, many times over, its ability to subvert even the most rigorous of analyses. However, if governments and private enterprises continue to avoid investing in environmental assessments in advance of project execution then we will continue to see projects located in areas inappropriate for intensive agriculture, or even in areas too swampy for healthful settlement.

The results of such misjudgments will lead governments and private enterprises to decide, after less than 5 years, whether the project is yielding acceptable economic results and for the agencies to abandon direct operation of the colonisation project. In some cases, the projects might be abandoned by settlers. In most cases, they will not be. Settlers will continue to try to make a living from the land but now without institutional support.

If governments and private enterprises fail to take account of research findings from colonisation areas elsewhere in the humid tropics, they will continue to execute plans that are at odds with field results. Similarly, they will attempt to carry out farming schemes rather than see each colonisation area as an experimental area, where only with time (c. 10 years) can one expect to see farming strategies begin to emerge that are sensitive to climatic, soil, socio-economic, and locational realities.

We can forecast that if governments and private enterprises continue to insist in believing that settlers to a new frontier area will be either the landless or the small farmer, colonisation projects will continue to surprise and disappoint planners with their heterogenous mix of settlers – many of whom will have little farming experience. This initial mix may be a blessing in disguise. The availability of craftsmen and professionals in most early frontier settlements provides a cadre of educated persons with skills needed in the development of a forested frontier. It is these persons who in other frontiers have become advocates of urban development, market links to the larger society, and leaders in the social and political life of frontier communities (Moran, 1988).

The question remains whether the relevance of environmental variation, bureaucratic expertise, and colonist heterogeneity will be both recognised and welcomed by government planners and private enterprises charged with colonisation scheme development. There is very little evidence of significant learning from past experiences in the execution of Amazonian settlement. I have shown elsewhere (Moran, 1985) that the execution of settlement in Rondônia repeated many, but not all, of the mistakes made in the Transamazon Highway Colonization Scheme. In this chapter, we have seen the similarity in outcomes between the Transamazon and a private colonisation scheme. Progress occurred in the creation of institutional capacity to build roads and other infrastructure, but expectations for the colonists were too high. Until such time as planners accept that it will take at least 10 years for a migrant population to learn about, and adjust to, a new environment they will continue to be disappointed and to judge many of the colonisation schemes failures.

What is success and what is failure in colonisation? As the term has been used by governments and private enterprises it tends to mean the achievement – within a 3- to 5-year period – of levels of agricultural output projected at the outset. Such projections are based upon either national or regional average yields per unit of land. The persistence of this standard of success or failure guarantees that most colonisation projects will be judged failures. This criterion fails to recognise the difference between colonisation (essentially a human adaptive process) and the construction of civil works, where one can evaluate within 3 to 5 years whether the infrastructure has led to a measurable improvement in the aggregate income of a region or to a notable increase in flow of traffic.

Colonisation fundamentally refers to the movement of people from one region to another one. Presumably, and it is certainly true for Amazonian colonisation, the migrants are unfamiliar with the new region. Their entire repertoire of knowledge has to be put to the test in the new region, then adjusted gradually and cautiously (because their biological survival is at stake) to the new environment and the differences in the social and economic circumstances. Only then will the population be able and willing to begin to take risks, to experiment with parts of their fields, and to begin to achieve average regional yields.

The first settlers in a newly-opened frontier face not only lack of knowledge about what works, but a severe lack of labour and capital. This is well as it should be. The lack of capital means that they will learn gradually and that mistakes made in farming will not be so devastating for them. The lack of labour means that they will develop forms of labour exchange to make up for the lack of capital to hire others, thereby creating social obligations that lay the foundation for social, religious and political organisation. When governments and private enterprises try to speed up the process of adaptation (which they cannot do) by providing credit at favourable rates during the early stages of settlement, the result is a greater cost to financial institutions and farmers for errors in management, greater rates of deforestation, higher rates of failure, and more rapid stratification of the population without an adequate period of time for community-based social, political, and religious organisations to emerge. In short, the probability of inadequate responses increases and with it the likelihood of failure from the point of view both of the farmer and of the creditors, who are unable to get a return on their investment. Adaptation to the frontier should take precedence over attainment of a given level of crop yield, over delivery of credit, and delivery of extension services. Both public and private agencies promoting colonisation fail to recognise the developmental characteristics of colonisation, fail to learn from past experiences, and fail to accept that frontiers reproduce the economic relations of areas of origin and that, as such, they do not *solve* the problems of access to land or inequitable distribution of income.

References

Bunker, S. (1985), *Underdeveloping the Amazon: Extraction, Unequal Exchange and the Failure of the Modern State*, Urbana: University of Illinois Press.

Butler, J. (1985), *Land, Gold, and Farmers: Agricultural Colonization and Frontier Expansion in the Brazilian Amazon*, Ph.D. dissertation, Department of Anthropology, Gainesville: University of Florida.

Crist, R. and Nissly C., (1973), *East of the Andes*, Gainesville: University of Florida Press.

Falesi, I. C. (1972), *Os Solos da Rodovia Transamazônica*, Belém: Instituto de Pesquisas Agropecuária do Norte (IPEAN).

Fearnside, P. (1985), *The Carrying Capacity of the Brazilian Rainforest*, New York: Columbia University Press.

IBRD (1979), *Brazil: A Review of Agricultural Policies*, Washington, D.C.: World Bank.

IPEAN (1974), *Os Solos da Rodovia Transamazônica: Trecho Itaituba–Rio Branco*, Belém: Instituto de Pesquisas Agropecuária do Norte.

Jordan, C. (1982), 'Rich Forest, Poor Soil', *Garden*, 6(1), pp. 11–16, New York Botanical Garden.
Kleinpenning, J. M. G. (1977), 'An Evaluation of the Brazilian Policy for the Integration of the Amazon Region', *Journal of Economic and Social Geography*, 67(5) pp. 345–60.
Margolis, M. (1973), *The Moving Frontier*, Gainesville: University of Florida Press.
Ministério da Agricultura (1972), *Altamira I*, Brasília: Ministério da Agricultura/INCRA.
Moran, E. (1976), *Agricultural Development along the Transamazon Highway*, Center for Latin American Studies Monograph Series, Bloomington: Indiana University.
Moran, E. (1977), 'Estrategias de Sobrevivência: O Uso de Recursos ao Longo da Rodovia Transamazônica', *Acta Amazônica*, 7(3) pp. 363–79.
Moran, E. (1979), 'Criteria for Choosing Successful Homesteaders in Brazil', *Research in Economic Anthropology*, 2, pp. 339–59.
Moran, E. (1981), *Developing the Amazon*, Bloomington: Indiana University Press.
Moran, E. (1983), 'Growth without Development: Past and Present Development Efforts in Amazonia', in E. Moran (ed.), *The Dilemma of Amazonian Development*: Boulder Col.: Westview Press, pp. 3–23.
Moran, E. (1984), 'The Problem of Analytical Level Shifting in Amazonian Ecosystem Research', in E. Moran (ed.), *The Ecosystem Concept in Anthropology*, Washington, D.C.: American Association for the Advancement of Science, pp. 265–88.
Moran, E. (1985), 'Colonization in the Transamazon and Rondônia, in M. Schmink and C. Wood (eds), *Frontier Expansion in Amazonia*, Gainesville: University of Florida Press, pp. 288–303.
Moran, E. (1987) 'A Produção Agrícola em um Projeto de Colonização em Carajás', in G. Kohlhepp (ed.) *Homem e Natureza na Amazônia*, Tübingen, Geographisches Institut, pp. 353–66.
Moran,E. (1988), 'Social Reproduction in Agricultural Frontiers', in J. Bennett and J. Bowen (eds), *Production and Autonomy: Anthropological Studies and Critiques of Development*, Washington, D.C.: University Press of America.
Nelson, M. (1973), *The Development of Tropical Lands*, Baltimore: Johns Hopkins University Press.
Posey, D. (1982), 'The Keepers of the Forest', *Garden* 6(1) pp. 18–24.
Schmink, M. (1981), *A Case Study of the Closing Frontier in Brazil*, Amazon Research Paper Series, 1, Gainesville: University of Florida, Center for Latin American Studies.
Schmink, M. (1982), 'Land Conflicts in Amazonia', *American Ethnologist* 9(2) pp. 341–57.
Seubert, C. E., P. A. Sanchez and C. Valverde (1977), 'Effects of Land Clearing Methods on Soil Properties of an Ultisol and Crop Performance in the Amazon Jungle of Peru', *Tropical Agriculture* 54(4) pp. 307–21.
Smith, N. (1982), *Rainforest Corridors*, Berkeley: University of California Press.
Tavares, V. P., Considera, C. M. and de Castro, M. T. L. L. (1972), *Colonização Dirigida no Brasil*, Rio de Janeiro: IPEA (Instituto de Pesquisas Econômicas e Sociais).

5 Future Hydroelectric Development in Brazilian Amazonia: Towards Comprehensive Population Resettlement[1]

Luc J. A. Mougeot

INTRODUCTION

Population resettlement is expected to continue to grow in the near future throughout the world, much as a result of water resource development (PEEM, 1986, p. 74). The Indira Sarovar Complex on India's River Narmada will require the re-accommodation of some 85,000 people from 253 villages, and impoundments projected in the River Plate basin, over 94,000 people (Goodland and Ledec, 1986, p. 28; Szekely, 1982, pp. 240–1). Thailand's Pa Mong Scheme is to displace some 400,000 individuals, with resettlement costs possibly absorbing 29 per cent of the Scheme's budget (Lightfoot, 1981, p. 97). China's proposed Three Gorges Dam Scheme might uproot some 1,400,000 people; this is more than were dislocated over a period of twenty years at 18 major reservoirs in Africa and South Asia (Goldsmith and Hildyard, 1984, p. 15; Mougeot, 1986, p. 402).

These prospects should cause deep concern in the light of appraisals of past hydroproject-related assisted resettlements, most carried out in developing countries over the last thirty years. They have usually been conceived as emergency measures, supported by inadequate baseline surveys. Too frequently, they have failed to perceive and respond to resettlees' stress, being poorly funded and administered and having little concern for the post-relocation fate of the affected populations. Administrators have often failed to communicate efficiently with relocated people and have not involved them in the planning and execution of resettlement. They have also failed to ensure proper interaction between resettlement projects and other, potentially enhancing or hindering, neighbouring developments. Contrary to what could be expected from the abundance of experience, there has been a considerable lag in the managers' willingness and/or ability to exploit the lessons of the past (Brokensha and Scudder, 1968; Chambers, 1970; UNEP, 1978; Barrow, 1981; Interim Mekong Committee, 1982; Canter, 1985).

Successful resettlement can to some degree be gauged by the extent to which the groups to be resettled, and others who might be affected, do not oppose resettlement plans or make their own unofficial arrangements. (Koenig, 1986, p. 29; PEEM, 1986, p. 74). Counter-initiatives normally result from specific causes and usually translate into specific spatial responses. The need a community perceives to transfer or be transferred from *A* to *B* brings into view for its members a potentially wide range of alternative voluntary movements. By weakening adherence to and interfering with the assistance scheme, these counter-movements might eventually threaten the scheme's viability. The risk of such a situation arising seems to grow when one or more of the following conditions are present: (a) the community at *A* is only partially entitled to assisted transfer to *B*; (b) *B* is a pre-existing settlement; (c) the region where *A* and *B* are located is subject to intense migrations; (d) various phases of the resettlement scheme suffer delays; (e) various transfers are required to attend simultaneously to the needs of several distinct communities; and (f) successive resettlements are needed to accommodate the phased expansion of the original development or the implementation of others in its vicinity. Success thus largely depends on resettlement managers' ability to anticipate and discourage the more likely counter-initiatives in each situation, by either forestalling, curbing or eliminating their causes. Otherwise there may not only be markedly increased costs to both resettlement officials and beneficiaries alike but also resettlement may be prevented from achieving even minimum expected returns (Mougeot, 1988, pp. 21–2).

There is growing recognition that successful assisted resettlement requires a more comprehensive approach than has been the case until now. Hydroproject-related resettlement has been comprehensively defined as one which enhances the habitability of the area, improves the resettlees' living standards, as well as contributing to sustainable local and national development. To be successful, comprehensive resettlement requires in-programme provisions, such as agricultural extension, mass media propagation of information, application of science and technology, social welfare and community development (Afriyie, 1973, p. 728). In order to be sustainable, resettlement needs to be deliberately planned and executed within an encompassing, long-term, area development strategy (Scudder, 1966, p. 20; Brokensha and Scudder, 1968, pp. 22–3).

But is it really worthwhile for a country like Brazil to invest in comprehensive hydroproject-related population resettlement in its Amazon region? The answer may be found by considering specific aspects of this question. Is Amazonian hydrodevelopment just a Brazilian fantasy in the first place, or is it consistent with long-term, global hydrodevelopment trends? After all, will planned hydroprojects disrupt only negligible numbers of people in the region? Are most people to be displaced likely to require only minimal resettlement assistance? If it should prove desirable for Brazil to invest in comprehensive resettlement, one then might ask: How should this be done?

Why should efforts be made to incorporate assisted resettlement within more comprehensive area development planning? What administrative framework might this require to become viable? Above all, how might assisted resettlement in this way contribute to improving relocation performance in the region?

AMAZONIAN HYDRODEVELOPMENT AND LONG-TERM GLOBAL TRENDS

Brazil's programme to expand its electricity supply by increasingly harnessing major Amazonian water courses is anything but isolated from world hydrodevelopment trends. Present and foreseeable market conditions are now making it technologically feasible and economically attractive, though environmentally problematic, to exploit the abundant potential still available in the world's subarctic and humid tropical frontiers. To be understood, the Brazilian decision deserves to be put into this perspective.

A very small proportion of the world's colossal hydroelectricity-generation potential has been exploited so far. Some 0.0087 per cent of earth's total volume of water (1.4 trillion cubic kilometres) is estimated to be available at any time in rivers, streams and natural lakes, but only 15.9 per cent of a corresponding hydroelectric potential of 2,342,639 megawatts (MW) was being used in the early 1980s (Petts, 1984, p. 1; Gordon, 1983, p. 21; Biswas, 1983, p. 28).

However late and limited has been their exploitation, hydraulic energy sources are now being rapidly tapped around the world; massive human interference in the natural environment is turning water courses into the single most important suppliers of renewable energy for development. Axial-flow generation has been applied only rarely; most hydroelectricity schemes now require some water storage. River damming for flood control, irrigation and, increasingly, hydroelectricity generation, has only expanded over the last thirty years or so. Yet by the early 1970s many of the world's major water courses had been modified by impoundment to some extent, with more than 12,000 dams of more than 15 metres in height impounding some 4,000 cubic kilometres of water, in reservoirs covering about 800,000 square kilometres (Petts, 1984, p. 4). Schemes were largely completed in the post-Second World War period, thanks to global economic recovery and progress in earthmoving and concrete technology; after 1950 many world regions staged dam construction activity, with man-made stream regulation now exceeding that of natural lakes threefold (Barrow, 1981, p. 135; Petts, 1984, p. 4–5).

Recent technical advances in hydroelectricity generation, as well as growing constraints on the use of other energy sources, have enhanced the cost competitiveness of hydrodevelopment. In spite of the rising costs of land, building materials and labour, the progress in planning, design and construction techniques and power station and transmission equipment, as well as routine environmental-effect accounting, have contributed to weaken cost increases compared with those of nuclear and coal-fired thermoelectric alternatives (Biswas, 1983, p. 28). Technology transfers to developing regions have been spearheaded by consultancy and construction firms of the developed countries, and supported by international aid agency programmes. In particular, countries with severe balance-of-payment deficits and facing mounting fossil-fuel import costs have preferred to develop reliable indigenous sources of renewable energy with an already well-established technology; schemes may also serve other development purposes, if adequately planned (flood control, irrigation, etc.). As a result, global generating capacity grew remarkably in the 1958–83 period, when some 31 countries installed an average of 10,400 MW a year; by the early 1980s hydroenergy was providing nearly 25 per cent of the world's electricity supply (Biswas, 1983, p. 25–6).

In the coming decades, the world's hydroelectric potential will almost certainly be harnessed at an increased rate, with dramatic changes taking place in the location, size, purpose, and impacts of future schemes. It has been estimated that the amount of energy produced globally by hydroelectricity facilities will multiply fourfold between 1980 and 2020. This growth will largely occur in developing countries where most of the available potential lies. By 1980, North America and Europe had tapped 59 per cent and 36 per cent of their potential, respectively, while Asia had by then developed only 9 per cent, Latin America 8 per cent and Africa 5 per cent (Biswas, 1983, p. 26; Flavin, 1986, p. 26). Altogether, slightly more than half of the global hydroelectric potential exists in developing countries, of which merely 10 per cent has been harnessed so far. According to a World Bank forecast, Third World hydroelectric capacity (excluding China's) should more than double, from 100,000 MW to 201,000 MW between 1980 and 1990; hydrocapacity could grow by as much as 300,000 MW by the year 2000 (World Bank, 1987, p. 7). Developing countries' current electricity consumption is so limited and there are so many potential uses that demand will probably expand, even when the economy does not, at annual rates expected to average 7 per cent during the 1990s, possibly more in many countries. Mexico, Brazil and Argentina have already raised their capacity by 43 to 58 per cent between 1978 and 1983 alone (Flavin, 1986, p. 26).

This locational shift in the current expansion of global capacity is associated with growth in the scale of generating schemes. The tendency to build large numbers of small and medium-sized dams for single purposes probably peaked in developed countries in the late 1960s (Petts, 1980, p. 326). The trend now is to construct fewer but much larger dams with greater generating capacities in developing regions. For instance, North and Latin America respectively claimed 21 and 11 of the world's 77 largest hydrodams completed by 1981; however, the former had only 14, whereas the latter claimed 25 of the 109 largest hydrodams then under construction (author's calculations based on Petts, 1984, p. 8).

A major share of developing regions' future hydropower development will take place in, or depend on the existence of, forested humid tropical watersheds. The hydroelectric potential of 93 developing nations surveyed by the World Bank in 1980 totalled 1,194,390 MW. No less than 42.5 per cent of this potential was concentrated in only 10 countries, between them holding 75.6 per cent of the world's tropical rainforest area; 63.4 per cent of the hydropotential was found in 30 countries containing 100 per cent of such forests (Biswas 1983, pp. 27–8; Guppy, 1984, p. 930). Until recently rivers in tropical, particularly equatorial, countries have remained relatively undisturbed. Considerable efforts are now being made to harness their huge hydropower potential (Petts, 1984, p. 4). In 1978, 17 important rainforest countries had dams either completed, under construction or planned; ten years later, however, these countries' share of all dams listed had doubled and their schemes' average rated capacity was well above the world average. Between 1978 and 1988 rainforest countries' share of the world's listed dams grew from 13.6 (30 out of 220) to 27.4 per cent (89 out of 325). Using figures on planned rated capacities available for 172 of the 325 dams listed in 1988, the average plant capacity was 2,065.5 MW for the world as a whole, but 2,300 MW for all rainforest countries possessing at least one such scheme, and 2,373.5 MW for the 10 major rainforest countries. Also, these countries were claiming nearly half of the world's hydrodams then on the planning boards or due to be completed beyond 1988. In fact, 59 of the 325 hydrodams in the world register were still at the planning stage or were due to be completed after 1988, of which 42.3 per cent were located in rainforest countries (author's calculations based on Mermel, 1978, pp. 43–54, and Mermel, 1988, pp. 47–55).

Developing countries are harnessing the hydroelectric potential of their humid tropical watersheds to fulfil many development needs. In regions where both hydropotential and mineral resources are abundant, the former is commonly being tapped to power local ore-processing or smelting plants, as in Indonesia, Venezuela's lower Caroni basin and Brazil's lower Tocantins basin (Goldsmith and Hildyard, 1984, p. 14; Siso, 1983, p. 15; Mougeot, 1986, p. 402). In others, hydroelectricity is also being generated on a large scale to be transferred, often over long distances, to energy-deficient regions. This is the aim of current projects in Colombia's Orinoco watersheads and on Brazil's lower River Xingú (Ospina, 1987, pp. 35–6; Duarte *et al.*, 1983, p. 23). In the humid tropics of Peru and Zaire, plans to exploit the hydroelectric potential through large schemes have temporarily been shelved, due to political, financial or physical constraints. In the meantime, short-run emphasis is being put on more numerous and smaller schemes to meet local communities' needs (Wicke, 1987, pp. 49–50; Anonymous, 1980, p. 52). In Tasmania, India's Western Ghats and Peninsular Malaysia, projects have been discontinued or abandoned, mainly for environmental reasons (Dragun, 1983, pp.198–200; Anon., 1980, pp. 15–16; Rachagan, 1983, pp. 44–6; Goodland, 1986, p. 26).

Brazil is representative of worldwide trends towards large-scale harnessing of the abundant hydropotential available in remote humid tropical watersheds. In 1980, 90 per cent of the country's electricity output came from hydraulic energy sources. Some 470 large dams and associated hydroelectric power plants were either in operation or under construction, three quarters of which were built during the post-Second World War industrial boom, with more than 100 initiated or completed in the 1970s alone. They are mainly located in or near the large industrial centres of the Centre–South, where most of the accessible potential has now been exploited (Budweg, 1980, p. 19; Budweg, 1982, p. 48).

Estimates of Brazil's hydroelectric potential were recently updated to 213,000 MW; 90 per cent of the nation has been surveyed and analysed, but only 14 per cent has so far been tapped. Until 1995, this potential will be harnessed at an expected annual rate of 11.3 per cent, well above economic growth (Biswas, 1983, p. 33). Installed capacity should rise from 34,035 MW in 1983 to 55,382 MW in 1990. By the year 2000, when electricity will still continue to be derived mostly from hydraulic sources, the country's capacity is likely to be between 77,600 and 83,200 MW, depending on growth of the electricity market, technology acquisition policies and financing arrangements (Holtz, 1985, pp. 17, 19).

Also consistent with the general global pattern, the average size of Brazil's hydroelectricity plants has grown steadily since the mid-1930s. Future growth in supply is expected to come from fewer schemes affording an unprecedented generating capacity. Fewer plants went into operation between 1976 and 1985 and between 1986 and 1990 than in the previous five-year periods. Average installed plant capacity has been rising since the mid-1930s: 13.19 MW (1936–45), 21.93 MW (1946–55), 83.66 MW (1956–65), 229.89 MW (1966–75), 642.48 MW (1976–85), and 2882.43 MW (1986–90). In fact, the few projects scheduled to be built between 1986 and 1990 have nearly as much generating capacity as that of 252 hydroelectricity schemes built in the country before 1975 (author's calculations based on Holtz, 1985, p. 17).

Most of Brazil's unharnessed hydropower potential is located in the Amazon region, which holds an estimated 97,800 MW. By late 1984 the federal power-utility holding company's subsidiary for northern Brazil, Centrais Elétricas do Norte do Brasil (ELETRONORTE), had identified some 63 sites for hydropower-generating reservoirs in the River Araguaia–Tocantins (AT) basin and along major tributaries of the River Amazonas; if all are exploited, these impoundments would flood about 100,000 square kilometres (Junk and Nunes de Mello, 1987, p. 367). Inventories now available on the AT and Xingu river basins suggest that more than half of the regional potential could be tapped through a few large schemes, with final costs competing favourably with those of smaller hydroplants and coal- or nuclear-fired thermoplants in the south-east (Monosowski, 1983, p. 11; Holtz, 1985, p. 16). It therefore seems inevitable that at least the larger planned schemes will go ahead, and that most flooding will be in the AT and Xingú river basins.

PLANNED HYDROPROJECTS AND THE DISRUPTION OF COMMUNITIES IN BRAZILIAN AMAZONIA

Large hydroelectricity-generating projects and their related infrastructure will probably be the single most disruptive development actions in Brazilian Amazonia in the near future. They will be significant both in terms of the population disturbed and in terms of water and land resources they will deny or make available to human communities.

Populations likely to be dislocated by future hydropower projects in the Amazon region, even those currently and permanently resident on the proposed reservoir sites, can hardly be reliably estimated at the present time. One reason is that much local ground surveying remains to be done. Only 4.5 per cent of Amazonia's hydroelectric potential had been tapped or was under construction in late 1983, while less than half of the amount still available had been inventoried to any extent (Holtz, 1985, p. 6). Once created, the actual pattern of impoundments could differ significantly from original plans in terms of numbers, location and areas submerged, as a result of deleting, adding, re-siting and re-scheduling schemes, in response to more accurate ground knowledge and changing market conditions.

With respect to schemes already under construction, the delimitation of areas to be evacuated depends on assessing topographic information and on defining maximum storage levels. This is particularly true for large reservoirs to be sited on the flat lower reaches of rivers. In most cases, as with the Altamira Hydroelectric Complex (AHC), storage levels cannot be established immediately and accurate topographic base maps also take time to compile (CNEC, 1985). For instance, assessment of the area to be submerged by a series of 6 or 7 reservoirs on the lower Xingú and Iriri critically depends on selecting one of two possible maximum reservoir levels for the downstream Babaquara Dam; it could be as large as 18,300 square kilometres (Duarte *et al.*, 1983, pp. 25–6). Accurate laser-levelling of future lakeshores is now feasible but at Tucuruí, as at Kariba (Brokensha and Scudder, 1968, p. 43), contours remained unknown until the lake filled; consequently some families were resettled in areas which were subsequently flooded and needed to be relocated once more. Even when reliable contours are known and census data are available, estimates could still be considerably affected by the re-programming of specific schemes and by intervening development actions in the vicinity of planned reservoirs.

Currently only crude provisional estimates can be attempted for most of the region, using available data on permanent residents in the reservoir areas. Assumptions on which calculations were made have been discussed elsewhere (Mougeot 1988, pp. 37–8). Estimates of disrupted populations for reservoirs proposed in the AT river basin and along major southern tributaries of the River Amazonas (which presently account for nearly 83 per cent of the regional hydropotential) suggest that, *if* the impoundments currently contemplated in these basins were all created, they would by 1985 have displaced between 85,000 and 156,000 residents (Table 5.1).

The upper estimate in Table 5.1 will almost surely have been exceeded manyfold by the time all likely impoundments are formed. The great majority of these are still on the drawing board; in the meantime, however,

Table 5.1 Estimates of population displacements in river basins of southern Amazonia

Estimates / *River basins*	*Projected population density per km² 1985* Total[a] (1)	Rural[b] (2)	*Estimated total reservoir area in km²* [c] (3)	*Population estimated in 1985 to be displaced by reservoirs* (1) × (3) (4)	(2) × (3) (5)
Araguaia–Tocantins	3.3863	1.9068	25,300	85,673	48,242
Xingú	0.3789	0.2436	21,000	7,957	5,116
Tapajós	1.3935	0.6683	19,200	26,755	12,831
Madeira	0.9834	0.4607	16,350	16,079	7,532
Total	1.9047	1.0444	81,850	155,900	85,484

[a] Total density ratio, calculated on basis of total area and total projected population for 1985, of all municipalities traversed or bounded by the basin's major rivers and their more important tributaries.

[b] Rural density ratio derived from the basin's total density ratio, assuming that rural density represents a proportion of total density which, in 1985, was equal to that calculated from 1980 census figures.

[c] Considering that the current potential of 98,680 MW for 15 river basins of Brazilian Amazonia holds a nearly one-to-one relationship with the corresponding number of km² of required reservoir area, each basin's total reservoir area was estimated to equal its current hydroelectric potential. This generalisation should be more valid for whole basins than for individual schemes; UHE Tucuruí is expected to generate more than 3 MW/km², whereas this ratio shall fall under 1 MW/km² at UHE Santa Izabel in the Araguaia–Tocantins.

Sources: Instituto Brasileiro de Geografia e História (IBGE), *Anuário estatístico do Brasil 1986* (Rio de Janeiro: IBGE, 1986), p. 59; IBGE, *IX Recenseamento Geral do Brasil 1980. Censo Demográfico – Dados distritais* (Rio de Janeiro: IBGE, 1982), vol. 1, tomo 3, No. 1, 3–4, 22–3, tab;e 1; IBGE, *VIII Recenseamento Geral do Brasil, 1970. Censo Demográfico* (Rio de Janeiro: IBGE), Vol. 1, tomo 1, III–IV, XII–XIII, table 53; Wolfgang J. Junk and J. A. S. Nunes de Mello, 'Impactos ecológicos das represas hidrelétricas na bacia amazônica brasileira', in *Homem e Natureza na Amazônia/Hombre y Naturaleza en la Amazonia*, ed. Gerd Kohlhepp and Achim Schrader (Tubingen: Geographisches Institut der Universitat Tubingen, 1987).

population densities are escalating. Between 1980 and 1985 alone total density ratios for the four basins concerned rose by 29.4 per cent, ranging from 20 per cent in the already densely peopled AT basin, to 49 per cent in the Madeira basin. In the state of Maranhão, densely populated municipalities bordered or crossed by the Rivers Tocantins and Manuel Alves Grande are to accommodate four schemes; their total density grew by at least 45.4 per cent in only five years. The numbers of those to be displaced from reservoir areas will easily surpass current estimates in the active colonisation areas where large impoundments are being or might be postponed. The planned Santa Isabel Reservoir on the lower River Araguaia, upstream of Tucuruí, was originally to submerge some 3,840 square kilometres in 1992, probably ousting a population well in excess of the 1980 estimate of 60,000. The scheme has now lost priority to the AHC on the lower Xingú. Meanwhile, however, an extensive state road-upgrading programme has been launched in its proposed vicinity, and the total density of the local Pará municipalities has grown by at least 80.1 per cent, from 2.29 to 4.13 persons per square kilometre between 1980 and 1985.

Measures only concerned with people permanently residing on reservoir sites may greatly underrate the real extent of local population disruption by hydroelectric development in Brazilian Amazonia. Estimates may vary according to the orders in effect and to the land tenure conditions. Populations directly ousted are those losing access to land and water-based investments and areas, natural resources and work opportunities, due to site requirements of reservoirs, construction camps, raw material procurement, roads and transmission lines. But people can also be uprooted indirectly (in space and time), by changes these actions bring about in the biogeophysical environment and in land valuation and utilisation patterns, as well as by new diseases, adverse social changes and economic behaviour which outsiders might introduce.

With respect to land tenure, local residents owning local land can be affected directly by project site requirements, as well as non-resident owners of local properties, and non-residents who, while not owning local land, might seasonally depend on access to river-based resources and services for their subsistence. These categories can be further affected indirectly, once they relocate in the vicinity of the new reservoir (including populations initially unaffected by project site requirements). When relocating because of the spatial impacts of the project, all these categories of people could themselves become agents of further project-triggered effects, by displacing one another or yet unaffected and more vulnerable groups of people.

It is well known that the poorer groups in Brazilian society compete with one another, in their struggle to gain a living in frontier areas where unclaimed land is increasingly in short supply. Hydroprojects could further increase pressure on these people. Native populations are particularly at risk. In 1981 at least seven hydroprojects planned or under construction throughout Brazil, each comprising between one and seven individual dams or sets of dams, were to submerge or require evacuation of at least 100,000 hectares in 32 to 34 separate indian areas, of which 22 to 24 were located in Amazonia (Aspelin and Coelho dos Santos, 1981, pp. 3–5). On the River Trombetas, the Cachoeira Porteira and Mapuera Reservoirs are to flood nearly 1,000 square kilometres of mostly dryground forest which, though environmental consultants claim they are not permanently settled, at least serve as hunting, fishing and fruit-gathering grounds for native peoples (communication of Guaraçí Sathler, Engerio, 11 June 1987). The Tucuruí Hydroelectric Complex (THC) reportedly encroached on three indigenous peoples' reserves and, more directly, affected at least six tribal groups totalling some 800 individuals (Monosowski, 1986, p. 196). The Parakanã and Gavião Reserves had already been reduced by developments prior to the THC, but were further partly expropriated or re-sited to make room for hydroproject needs. The indians have in this way lost direct access to the Tocantins waterway and its resources; they also have become exposed to disease risks and resource-base destruction by non-indian poachers (*O Liberal*, 23 April 1987, p. 3; 7 August 1987, p. 7).

RESETTLEMENT ASSISTANCE REQUIREMENTS

In principle, various degrees of State assistance may be offered to populations displaced by hydroprojects. In Amazonia, however, communities that will be dislocated are mostly rural and, because of their social and cultural vulnerability, are likely to require substantially upgraded resettlement support, even to merely maintain their standard of living after relocation.

Populations disrupted by development actions should be consulted on how they prefer to relocate, either as communities, individuals, or both. This will mainly depend on the prevailing degree of community cohesion. Pioneer crossroad-towns of migrant nuclear families with diverse cultural origins are more likely to favour self-relocation than traditional, riverine villages of sedentary households with extensive local kinship ties. Assisted resettlement can in fact provide the opportunity to strengthen a sense of community which may have been or might be undermined by other, adverse, developments. Each official assistance programme may vary in degree and duration, from minimally assisted self-relocation to fully-fledged resettlement (Dasmann *et al.*, 1973, p. 220; Lightfoot, 1981, p. 113). Most experiments worldwide have adopted solutions between these extremes, with varying degrees of government funding, provision of infrastructure and housing, and assistance to modify existing social attitudes and economic practices. Populations moved as communities normally require more elaborate resettlement assistance than individuals (Lassailly-Jacob, 1983; Barrow, 1981, p. 141).

When community preservation is at stake, the perceptions of the State and the population should coincide as much as possible, as to the most appropriate form, degree, and duration of assistance. Experience suggests that national ethnic minorities and rural-based groups, living at or near subsistence levels, are usually better aided by fully-fledged resettlement solutions, while the more mobile, urbanised, better educated and wealthier cope better with self-relocation (Scudder and Colson, 1982, p. 277). However preferable one or another alternative might appear to be, its actual adoption may critically depend on people's confidence in governmental competence, as much as on the local elite's influence on resettlement authorities. The more needy's willingness to participate in sponsored projects is likely to be tempered by their perception of the government's ability to carry out timely and effective implementation of community resettlement (Scudder and Colson, 1982, p. 278; Koklu, 1982, p. 6). On the other hand, particularly in closed communities, resourceful individuals are prone to self-relocate themselves and, when doing so, may deprive their communities of the crucial leadership required to overcome resettlement stress. In a Mexican frontier region, 'open' communities facing relocation tended to draw inward and develop a guarded attitude, thus preventing fragmentation and out-migration of the fittest. In more closed and rigidly structured groups, divisions along latent factional lines were precipitated when risk-takers seized opportunities previously unavailable to them (Partridge *et al.*, 1982, p. 260).

In Brazilian Amazonia, communities likely to be displaced are largely composed of the more vulnerable human groups: native indigenous people and subsistence peasants of mixed-blood, as well as small-scale immigrant farmers. International experience strongly suggests that, in most cases and instead of individual cash compensation, these populations are more likely to require assisted resettlement, in order at the very least to maintain previous living standards. The potential benefits of assisted resettlement to these people and to hydropower projects themselves will essentially depend on the extent to which resettlement is part of a watershed or basin development strategy, as explained in the next section.

ASSISTED RESETTLEMENT AS PART OF INTEGRATED AREA DEVELOPMENT

The weight of evidence of social and environmental costs gathered from various resettlement experiments is at last changing official understanding of resettlement's requirements and opportunities. Both within countries and on the international scene, the emphasis seems to be clearly shifting from limited involvement outside strictly defined resettlement to solutions integrated with overall local development, and concerned with shortening the more stressful stage of resettlement (Butcher, 1971, p. 6). International development agencies are now recognising that social aspects of hydroelectric projects may be as expensive, if not as time-consuming and complicated, as engineering and construction components; indeed, resettlement is often the most costly item (Goodland and Ledec, 1986, p. 60).

As a result the World Bank has issued guidelines and has been attempting to quantify various social impacts, including forced resettlement and associated problems (unwanted migration, health risks, protection of minority peoples and natural resource conservation) (World Bank, 1980; Olivares, 1986, pp. 135–6). The FAO has recently devoted attention to the environmental impacts of resettlement itself (PEEM, 1986). More importantly, as resettlement costs are more fully appraised and internalised by hydropower project planners, efforts will be made to limit the need for displacement. This may be unavoidable, however. In such instances and in order to be approved by the World Bank, projects are now required to submit a satisfactory resettlement proposal, the planning and implementation of which become an integral part of the development project (Olivares, 1986, p. 138). Some have already been approved, for example, India's Indira Sarovar Irrigation and Power Complex in the Narmada river basin, where relocated people are to be given priority in project-created employment and training will be provided to assist them where necessary (Goodland and Ledec, 1986, pp. 28, 33).

In many countries, the understanding of what should be the more appropriate approach to resettlement is changing. The now famous Nam Pong case study illustrated that resettlement should be considered as but one component of water resource management, along with and in interaction with reservoir shoreline and upstream watershed development (Srivardhana, 1987). Asia's extensive experience stresses that there is certainly room for more encompassing approaches to water resource management itself, in terms of the resource base contemplated, perceived potential benefits and measures required. Planning needs to account for more than just engineering and financial aspects: it needs to value the understanding, support and participation of local populations in the management system (Bower and Hufschmidt, 1984, pp. 351–2). In northern Thailand watershed management through agroforestry has been successfully linked with reservoir-related population resettlement (Perry and Dixon, 1986, p. 37).

In Brazilian Amazonia the prospect of resettling large numbers of culturally distinct, often environmentally adapted and territorially attached, rural communities represents a major opportunity to involve local populations in water resource management. Long recognised by students of African reservoirs built between 1958 and 1981, this potential has been overlooked by most advocates and many opponents of large hydropower projects. These have been justifiably criticised for benefitting wealthy minorities, favouring labour-saving export activities, increasing water-borne diseases, damaging the natural environment, displacing indigenous people and destroying their culture (Goldsmith and Hildyard, 1984, pp. 231–8). But these problems are not the sole province of hydroelectricity schemes; in rapidly developing regions such as Amazonia, they have continued and will probably continue unabated, regardless of whether or not hydropower projects are developed. On the other hand, project cost-benefit appraisals by funding and aid agencies are becoming more comprehensive, at the same time that socially marginalising and environmentally destructive trends are really threatening the massive investments involved. This developing situation should lead governments to try and intervene more than ever before, in order to protect and sustainably exploit the riches of their country's river basins. In many instances hydroelectric development might just provide the last opportunity to do so.

Deforestation is probably the single most important threat to the sustained exploitation of major tropical river basins' hydropotential and related uses. It is being encouraged by inequitable land distribution, low agricultural productivity, poor land-use policies, inappropriate timber and forest product development, weak institutions and rapid population growth. Open and closed tropical forests (2,970 million hectares in 1980) were cleared at a rate of 11.3 million hectares a year between 1976 and 1980, when an average of 18 to 20 million hectares were being modified from their original state every year (Hadley and Lanly, 1983, p. 11; FAO *et al.*, 1987). Closed tropical forests represent 1,188 million hectares and often are being logged as part of commodity-export programmes (Guppy, 1984, pp. 938, 950). In Thailand, the Philippines, Peninsular Malaysia and parts of Indonesia, forest assets have been depleted beyond the point of domestic self-sufficiency; regions with the more extensive forest areas now remaining are being felled rapidly. These regions include Indonesia's Kalimantan and central Brazilian Amazonia. In Indonesia forests have been disappearing at a rate of 1 million hectares a year and in the Brazilian Amazonia, at about 3 million hectares a year, mostly as a result of highway and cattle-ranching development (Kartawinata *et al.*, 1981, pp. 115–19; Salati, 1987, p. 64; Guppy, 1984, p. 941).

In Brazilian Amazonia, deforestation has increased exponentially, with much of the clearance occurring in the upper reaches of the rivers that hold most of the region's hydroelectric potential: for example in the transitional rainforest-cerrado zone in Mato Grosso, in southern Pará, Maranhão, Goiás and in eastern Pará. Here, clearing is increasing even without further considerable immigration, largely as a result of ownership transactions and conversion into speculative pasturelands (Fearnside, 1986a, pp. 74–81).

In tropical watersheds with intense seasonal rainfall regimes, extensive forest clearing encourages soil erosion and depletion of fertility. Seepage is hindered and runoff is accentuated. Seasonal river flow amplitudes are magnified and stream sediment loads increased. These curtail flood regulation and power generation and reduce the live storage capacity of reservoirs. Thermal and physico–chemical changes in the water quality may take place, up and downstream of the dam; these changes can affect water-dependent biota and, ultimately local human populations (Russell, 1981, pp. 942–3).

Destructive floods have been related to the denuding of Indonesia's Barito and Kahayan river catchments and of the River Ganges' Yamuna sub-basin in northern India (Guppy, 1984, pp. 942–3; Jayal, 1985, pp. 95–8). In Peninsular Malaysia, water-flow into the Muda and Pedu Reservoirs is now so low in some dry seasons that costly cloud-seeding has been attempted, in order to try and maintain acceptable levels for irrigation and power generation (Goh Kim Chuan, 1984, p. 28). Deforestation-related sedimentation now threatens the life span of the Karangkates Reservoir in Java's Brantas river basin, and India's Nizamsagar and Tehri Dams (Suryono, 1987, pp. 27–8; Dogra, 1986, pp. 201–2; Goodland and Ledec, 1986, p. 57). In order to save China's Samsen Gorge Project from excessive siltation, its dam and power plant had to be redesigned and rebuilt, with a sizeable generating loss compared with the originally planned installed capacity (Xiutao, 1986, pp. 23–4). In Kenya cattle are now grazing totally silted-up reservoirs on the Rivers Thiba and Chania (Pereira, 1981, pp. 7–8).

THE INSTITUTIONAL FRAMEWORK OF ASSISTED RESETTLEMENT INTEGRATED AREA DEVELOPMENT PROGRAMMES

Brazil's traditionally limited concern with integrated area development is somewhat surprising, given the numbers and potential of its river basins. This needs to alter rapidly to maximise the benefits of individual development projects in Amazonia. The current situation has largely been the cause and effect of the institutional and legislative vacuum at national and regional levels. There has been little interest in creating high-level management structures, with broad-based societal representation and sufficient jurisdictional and financial powers for satisfactory reservoir or watershed management, not to say basin development. Streams and rivers are federal property and their exploitation has customarily been assigned to private, then federal, hydropower enterprises. Without any legislation requiring dam planning to consider other possible uses, most reservoirs in Brazil have been created for single, mainly electricity-generation, purposes and have seldom been adapted to additional uses (Budweg, 1982, p. 48).

Admittedly, federal power holding-companies and their subsidiaries face a very difficult task in regions lacking a regional planning tradition. These companies are being challenged to assume developmental functions clearly beyond their conventional expertise and jurisdiction, and to cope with local and regional claims not always agreeable to federal priorities. In Brazil, recent environmental legislation and rising construction costs in remote areas might incline dam owners and designers to attempt integrated reservoir planning (Budweg, 1983, p. 30). But other countries' experience suggests that these have been little able to enforce multiple utilisation and environmental preservation beyond the limits of their project areas (Scudder, 1966; Pereira,

1981; Babcock and Cummings, 1984; Koenig, 1986; Srivardhana, 1987).

A Brazilian committee on integrated watershed studies was set up in 1979, to coordinate federal water resource development and establish executive committees for the various basins, in order to promote regional planning and the multipurpose development of streams and rivers (Budweg, 1982, p. 49). These executive committees' ability to establish river basin authorities will determine the extent to which, beyond reservoir-confined management, planning will contribute to local and regional development needs and opportunities. In Amazonia the Araguaia–Tocantins Integrated Development Project is presently the most serious effort towards integrated basin development (PRODIAT, 1982). But PRODIAT has been charged only with formulating guidelines and objectives for coordinating governmental plans and activities in support of socio-economic development in the AT basin. Beyond PRODIAT, the creation of an effective management structure for integrated and sustained basin development basically depends on the federal government being committed to such development.

Asia's more extensive experience in comparable geographical contexts shows some of the steps and requirements Brazil probably needs to address, in order to make river basin planning work in Amazonia. The Philippines' Bico river basin is comparable to the AT basin, in terms of its inequitable income distribution, poor transportation, natural hazard risks, inequitable land tenture, rapid population growth, and institutional deficiencies. Following studies by a planning board with advisory functions comparable to PRODIAT's, a development company with executive and financial powers was created in 1966, then replaced in 1973 with a River Basin Council. This coordinated and produced substantial feasibility studies for domestic and foreign development financing. However, due to inadequate jurisdictional and budget autonomy and ambiguous linkages with other sectors of the national government, the Council gave way in 1976 to the current River Basin Development Programme.

An appraisal of this programme's results after 10 years stresses that, in order to achieve integrated and sustained area development, the management structure needs to involve a broadly representative cross-section of basin society, as well as a broadly-based constituency at the national level. Participatory management is essential because sustainability requires project users' continuing commitment to utilise, maintain and further develop early investments. However, in order to flourish, local participatory initiatives need support and commitment from the central government itself. Extensive forms of coordination and consultation can do little when local initiatives are constrained by central governmental prescriptions and by powerful lobbying on the part of dissenting regional or local interest groups (Koppel, 1987). Chinese water resource development is effectively related to a viable administrative framework; it is based on centralised planning of large-scale projects, with river commissions and research institutes supporting local community-run managements units (Gustafson, 1987).

Certainly it would take time for functional basin authorities to become established, not to say basin development plans to be approved, in those Amazonian river systems which will be most disrupted by hydroelectric reservoirs. In the meantime, however, provisions are urgently needed to improve the planning of forthcoming reservoir-associated resettlement programmes. In order to be viable in the short run measures should make use to the largest possible extent of the current sectoral planning framework, institutional structure and domestic legal system.

Official guidelines for environmental planning during the hydroproject cycle currently assign plan-level priority to four major areas of concern (baseline surveys, environmental control, multiple utilisation, and expropriation). In each area, actions are to be developed from the plan to the project level, as the hydroproject itself evolves from the feasibility to the operational stage. Most actions are to reach their project-level definition at the basic-hydroproject stage (MME/ELETROBRÁS, 1986, Table I).

These guidelines have serious implications for forthcoming resettlement planning. First, although nearly half of the manual for hydroelectric projects is devoted to the baseline plan at the feasibility stage (a checklist of some 47 possible effects is provided), it is not clear how, why and by whom the aspects to be considered in baseline studies are selected. Secondly, resettlement *per se* (here including compensation, relocation, rehabilitation) is only one of many provisions to be covered by the environmental control plan, but no reference is made as to how resettlement might interact with other provisions of this plan. Resettlement appears to be conceived essentially as a mitigation and only economic activities to be directly affected are to be rehabilitated, although baseline studies should encompass the area of influence of the project. There is no accounting for how resettlement might affect or be affected, benefit or be benefited, by the master plan for reservoir development, for instance. Last but not least, the development cycle of resettlement is intimately associated with that of the hydroproject; it should reach programme-level definition at the feasibility stage and project-level definition at the executive-project level. No monitoring or evaluation of resettlement projects are mentioned at the operational stage.

The hydroelectric sector claims that project manual guidelines do nothing but consolidate the experience of past practice in the country. However, these guidelines present little which convincingly demonstrates that hydrodevelopers are capable of the timely and coordinated implementation, monitoring and evaluation of environmental actions, including resettlement. If these are the result of experience, then it is not difficult to understand why at Tucuruí, baseline surveys were conducted only at the implementation stage of the hydroproject, and were essentially concerned only with the area of specific intervention (future reservoir area). There was little concern to link resettlement baseline surveys and their results with surveys and findings concerning other related physical and biotic aspects, as well as resettlement planning with environmental control and multiple utilisation/reservoir development planning. Resettlement projects were all executed well before any master plan for reservoir development had been approved. Original resettlement proposals and the timetable were greatly disturbed by financial cutbacks and delays in the execution of the hydroproject itself. Clearly, current guidelines give little reason to believe that such deficiencies will be eliminated to any significant extent in future schemes in the country as a whole, and in Amazonia in particular.

Viable resettlement should re-create a complex social and environmental system at a new location and provide options for future development in time. If the latter is not possible, then resettlement managers should indicate alternatives and, if resettlees agree, provide the support for them to exploit such alternatives. These objectives require integrated planning and control over development in the area of influence of the reservoir. Where such requirements are generally not satisfied, as in Brazilian Amazonia, two basic provisions should be taken. First, the environmental control plan should be merged with the multiple utilisation plan, giving top priority to the definition and protection of the resettlement sites for the displaced communities. Second, the phasing of plans throughout the hydroproject cycle should be enforced. Currently, the cycle is subject by law to a multi-stage licensing procedure which requires the preparation of an environmental impact report, including recommendation of mitigation and enhancement measures, as well as their implementation once approved. Additionally, this licensing should also be subject to progress reports on the execution of the environmental control and utilisation plans.

In order to implement these provisions the hydrodeveloper should be granted sufficient powers to execute whatever provisions are required for project licences to be secured and renewed. As to the review on plan progress, it may take time to improve the operational capabilities of the competent authorities. Meanwhile, an advisory panel-type system could be used, involving community representatives and non-governmental organisations. Support for such a system already exists, both in international development agencies and in the hydroelectric sector in Brazil. In this way, and until functional river basin authorities can be established, resettlement would continue under the responsibility of hydrodevelopers. However, it would benefit from more comprehensive conception and timely implementation, which would be controlled by an external body and which would be required for the multi-stage licensing of the hydroproject itself.

THE CONTRIBUTION OF COMPREHENSIVE RESETTLEMENT PLANNING

One of the more difficult tasks assigned to power utility companies exploiting Brazil's rivers has been the resettlement of populations displaced by their power-generating schemes. Despite lessons of previous Brazilian resettlement projects (Augel, 1982; Germani, 1982; Andrade, 1983; Sigaud, 1984), at the Tucuruí Reservoir on the lower River Tocantins no basin or resettlement authority was designated or created which, given appropriate jurisdiction and financial autonomy, could have coordinated a more comprehensive strategy.

In many developing countries, even economically less advanced than Brazil, such as the Ivory Coast, Ghana and Nigeria, non-hydropower authorities have been specifically charged with resettling comparable or smaller population numbers than those displaced by the Sobradinho Reservoir on the River São Francisco or at Tucuruí. The Niger Dams Development Authority re-accommodated some 44,000 people relocated from the site of the Kainji Reservoir quite satisfactorily (Lassailly-Jacob, 1983, pp. 49, 50, 55). At Tucuruí, however, as at Sobradinho and other reservoir developments in Brazil, the owner of the hydroelectricity scheme was delegated the prime responsibility of planning, funding, and conducting resettlement.

In this way, resettlement is dependent on a hydroproject's timing and its spatial and financial requirements. Not only can this hinder resettlement from realising the potential of a comprehensive approach to water-based resource development, it may even jeopardise successful implementation of narrowly planned resettlement. The Tucuruí Resettlement Programme (TRP), for instance, the first resettlement programme in Brazilian Amazonia, is unquestionably a late addition to the global record. As such, it entailed costs which could well have been avoided and missed opportunities that could well have been seized, if a comprehensive approach had been adopted.

The significance of the TRP is considerable not only because international and local information is now available to assess it, but also and mostly because it is already acting as a reference for other reservoir-related proposals in Brazilian Amazonia. The following sections discuss selected components of resettlement planning which may be enhanced by such proposals, when conceived within an integrated approach to area development. Baseline studies, resettlement siting, and opportunities for economic and social development, are of particular concern.

Baseline Studies and Resettlement Planning

Baseline surveys can become powerful tools for sound resettlement proposals, when timely and properly carried out by social and natural science specialists (Eidem, 1973, p. 735). Unfortunately, in contrast to the engineering aspects of hydroprojects, there has been notably less concern to subject resettlement proposals to internationally recognised goals and standards. Also, their implementation and monitoring has seldom been submitted to the impartial scrutiny of experts (Lightfoot, 1981, p. 96). At Tucuruí, the initial report on the environmental risks was prepared by a single World Bank consultant (Goodland, 1978), but socio-economic surveys were done much later by a consultancy from Southern Brazil which specialised in construction and topographic surveys (BASEVI, 1979).

When baseline studies are assigned a relatively low priority their quality and ability to guide resettlement planning are usually severely curtailed. At Tucuruí, the stated objective of the studies was to predict the effects of the hydroproject, mainly its reservoir, on the population in its area of influence, and to set forth guidelines for site selection and urban resettlement (BASEVI, 1979, pp. 12, 13, 15). However, the studies present a number of important limitations, with respect to survey timing and approach, study universe, sample design, data collection, processing and analysis. Such limitations could explain why, when making decisions and choices during resettlement planning and execution, important disruptive effects were overlooked.

Survey Timing and Approach

At Tucuruí, there was little time for the survey of the area to aid planning, indemnification, transfer and resettlement. Field enquiries were carried out some three years after the beginning of hydroproject construction and only two years before the reservoir was scheduled to fill. The surveys came after a painstaking two-year analysis of local land tenure. They were possibly hastened through pressures by local municipal councils on various levels of public administration. Originally, the reservoir waters were to reach a height of 32 metres in 1979, which thus would have flooded two important villages; a leaflet advising people not to acquire any land within a mapped area had been distributed. However, the municipalities remained dissatisfied with the information given to them and, in late 1977, joined together to press for greater details on resettlement prospects. Shortly after, the hydroproject owner hired a consultancy firm to undertake baseline surveys and prepare a report.

A multi-stage approach carried out over at least one year would provide thorough and convincing documentation of riverine communities' seasonally changing relationships with their resource base. Instead, at Tucuruí a one-shot survey was attempted in the late rainy season (March 1979). This period typically affords limited insights into the real extent of river-related resource uses. Rural populations assemble at highground settlements, as local agrarian activities become more limited. The low prevalence of fish consumption reported in the study is probably due to question formulation and survey timing (low incidence), as it contradicts an earlier survey (Hidroservice,

1977), personal field interviews, and the observed increase in post-impoundment reservoir fisheries. Because surveys were scheduled for the wet season travel was difficult and more people were present in the survey region than would be the case at other times of the year. It is likely that the timing of the survey prevented field officers from verifying the actual extent and nature of land and water-based resource utilisation by local people.

Also, as a result of inappropriate timing, a survey may fail to detect a community's vulnerability to periodic environmental stress. As noted at other tropical impoundments (Koenig, 1986, p. 13; Reining, 1982, pp. 201–5), the high-water season at Tucuruí is very likely to be a period of the year when local people's savings have mostly been spent, food shortages are more likely, and morbidity and mortality are particularly high (Koenig, 1986, p. 13). A multi-stage survey could have supplied the evidence to show the real risk involved with programming relocation during the harvest period, as opposed to the post-harvest period. The survey strategy adopted prevented the identification of such risk, but secondary-source evidence could have been retrieved and used to prepare the report. At Tucuruí the evacuation and population transfer to new settlements largely took place during the rainy season of 1984.

Study Universe and Sample Design

When the study universe is ill-defined and the sample design defective, unnecessary survey costs may arise, as well as statistical bias in analyses. At Tucuruí, questionnaires were applied to 1,128 heads of household at 825 homes, representing very large settlement-specific and overall population percentages. This is quite unjustified by the details of reported analyses. Criteria which guided the selection of 11 settlements were not stated. For example, a major riverine town, Itupiranga, which was to be unaffected by flooding, *was* included in the survey, whereas important villages that were to be affected were omitted. Repartimento Central, the only dryground crossroad, was excluded (BASEVI, 1979, pp. 18, 22). Also, at each settlement, and contrary to what was argued by the report, sample stratification would have been feasible, if SUCAM field data had been duly exploited; this would also have permitted smaller sample sizes to afford equally or more representative population estimates.

Issues in Data Collection, Processing and Analysis

At Tucuruí, questionnaires and interviews covered a wide range of items, though the report only stresses a few aspects useful to planning. However, in spite of the detail, the survey failed to collect relevant information, or exploit the data base through meaningful cross-tabulations. Nor did it fully interpret even the simpler data displays or make practical recommendations based on them.

The study did record a high prevalence of local kinship ties amongst most interviewees, as well as generally strong community cohesion and attachment of people to their settlements; people preferred to be transferred along with their neighbours, to areas near water courses or at crossroads (BASEVI, 1979, pp. 264, 288, 290, 298). The study mentions the need to instal water treatment facilities and to undertake health education campaigns at new settlements; it also warns that relocated communities might be adversely affected by influxes of outsiders and by the outmigration of young adult relocatees (BASEVI, 1979, pp. 35,136).

On the other hand, the survey's questions on migration failed to address some aspects extremely relevant to resettlement assistance and siting. Data on interviewees' previous places of residence, and on their relatives and occupations held at these places, would have shown the location and importance of displacees' possible destinations. Additionally, data on local spatial mobility and resource utilisation at current place of residence would have indicated the river-dwellers' activity space, and the extent and frequency with which they resorted to river-based opportunities (marginal-lake fishing and floodland cropping). Both series of data could have suggested more sensible, possibly more economical, site alternatives at the time resettlement was being planned, rather than some relocation site being adopted in a spontaneous and untimely manner by resettlees themselves (as would later be the case).

If it had been available, the aforementioned information could have been correlated with that which was collected on income, household and community sizes. For instance, the average household size was found to be larger, average household income lower and the settlement-specific income range narrower in hamlets than in large villages (BASEVI, 1979, pp. 63, 66, 83). Poor rural families were, it appeared, usually large and their subsistence dependent on many individuals' ready access to local natural resources and work opportunities. Households' activity space is therefore not only necessarily confined but primarily defined by natural attributes of the residential site and its surroundings. Rural subsistence communities are not distributed randomly in space. Because so many poor families depend daily on such activity spaces, resettlement sites lacking one or more of these attributes can have a very negative effect on large numbers of relocated people.

Rather than elucidating this point, the report argued that since most interviewees were of rural origin, these would be unlikely to resettle in rural areas (BASEVI, 1979, p. 43). Residence however is undefined, and numbers of 'urban' residents at the time of surveys could partly reflect seasonal incidence. The study does not recognise urban residents' dependence on agrarian resources or on rural kin, in order to maintain themselves at settlements. Concern with collecting data on access to agrarian resources was limited to measuring landownership; even so this was not negligible, comprising 39 per cent (BASEVI, 1979, p. 117). Other potential indicators of urban residents' agrarian dependence were documented but overlooked by the report. For example 63.2 per cent of interviewees used charcoal and firewood as cooking fuel; and 46.3 per cent resorted primarily to home medicine (BASEVI, 1979, pp. 123, 145, 198). This medicine is available only to those with access to locally grown or naturally available medicinal plants. In developing tropical regions, rural families' nutritional well-being, and female-work contributions to it, largely depend on easy access to these natural commodities, plus such necessities as drinking water and wild food products (Koenig, 1986, pp. 29–32). At Tauiry near Tucuruí, the extraction by women of oil from palm nuts was important to the household economy of local residents (Biery-Hamilton, 1987, p. 20).

In the light of data on income, household and community size, it could further have been argued that poor families, the majority of the population, would risk becoming marginalised. This would happen even when simply relocating from small, rather homogeneous hamlets, to larger, more class-differentiated centres, without or within assisted resettlement projects. This might occur even with families which maintain former cash-income levels at new settlements. The report could have proposed occupational training strategies aimed at specific target groups in new settlements. Instead, high school absenteeism at old villages is emphasised and parents' desire for their children's betterment held as an endorsement for further schooling (BASEVI, 1979, p. 268). But the report disregarded youngsters' important role in poor families' survival strategies and how increased school attendance or outmigration might affect these communities' labour supply and future livelihood. If it had done so, it could have raised the potential for local training programmes to make use of local resources, in order to retain young people in rewarding occupations and improve their families' lot. At Tucuruí pre-reservoir communities were resettled in fewer and larger urban centres; traditional labour practices have evidently been affected (Biery-Hamilton, 1987, pp. 22–3) yet, four years later, no training programme had been implemented to address the problem.

Collective Resettlement Versus Individual Cash Compensation

Resettlement managers' perception of the area to be affected by a hydroproject, of population numbers to be disrupted, and of desirable indemnification procedures, largely depends on whether resettlement is considered more as an emergency capital cost or as a development opportunity. When charged with the resettlement the hydropower developer has tended to view it as self-contained social assistance, not as a highly interactive component of integrated area development. For obvious financial and jurisdictional reasons under this management, the real area affected by hydroprojects is usually misrepresented. The Tucuruí baseline study, for instance, did not define what was meant by 'area of influence' of the hydroproject to which it applied. Yet the fact that 10 out of 11 localities surveyed were situated within the actual reservoir areas, and that all were the larger settlements, suggests a very conservative understanding of what the disrupted area really might have been.

Narrowly defined hydroproject impacts, in turn, directly constrain the enumeration of people to be dislocated by the reservoir flooding. At Tucuruí, all estimates referred solely to those subject to evacuation from the reservoir: authorities were essentially concerned with people who would be legally entitled to indemnification. Even so estimates varied considerably over a short period of time (1976–9), from 4,000 to 17,000; all were probably lower than real numbers facing evacuation at the time estimates were made (Hidroservice, 1977, I, pp. 32–6; Goodland, 1978, p. 39; BASEVI, reported in *O Liberal*, 21 February 1984; communication of Martin Coy, 12 April 1984). Later estimates were attempted, based on SUCAM field counts, which suggest an on-site population of 25,000 to 35,000 in 1980 (Mougeot, 1986, p. 406).

However, none of these estimates considers the people living outside the flooded area but near the planned reservoir, who could have been seasonally dependent on river-based resources and employment for their subsistence. Some of these were probably seriously affected, even being forced to change residence as a result of impoundment, as observed at Sobradinho Reservoir in Brazil's North-East (Sigaud, 1984, pp. 46–51). At Tucuruí before impoundment, riverine islands were periodically submerged by the Tocantins and were being planted with subsistence crops by temporary settlers in the low-water season. The high-water baseline survey was confined to riverbank settlements; it does not seem to have taken into account those users' needs. A section of the west-bank Parakanã area which was expected to be flooded, was expropriated and evacuated; however, only part of that section was effectively flooded and remaining tracts, instead of being restituted, were retained to resettle colonists, thus denying the indians their previously direct access to the river.

Aspelin and Coelho dos Santos (1981, pp. 13–14) suggest a 20 kilometre-wide lake perimeter as the area likely to be most affected, directly and indirectly, by reservoirs in Brazil. Remarkably, at Tucuruí all preferred resettlement localities, except one, were located within 15 kilometres of major settlements that were expected to be flooded; these sites were possibly preferred because they were well known to the people and part of their activity space. Of course the quoted radius only affords a crude yardstick of the immediate area which might be affected in any particular situation. However, the order of magnitude is of interest and in any case this should be verified through sampling surveys on the activity space of representative communities within 15–20 kilometres of the margins of future reservoirs.

When resettlement is viewed as a hydroproject responsibility, the population effectively disrupted is likely to be undercounted. The real displacement costs and opportunities are also likely to be underrated or overlooked. The probable result is that large numbers of officially enumerated displacees end up being poorly benefited by resettlement; the more destitute groups are excluded through such a legally correct but inappropriate approach. This approach further selects from enumerated displacees those entitled by law to some compensation, and then from these, those with a right to assisted resettlement. Particularly in underdeveloped regions, as at Sobradinho (Sigaud, 1984, p. 22) and Tucuruí, groups have in this way been excluded from fair resettlement and make up a large share of the officially enumerated population.

This is because indemnification criteria and procedures, though legally consistent, can be notably unsuited to the locally prevailing land tenure and agrarian economy. In the Tucuruí region, for instance, only 20.8 per cent of the 4,334 properties surveyed by Hidroservice (the consultancy firm) had any property title and 44 per cent were *minifúndios* (small agricultural holdings). Non-proprietors were granted cash compensation for buildings and land investments. On the other hand, proprietors were given the option of either some cash compensation for possessions with rights to a new house or tract of land at an official resettlement, or a larger cash compensation without assisted resettlement. These procedures assumed that: (1) land is a private good for which the official permanent-ownership title is the necessary and sufficient proof entitling its bearer to land compensation; and (2) a one-time market value assessment of investments into, and returns from, land privately occupied or owned fully accounts for losses incurred by expropriation.

Currently in Brazilian Amazonia such assumptions poorly reflect the reality of riverine settlements. Here, an agrarian economy still prevails, characterised by social control but which is largely devoid of legal documentation. For instance at Tucuruí, Tauiry inhabitants cultivated in common some 260 hectares of land (Biery-Hamilton, 1987, p. 20). But the Tauiry system does not limit itself to dryground croplands. It also embraces: turtle-breeding beaches, seasonally-silted island, levee and marginal floodlands, dryground wild-fruit and nut stands, timber and firewood-supplying gallery forest, placer-mining sites, hunting and fishing grounds along marginal lakes, marshlands and tributaries. It depends heavily on a series of natural products seasonally available in land, air and water environments, largely exceeding individually or collectively farmed tracts of land.

As could be expected from the indemnification procedures originally applied, most of the reportedly 4,000 families displaced by the Tucuruí Reservoir were granted limited individual cash compensations only. Even in these terms fair indemnification was hindered by untitled properties, falsification, deficiencies and delays in updating records, and land squatting by outsiders in the early 1970s. Measures to normalise land tenure, in years preceding indemnification, were also evidently phased down in pace and extent in order to limit disbursements. The acquisition of permanently titled estates, as opposed to provisionally or untitled land, entailed sharply higher costs for the hydroproject owner; numbers and extent of the former would very much affect the amount of resources to be set aside for cash compensation and resettlement. As a result and despite longstanding occupancy, most tracts on the western margins of the Tocantins, to the north and between 110 and 153 kilometres of the BR-230 highway, had not been issued with permanent ownership titles by the time land assessment surveys were carried out. This, in part, was to limit expropriation expenses and contradicted an early decision, reported in Goodland (1978, p. 39) that the right of possession would be respected (Mougeot, 1986, p. 407).

Collective resettlement should be more appropriate than individual cash compensation in Brazilian Amazonia, where socio-economically vulnerable groups predominate. For one thing, fair cash compensation rates and policies have been difficult to establish, due to valuation factors involved and conflicting opinions on valuation. Misreporting can also occur, and it is difficult to verify when this has occurred. When based on assessed values, compensation rates usually undercut real market prices, and even when they do not, seldom enable evacuees to acquire land or livelihood of comparable value. More often, the compensated fail to recover their former living standards without loss (Lightfoot, 1981, p. 101). Also, externalities unforeseen or difficult to quantify are not taken into account, such as mass relocation-induced price inflation of land and other goods and services, maintenance expenses caused by delays in compensation payments or reaccommodation, and misappropriation of benefits.

Low-income groups tend to be particularly affected by such externalities. It was observed at Thailand's Huai Luang Reservoir (Lightfoot, 1981, p. 103) and at Tucuruí that expropriatees with more personal savings and/or larger indemnities were able to anticipate land price rises and secure the better and more accessible tracts of land, in the Tucuruí case at the Nova Jacundá resettlement (Mougeot, 1986, pp. 412–13). On the other hand, the majority who are ineligible for assisted resettlement receive small cash compensation. When seeking new accommodation or subsistence they became particularly vulnerable to profiteers. This risk can also be faced by expropriatees entitled to assisted resettlement, when there are long delays between cash indemnification and actual relocation; the risk grows when no food relief or payment adjustment are implemented, as was the case at Tucuruí. Here, as at the Huai Luang Reservoir (Lightfoot, 1981, pp. 101–2), clandestine trading of compensation and resettlement rights at discount prices was already considerable three years after compensation, at rural and urban resettlements released in 1981 (Mougeot, 1986, p. 413).

Integrated Area Development Planning and Resettlement Siting

In Brazilian Amazonia, the appropriate siting of sufficiently large areas of landholding granted as compensation will be crucial to the readjustment of rural populations which depend on agrarian activities, as opposed to non-farming populations (Gajaran, 1970). Total areas required will vary according to the numbers to be resettled, but unplanned, would-be resettlees could exert pressure on the carrying capacity of areas reserved for those officially entitled. However, as part of integrated area planning, provisions can be made to anticipate and curb unwanted immigration at resettlement sites (Leistritz and Murdock, 1986). The size of individual lots should account for various household sizes (Lightfoot, 1981, p. 111; Koklu, 1982, p. 19). Also, the need for contiguous expanses of land, mainly to preserve community cohesion, should be verified rather than assumed, since acquisition might otherwise entail unnecessary expropriation costs. Typically, however, riverine communities displaced by impoundments prefer alternative locations with similar locational and resource-base qualities and opportunities to those they are required to evacuate (Eidem, 1973, p. 735; Afriyie, 1973, p. 727; Lightfoot, 1981, p. 105; Mills-Tettey, 1986, p. 31). Failure to account for people's preferences may threaten post-resettlement community resilience or stimulate counter-productive self-resettlement initiatives. Populations self-relocating might face the need to relocate again, as will be the case at Tucuruí, where it is planned further to raise the reservoir's current storage capacity.

Resettlements have rarely secured sufficiently large areas at those places initially preferred by communities as relocation sites. The main reasons include scarcity of land available at affordable expropriation costs, land-use and health risks posed by post-impoundment environmental changes, and conflict with expected site requirements and areal impacts of other planned developments. Due to land scarcity in densely occupied regions, long-distance resettlement of people uprooted by reservoirs is likely to increase. Some 10,000 families displaced by West Java's Saguling Hydroproject might need to be 'transmigrated' to Sumatra, Sulawesi or Kalimantan (Otten, 1986, p. 231). In Brazil, the same problem has already hindered sensible solutions outside Amazonia, as epitomised in Paraná state by the need for many Itaipú Dam-displaced families to relocate to one of their cooperative's colonisation schemes in Amazonia (Branford and Glock, 1985, pp. 90–1). The problem is growing in Amazonia as well.

With a view to integrated area development, timely feasibility studies may not only permit resettlement managers to prevent the problems just mentioned and better to account for people's preference than is currently the case. They can also prevent planners from devising seemingly responsive but misinformed solutions, which might prove detrimental later on. More importantly, such studies can suggest opportunities both to planners and prospective resettlees; by articulating resettlement site requirements with those of other compatible local developments, costs might be curbed and benefits expanded.

Resettlement proposals which neglect this potential not only miss opportunities but also may find their implementation blocked as well. At Tucuruí, communities consulted during baseline surveys quite expectedly had suggested resettlement solutions which would preserve their territorial cohesion and access to river resources. Accordingly, the proposed network included ten new sites, of which eight were rural and were located upstream of the Dam near the projected reservoir shoreline and close to existing communities. But this proposal in reality was altered substantially, in terms of numbers, location and execution time. Fewer rural sectors (five) and more urban centres (six) have actually been created, most well away from the reservoir shoreline. Most rural sectors have been partly or mostly settled by families not displaced by the Reservoir, while most of the six urban sites have been located at pre-existing settlements. More importantly, tracts of farmland were made available to resettlees much later than were the urban plots, and generally were located at some distance from the latter. The pattern which finally emerged has thus concentrated formerly fairly isolated, small, riverine rural communities practising a subsistence economy, into fewer and larger roadside towns, located in dryground areas remote from the lake and suitable farming plots, in regions now undergoing intense population growth.

The key obstacle to the execution of the proposed siting stems from the way in which resettlement has been subordinated to the hydroproject's own funding, jurisdictional and timing priorities. These priorities did not always coincide with those of the siting proposal and, in fact, modified it substantially. One problem was the then (1976) underdeveloped road network within and around the future lake. And this is likely to be a recurrent problem with future reservoirs in Brazilian Amazonia. At Tucuruí, many road sections would have had to be built or restored on both margins of the Tocantins to meet plan needs. But road construction has actually proceeded to suit primarily the needs of the hydroproject's engineering works. One contributing factor was restrictions imposed during project implementation by external funding agents on the amount and rate of investment originally scheduled for the hydroproject itself. This induced the power company to reallocate spending to further productive investments while downgrading and putting off assistance items such as resettlement.

Consequently, on the site of the future reservoir, highway maintenance not considered vital to the hydroproject was phased out. This affected directly the more settled, western river margin, where few access roads were opened between established communities and existing roads and localities where the communities were to be evacuated. At the same time, service roads were being built on the less-peopled north-eastern river margin to permit selective clearing of timber from the area to be flooded. On future reservoir margins, road detours required on the western bank by the resettlement proposal and needed to support local trade were built long after those on the eastern bank had been extended and upgraded in order to transport supplies to the dam site and to service construction and maintenance of high-voltage transmission lines.

The overall result was that rural and urban resettlement sites originally proposed to be near the reservoir, mostly on its western shores, were finally sited further away and to the east of it on unused tracts of dryground estates already expropriated for the passage of transmission lines and roads. Together with the downward reassessment of probable reservoir flooding levels and late road opening in the recovered lakeshore area, these alternative sites reduced resettlement costs by permitting fuller occupancy of acquired lands. As a result, land on the north-western lakeshore in convenient reach of the new roads, the lake and the town of Tucuruí also was released to large ranches.

The siting of resettlement should form part of a plan for the multiple utilisation of the watershed, or even the reservoir region. At Tucuruí, this could have been done. Instead, studies for each purpose were conducted by different consultancy firms at different times. In 1985, when most of the resettlement had been largely implemented, ground surveys designed to assist in resettlement were still under way and no plan was to be approved until late 1987. Consequently, resettlement managers had little assistance in identifying and evaluating alternatives, nor in adjusting those implemented to interact with other local developments. Possible compatible or competing local developments include official agricultural colonisation and land reform schemes on the western margin of the reservoir and upstream at Carajás, industrial estates at the southern and northern extremities of the Reservoir, private logging and cattle ranching concerns in its vicinity, lake fisheries and related activities in lakeshore towns, drawdown and dryland agriculture. Other labour-intensive and profitable activities also exist, for which training and management could be provided in order sustainably to utilise, protect, and monitor the new lake environment.

Resettlement planning has been poorly integrated with other local developments and when this occurs it can generate avoidable costs and losses, and fail to attract potential benefits. At Tucuruí, one unnecessary loss was the partial flooding and, subsequently, entire deactivation of INCRA's Marabá Integrated Colonisation Project (PIC). One federal body thus was allowed in the late 1960s to plan and, throughout the 1970s, implement a costly colonisation scheme precisely in a region which other federal agencies knew or strongly suspected would be at risk of being disrupted by a reservoir. Hydrological surveys had been carried out in 1954, the THC was being planned at least as early as 1968 (well before the Marabá project was established) and was approved in 1974. Lack of intra-governmental communication has caused unnecessary spending in order to reroute highways and indemnify and resettle rural proprietors, and also precipitated the abandonment of the INCRA project.

On the western margins of the Tucuruí Reservoir, the flooding of the highway caused at least 621 colonists, possibly as many as 800, to be relocated. These people used to farm 100-hectare plots of land between 80 and 180 kilometres of the Marabá–Altamira stretch of the BR-230 highway. INCRA began withdrawing from the scheme nearly two years before indemnification and a five-year delay followed before entitled families could actually be relocated. Land titling procedures were halted in 1978, and in late 1979 the colonisation project was released from INCRA's jurisdiction, though the reservoir would begin to fill only in late 1984. Originally, only farmsteads likely to be flooded lost INCRA's support but, by 1980, all the Marabá colonisation project's 2,484 rural properties and 70 urban plots had been released (i.e., abandoned without support) along the BR-230 to the east and west of Marabá township, and at the *agrovilas* (planned rural service centres) Castelo Branco and Cajazeiras. In this way, the colonisation scheme lost access to POLAMAZONIA funds, which had proved vital to defray road maintenance, basic social services and land titling procedures. Also during that period, practically all PIC administrative staff were transferred from INCRA regional headquarters at *agrópolis* Amapá, in Marabá, to a local branch unit of the Grupo Executivo de Terras do Araguaia–Tocantins (GETAT), posted at Tucuruí. Not only losses would be incurred in dismantling this federal project to make room for another.

Development Opportunities for Future Comprehensive Resettlement

Opportunities can be neglected from the onset, or they can be seized or created through compatible land and water use planning for the future reservoir region. At Tucuruí, those farmers who were to be disrupted could, at the very least, have been informed and, if they so wished, should have been reaccommodated at other colonisation schemes in the region. In this way, costs would have been spared to both hydroproject owner and resettlees. At Tucuruí, the governmental body in charge of executing rural resettlement, GETAT, was also engaged in colonisation schemes elsewhere in the region. Its social funds were then aimed primarily at the *Companhia Vale do Rio Doce* (CVRD) colonisation projects at Serra dos Carajás. However, not only did rural resettlements at Tucuruí receive comparatively less assistance than CVRD schemes but also no support or encouragement was formally provided to displaced families to help them resettle at other agricultural projects. In early 1984, when evacuation of the Reservoir was in full swing, GETAT reported that about 10 per cent of the 520 colonist families at the CVRD's schemes, 125 kilometres south-west of the PIC Marabá, had come from the Tucuruí reservoir area but they had been admitted on their own initiative, without any assistance from the TRP.

Above all, reservoir-related resettlement can benefit greatly from closer interaction with wetland farming and fishing schemes. As is often the case in Brazil and elsewhere, resettlement is primarily in the hands of land settlement agencies and tends to be modelled on dryground agricultural colonisation schemes (Lightfoot,1981, p. 104). More often, it fails to interact with other agrarian opportunities and needs. Similarly narrow-minded resettlements were executed in the Ivory Coast, where land-use planning emphasised sites dissociated from the Volta Lake and where, as a result, various conflicts emerged between wetland farmers and fishermen (Thomi,1984, p. 11). More appropriate lake-based developments were implemented on the Ghanaian margins of the same lake and at Kainji, Nigeria, where irrigated fields, experimental farms and mechanisation projects were gradually introduced, to upgrade traditional drawdown-cropping practices (Lassailly-Jacob, 1983, p. 55). At Tucuruí, however, increased road accessibility has encouraged indiscriminate marginal and drawdown-logging, even in areas set aside for conservation, whereas former subsistence river users have been denied access to appropriate drawdown-cropping and lake-fishing practices (Biery-Hamilton, 1986, p.34; Mougeot, 1986, p. 411).

In Brazilian Amazonia, occupancy trends which are gaining momentum are likely to seriously hinder, if not overwhelm, comprehensive reservoir, not to say watershed, planning and management. At Tucuruí since 1975, land concentration and forest clearance have proceeded rapidly on the reservoir margins, with deforestation gaining momentum since 1980. Since 1983, further clearings have been prompted by additional highway construction to the west, and by resettlement-related feeder road development to the east of the Reservoir (Barrow and Mougeot, 1987, pp. 22–7). In combination with mining in tributary watersheds, expected massive industrial-fuelwood cutting and agrochemical-dependent agricultural development upstream of Tucuruí, deforestation probably will increase soil erosion, reservoir siltation and pollution (Mallas and Benedicto, 1986, pp. 248–9; Fearnside, 1986b, pp. 407–10). These trends, in turn, may pose serious economic and health risks to reservoir-dependent animal populations and human communities.

The most serious health risk to human communities in the reservoir region is the possibility that *schistosomiasis mansoni* will become endemic. The growth, drowning and rotting of vegetation over as much as 57,000 hectares (between lake levels of 66 and 72 metres) along some 3,700 kilometres of mostly dissected, gently sloping and shallow lake shoreline, has been observed to recur seasonally since 1985 (Junk and Nunes de Mello, 1987, p. 379). The natural acidity of lake waters could be reduced by this, by agrochemical and lime-contaminated runoff from marginal farmlands, and by urban industrial effluents. On the reservoir margins, the application of lime and chemical soil correctives, insecticides, fungicides and pesticides is a recent practice but one which is spreading very rapidly and at an increasing pace due to commercial cash-cropping and cattle-ranching. Though no firm census data are available, the pace of adoption of lime and agrochemicals suggests that agrochemical pollution and acidity reduction are almost certain to occur in the Reservoir and its tributaries (Barrow and Mougeot, 1987, pp. 27–30).

If the natural acidity of lake waters were reduced locally and seasonally, the already very serious risk of *schistosomiasis* becoming endemic could materialise. This is because a number of other required conditions are already present at Tucuruí. Extensive growth of potentially snail-harbouring macrophytes (*Salvinia auriculata*) has been recurrent during rainy seasons at Tucuruí. Potentially miracidia/ceracariae-carrying snail species, *Biomphalaria staminea* and *B. glaberata*, which exist in north-eastern and southern Brazil, have been found in wetlands and tributaries associated with limestone outcrops to the east, west, north and south of the reservoir (Goodland, 1978, pp. 23–4). *B. staminea* individuals were collected recently in the reservoir area by ELETRONORTE-contracted INPA–MPEG researchers (Mougeot, 1987, p. 3). Most migrant workers attracted by dam construction activities and migrant settlers previously established in the reservoir region before its creation are known to come from the North–East, where *schistosoma haematobium* and *s. mansoni* are widespread (Weil and Kvale, 1985, p. 193). They could introduce the latter at Tucuruí to infect snails; this is suspected to have been the case at Lake Kariba (Weil and Kvale, 1985, p. 204).

If *schistosomiasis* were to become endemic, it would primarily affect users of the drawdown and litoral zones of the lake environment. Evidence suggests that infection rates would be particularly high amongst children, and especially at the end of the rainy season when water levels are high and snail host populations larger, the risk being greater at midday when peak shedding of cercariae by snails occurs (Weil and Kvale, 1985, pp. 198–9).

The success of programmes to prevent, treat, monitor, and control the disease in the reservoir region, would very much depend on local community participation, as shown by Chinese experience and suggested by recent studies in Kenya (Szekely, 1982, p. 239; Toomey, 1986, p. 8). In the near future, programme effectiveness might be enhanced, notably through eradication of the intermediate hosts, the snails, using natural molluscicides such as the African endod plant, *Phytolacca dodecandra* (Mkamwa, 1986, p. 9). Proper siting of resettlement at minimal distance from the lake shoreline can also reduce infection risks, as observed at Ghana's Kpong Reservoir (Derban, 1984, p. 15). Most crucial would be adequate sanitation and public education. Other diseases which might be introduced or whose incidence might be increased and affect local populations, resettled or not, include, yellow fever, Chagas disease, river blindness, and bubonic plague (Barrow and Mougeot, 1987, pp. 30–4). Also, specimens of five potential phlebotomous vectors of *leishmaniosis* have been collected in the reservoir region by INPA–MPEG researchers (Mougeot, 1987, p. 3).

Agricultural pollution may affect the diversity and productivity of farming and fisheries. Research in Nigeria on pesticide application shows that it can considerably harm soil fauna, particularly detritivores which play a key role in soil dynamics (organic matter decomposition and nutrient cycling). This could lead to decline in crop yield and, in the long run, might jeopardise the soil system's capacity to recover productivity under fallow (Perfect, 1980, p. 21). There was a real risk that fish stocks would be decimated at the Sobradinho Reservoir in north-eastern Brazil, at the beginning of the rainy season in late 1987, as a result of indiscriminate applications of pesticides and other agrochemicals used to cultivate drawdown-area onion crops (*O Liberal*, 28 November 1987, p. 24).

When developed within integrated reservoir or watershed management, rural resettlement can further the local communities' well-being and a hydroproject's profitability in many ways. Intensive tropical forest management became urgent worldwide in the 1960s, and was revived in the late 1970s (Spurgeon, 1987, p. 6) as part of a diversified approach to watershed management in many tropical river basins (Russell, 1981, pp. 13–15). This now involves technically possible and economically sound labour-intensive land-use systems, ranging from natural forest management to productive afforestation of degraded lands, as in Ghana (FAO *et al.*, 1987, p. 17). These approaches can also make use of slightly modified traditional shifting cultivation (taungya), afforestation with indigenous and exotic species, combined with perennial crops or nut and fruit trees, or with carefully managed, short-season arable crops (Kunkle and Dye, 1981, pp. 104–5). In Brazil, some agroforestry practices have been encouraged on the eastern shores of the Itaipú Reservoir. However, as in any watershed undergoing rapid settlement, programmes need to be implemented swiftly and extensively to curb erosion and siltation to any significant extent. It is only through massive programmes that floods have been moderated, sedimentation rates lowered, and water availability and crop yields increased in India's Damodar Valley, and that the life span of Pakistan's Tarbella and Mangla might be extended (FAO *et al.*, 1987, pp. 9–10).

In reservoir areas, possible water-based alternatives include wetland and drawdown agricultural uses, through chinampa-like systems of flood-tolerant cropping, suitable for dryground-crop manure and compost, and highly productive, high protein-content and disease-resistant aquatic and semi-aquatic grasses and herbs for cattle fodder. Fish-farming and the productive management of aquatic fauna have also been attempted within such drawdown areas (Klinge *et al.*, 1981, pp. 26–8). Now widely used in many tropical countries, enclosure aquaculture can cost less than stocking or managing large hydroelectric reservoirs with wild fish stock; some suitable species are known to be little disrupted by seasonal reservoir-level fluctuations (Powles, 1986, p. 18). At Kenya's natural Lake Victoria, fish-catch marketing and processing provide important work opportunities for women and children; more equitable local income distribution is being sought through increasing the involvement of fishermen cooperatives in such activities (Prazmowski, 1987, p. 13). Here, as at man-made reservoirs, processing and transportation to markets can be upgraded to increase fish-product value and sales, but at reservoirs some costs can further be curbed through appropriate resettlement siting. At reservoirs with extensive drawdown areas, as will be the case in the lower reaches of Amazonian rivers, large quantities of invertebrates stranded by receding waters can provide a rich source of protein for productively-managed, drawdown-feeding wildlife; this might be preferred to intensive cattle-grazing. By releasing nutrients (phosphate, ammonia) through defecation and urination, the cattle might boost flood-related phyto and zooplankton growth, thereby increasing waterborne-disease risks, as observed at Lake Kariba (Ramberg *et al.*, 1987, pp. 314, 319).

In areas where land- and water-based uses just discussed have been implemented, the participation of local populations has proved to be crucial. This is likely to be the case in Brazilian Amazonia as well. Concerning agroforestry, low-cost nurseries run by local residents where trees need to be planted have been essential to progress in Thailand, India, Nepal and Peru. Different users can and should be encouraged to join in these efforts in order to enhance results. In Mexico, private estate owners and subsistence farmers collaborate to protect reserves and cooperate with scientists to improve land-uses in their vicinity. Measures can be taken to finance and encourage the adoption of such practices. In Nepal, wages are being paid to tree planters over an initial period; in Colombia, revenues from a tax on the sale of electric power from large hydroprojects are used to promote watershed management; in Zambia, workers on an industrial timber plantation are given forest and sawmill residues which they use to produce charcoal (Perry and Dixon, 1986, pp. 43–4; FAD *et al.*, 1978, pp. 9–10, 14, 19).

Indigenous populations can play an invaluable role in integrated area development, as much in Brazilian Amazonia as they have done elsewhere in the world. Panama's Cuna Indians have been working with national and international agencies to develop multiple-use areas and tropical forest reserve management; they also have been engaged in environmental education programmes and in research on traditional agroforestry (FAO *et al.*, 1987, p. 22). Native Amazonian people can indicate to planners hitherto unknown, low-cost, and environmentally sound methods to exploit specific environments. The Kayapó of Southern Pará are now known to possess a remarkable ability to create and manipulate microenvironments, within and without natural ecozones, in order to increase biological diversity and sustained productivity; techniques include cultivation of forest islands and use of pest biocontrols, among others. Also, the Kayapó's reliance on herbal medicine has led them to develop very elaborate pharmacological classifications of plants and of the diseases against which these are aimed. The indians' categories and treatments of fuelwood species are equally complex (Posey, 1987, pp. 98, 100–3). This practical knowledge is obviously pertinent to the sustained and socially beneficial utilisation of reservoirs and their vicinity. Participatory strategies to impoundment-related assisted resettlement can not only better respond to traditional strengths and capabilities of local populations; they can also avoid destroying their productive ecological adaptations and ways of life (Babcock and Cummings, 1984, p. 23).

CONCLUSION

Over the next decades, Brazil will probably experience some of the largest population dislocations ever associated with hydrodevelopments in the humid tropics. A strong case can be made for the comprehensive resettlement of populations which will be disrupted by future hydroprojects and related actions in Brazilian Amazonia: (a) the country possesses a well-established programme to tap the energy potential of this region's river system; (b) this programme will predictably displace not only large population numbers but communities requiring differentiated, assisted resettlement; (c) the programme is the first to make use of Amazonia's natural geounits for spatially integrated development: its river basins. Comprehensive assisted resettlement could ensure that local communities are satisfactorily involved in developing the resource potential created by hydroelectric schemes; it could also ensure nothing is overlooked that will be required for the proper functioning of these projects. When devising a strategy appropriate to regional needs, Brazil can benefit much from similar experiences in other tropical and humid tropical regions, where sizeable populations have been displaced by large reservoirs.

Note

1. Research for this paper was made possible by a CAPES–British Council postdoctoral fellowship. The author is grateful to Chris Barrow, Ron Bisset, Anthony Hall, and David Goodman for their very useful comments on earlier drafts and related papers. Errors and opinions still present in this version are the author's sole responsibility.

References

Afriyie, E. K. (1973), 'Resettlement Agriculture: An Experiment in Innovation', in W. C. Ackermann, G. F. White and E. B. Worthington (eds), *Man-Made Lakes: Their Problems and Environmental Effects*, Geophysical Monograph 17. Washington, D.C.: American Geophysical Union, pp. 726–9.

Andrade, M. C. de (1983), 'Nordeste, progresso e pobreza', *Revista Brasileira de Tecnologia*, 14(1) pp. 3–10.

Anon. (1980), 'Zaire's Hydropower', *International Water Power & Dam Construction*, 32(2) (February), pp. 50–2.

Aspelin, P. L. and Coelho dos Santos, S. (1981), 'Indian Areas Threatened by Hydroelectric Projects in Brazil', *IWGIA Document*, 44, Copenhagen: International Workgroup for Indigenous Affairs.

Augel, J. (1982), 'Human Settlement Problems in Brazilian Development', *Ekistics*, 49(292) pp. 31–6.

Babcock, T. G. and Cummings, F. H. (1984), 'Land Settlement in Sulawesi, Indonesia', *Malaysian Journal of Tropical Geography*, 10 (December) pp. 12–25.

Barrow, C. J. (1983), 'The Environmental Consequences of Water Resource Development in the Tropics', in Ooi Jin Bee (ed.), *Natural Resources in Tropical Countries*, Singapore: Singapore University Press, pp. 439–71.

Barrow, C. J. (1981), 'The Health and Resettlement Consequences and Opportunities Created as a Result of River Impoundment in Developing Countries', *Water Supply and Management*, 5(2) (1981) pp. 135–50.

Barrow, C. and Mougeot, L. J. A. (1987), 'Natural Resources Exploitation in Eastern Amazonia: The Araguaia–Tocantins River Basin', Swansea and Belém: Centre for Development Studies and Núcleo de Altos Estudos Amazônicos (mimeo).

BASEVI, C. e T., SA (1979), *Estudo das condições sócio-económicas da área de influência do reservatório da UHE de Tucuruí, PA*, 3 vols, Brasília, D.F.: Centrais Elétricas do Norte do Brasil, SA.

Biery-Hamilton, G. M. (1987),'The Impact of the Tucuruí Dam on a Small Riverside Town', Centre for Latin American Studies, Gainesville: University of Florida (mimeo).

Biswas, A. K. (1983), 'World Status Report on Hydroelectric Energy', *Mazingira* 7(3) pp. 24–34.

Bower, B. T. and Hufschmidt, M. M. (1984), 'A Conceptual Framework for Analysis of Water Resources Management in Asia', *Natural Resources Forum* 8(4) (October) pp. 343–56.

Branford, S. and Glock, O. (1985), *The Last Frontier: Fighting Over Land in the Amazon*, London: Zed Books.

Brokensha, D. and Scudder, T. (1968), 'Resettlement', in N. Rubin and W. M. Warren (eds), *Dams in Africa: An Interdisciplinary Study of Man-Made Lakes in Africa*, London: Frank Cass, pp. 20–61.

Budweg, F. M. G. (1980), 'Environmental Engineering for Dams and Reservoirs in Brazil', *International Water Power & Dam Construction*, 32(10) (October) pp. 19–23.

Budweg, F. M. G. (1982), 'Reservoir Planning for Brazilian Dams', *International Water Power & Dam Construction*, 34(5) (May) pp. 48–9.

Budweg, F. M. G. (1983), 'Water Resources and the Environment: Development Planning in Brazil', *International Water Power & Dam Construction*, 35(7) (July) pp. 23–30.

Butcher, D. A. P. (1971), *An Operational Manual for Resettlement: A Systematic Approach to the Resettlement Problem Created by Man-Made Lakes, with Special Reference for West Africa*, Rome: FAO.

Canter, L. (1985), *Environmental Impact of Water Resources Projects*, Chelsea, Michigan: Lewis Publishers.

Chambers, R. (ed.) (1970), *The Volta Resettlement Experience*, London: Pall Mall.

CNEC (Consórcio Nacional de Engenheiros Consultores, SA) (1985), 'Viabilidade do

complexo hidrelétrico de Altamira: avaliação preliminar dos impactos na qualidade de água face à vegetação inundada', Rio de Janeiro (mimeo).
Dasmann, R. F., Milton, J. P. and Freeman, P. H. (1973), *Ecological Principles for Economic Development*, London: John Wiley.
Derban, L. K. A. (1984), 'Health Impact of the Kpong Dam in Ghana', *International Water Power & Dam Construction*, 36(10) (October) pp. 13–15.
Dogra, B. (1986), 'The Indian Experience with Large Dams', in E. Goldsmith and N. Hildyard (eds), *The Social and Environmental Effects of Large Dams*, vol. 2, Camelford: Wadebridge Ecological Centre, pp. 201–8.
Dragun, A. K. (1983), 'Hydroelectric Development and Wilderness Conflicts in South-West Tasmania', *Environmental Conservation*, 10(3) (Autumn) pp. 197–204.
Duarte, A. T., Pettena, J. L. and Paes Filho, J. da Rocha (1983), 'The Altamira Hydro Complex in the Amazon Region', *International Water Power & Dam Construction*, 35(10) (October) pp. 23–8.
Eidem, J. (1973), 'Forced Resettlement: Selected Components of the Migratory Process', in W. C. Ackermann, G. F. White and E. B. Worthington (eds), *Man-Made Lakes: Their Problems and Environmental Effects*, Geophysical Monograph 17, Washington, D.C.: American Geophysical Union, pp. 73–7.
FAO (Food and Agriculture Organisation of the United Nations), International Wildlife Research Institute, World Bank, and United Nations Development Program (1987), *The Tropical Forestry Action Plan*, Washington D.C.: FAO.
Fearnside, P. M. (1986a), 'Spatial Concentration of Deforestation in the Brazilian Amazon', *Ambio*, 15(2) pp. 74–81.
Fearnside, P. M. (1986b), 'Os planos agrícolas: desenvolvimento para quem e por quanto tempo?' in J. M. Goncalves de Almeida, *Caraás: desafio político, ecologia e desenvolvimento*, São Paulo: Brasiliense; Brasília, D.F.: Conselho Nacional de Desenvolvimento Científico e Tecnológico, pp. 362–418.
Fels, E. and R. Keller (1973), 'World Register on Man-Made Lakes', in W. C. Ackermann, G. F White and E. B. Worthington (eds), *Man-Made Lakes: Their Problems and Environmental Effects*, Geophysical Monograph 17, Washington, D.C.: American Geophysical Union, pp. 43–9.
Flavin, C. (1986), 'Trends in Third World Hydro Development', *International Water Power & Dam Construction*, 38(10) (October) pp. 26–8.
Gajaran, C. S. (1970), *Planned Rehabilitation and Economic Change: A Case Study of Tungabhadra River Project Rehabilitation Colonies at H.B. Halli*, Poona: Gokhale Institute of Politics and Economics.
Germani, G. (1982), 'Os expropriados de Itaipú. O conflito: Itaipú x colonos', *Cadernos do PROPUR 3*, Porto Alegre: Universidade Federal do Rio Grande do Sul.
Goh Kim Chuan (1984), 'Water Resources Management in Malaysia', *Malaysian Journal of Tropical Geography*, 9 (June) pp. 28–38.
Goldsmith, E. and N. Hildyard (1984), *The Social and Environmental Effects of Large Dams*, vol. 1, Camelford: Wadebridge Ecological Centre.
Goodland R. (1978), *Environmental Assessment of the Tucuruí Hydroproject, Rio Tocantins, Amazonia, Brazil*, Brasília, D.F.: Centrais Elétricas do Norte do Brasil, SA.
Goodland, R. (1986), 'Hydro and the Environment: Evaluating the Tradeoffs', *International Water Power & Dam Construction*, 38(11) (November) pp. 25–33.
Goodland, R. and Ledec, G. (1986), 'Environmental Management in Sustainable Economic Development', *Impact Assessment Bulletin*, 5(2) (Summer) pp. 50–81.

Gordon, J. L. (1983), 'Recent Developments in Hydropower', *International Water Power & Dam Construction*, 35(9) (September) pp. 21–3.

Guppy, N. (1984), 'Tropical Deforestation: A Global View', *Foreign Affairs*, 62(4) (Spring) pp. 928–65.

Gustafson, J. E. (1987), 'Some Reflections on the Chinese Water Resources Development Experience', *Water Resources Development*, 3(1) pp. 63–72.

Hadley, M. and J. P. Lanly (1983), 'Tropical Forest Ecosystems: Identifying Differences, Seeking Similarities', *Nature and Resources*, XIX(1) (January–March) pp. 2–19.

Hidroservice (1977), *Formulação de programas de desenvolvimento no vale do Rio Tocantins*, vol. 1, São Paulo: Ministério do Interior/Superintendência de Desenvolvimento da Amazonia.

Holtz, A. C. T. (1985), 'The Future of Hydropower in Brazil', *International Water Power & Dam Construction*, 37(1) (January) pp. 16–19.

Interim Mekong Committee (Interim Committee for Coordination of Investigations of the Lower Mekong Basin) (1982), *Environmental Impact Assessment: Guidelines for Application to Tropical River Basin Development*, Bangkok: Mekong Secretariat.

Jayal, N. D. (1985), 'Destruction of Water Resources – The Most Critical Ecological Crisis of East Asia', *Ambio* 14(2) pp. 95–8.

Junk, W. J. and Nunes de Mello, J. A. S. (1987), 'Impactos ecológicos das represas hidrelétricas na bacia amazônica brasileira', in G. Kohlhepp and A. Schrader (eds), *Homem e Natureza na Amazonia / Hombre y Naturaleza en la Amazônia*, Tübingen: Geographisches Institut der Universität Tübingen, pp. 367–85.

Kartawinata, K., Adisoemarto, S., Riswan, S. and Vayda, A. P. (1981), 'The Impact of Man on a Tropical Forest in Indonesia', *Ambio* 10(2–3) pp. 115–19.

Klinge, H., Furch, K., Irmler, V. and Junk, W. S. (1981), 'Fundamental Ecological Parameters in Amazonia in Relation to the Potential Development of the Region', in R. Lal and E. W. Russell (eds), *Tropical Agricultural Hydrology: Watershed Management and Land Use*, Chichester: John Wiley, pp. 19–36.

Koenig, D. (1986), 'The Manantali Resettlement Project', *Working Paper*, 29, Institute for Development Anthropology, Binghampton, N.Y.: State University of New York.

Koklu, R. N. (1982), 'Forced Relocation of Reservoir Inhabitants: Their Resettlement and Rehabilitation with Reference to the Keban Dam in Turkey', Master's dissertation, Centre for Development Studies, Swansea: University College of Swansea (mimeo).

Koppel, B. (1987), 'Does Integrated Area Development Work? Insights from the Bicol River Basin Development Program', *World Development* 15(2) (February) pp. 205–20.

Kunkle, S. H. and Dye, A. J. (1981) 'The Effects of Forest Clearing on Soils and Sedimentation', in R. Lal and E. W. Russell (eds), *Tropical Agricultural Hydrology: Watershed Management and Land Use*, Chichester: John Wiley, pp. 99–109.

Lassailly-Jacob, V. (1983), 'Grands barrages africains et prise en compte des populations locales', *Espace géographique*, 12(1) pp. 46–58.

Leistritz, F. L. and Murdock, S. H. (1986), 'Impact Management Measures to Reduce Immigration Associated with Large-Scale Development Projects', *Impact Assessment Bulletin*, 5(2) (Summer) pp. 32–49.

Lightfoot, R. P. (1981), 'Problems of Resettlement in the Development of River Basins in Thailand', in S. K. Saha and C. J. Barrow (eds), *River Basin Planning: Theory and Practice*, Chichester: John Wiley, pp. 93–114.

Mallas, J. and Benedicto, N. (1986), 'Mercury and Goldmining in the Brazilian Amazon', *Ambio* 15(4) pp. 248–9.

Mermel, T. F. (1978), 'Major Dams of the World', *International Water Power & Dam Construction*, 30(8) (August) pp. 43–54.
Mermel, T. F. (1988), 'Major Dams of the World – 1988' *International Water Power & Dam Construction Handbook 1988*, pp. 47–55.
Mills-Tettey, R. (1986), 'New Bussa: The Township and Resettlement Scheme', *Third World Planning Review*, 8(1) (February) pp. 31–49.
Mkamwa, J. (1986), 'Plants versus Snails', *IDRC Reports*, 15(3) (July) p. 9.
MME/ELETROBRÁS (Ministério de Minas e Energia/Centrais Elétricas do Brasil, SA) (1986), *Manual de Estudos de Efeitos Ambientais dos Sistemas Elétricos*, Brasília, D. F.: MME/ELETROBRÁS.
Monosowski, E. (1986), 'Brazil's Tucuruí Dam: Development at Environmental Cost', in E. Goldsmith and N. Hildyard (eds), *The Social and Environmental Effects of Large Dams*, vol. 2, Camelford: Wadebridge Ecological Centre, pp. 191–8.
Monosowski, E. (1983), 'The Tucuruí Experience', *International Water Power & Dam Construction*, 35(7) (July) pp. 11–14.
Mougeot, L. J. A. (1986), 'Aménagements hydro-électriques et réinstallation de populations en Amazonie: les premières leçons de Tucuruí, Pará', *Cahiers des Sciences humaines ORSTOM*, 22(3–4) pp. 401–17.
Mougeot, L. J. A. (1987), 'Síntese da produção científica (1985–7) do projeto sobre riscos e impactos ambientais decorrentes da formação do reservatório de Tucuruí, Pará, Brasil', Report presented at Seminário de Avaliação SEPEQ/NAEA, Belém (5 November 1987) (mimeo).
Mougeot, L. J. A. (1988), 'Forced Population Resettlement in Brazilian Amazonia: Past and Current Policies and Future Needs', Centre for Development Studies, Swansea: University College of Swansea (mimeo).
Olivares, J. (1986), 'The Assessment of Water Resources Projects Involving Economic, Social and Environmental Costs and Benefits: The World Bank's View', *Natural Resources Forum*, 10(2) (May) pp. 133–43.
Ospina, C. S. (1987), 'Hydroelectric Power in Colombia', *International Water Power & Dam Construction*, 39(7) (July) pp. 33–8.
Otten, M. (1986), *Transmigrasi: Indonesian Resettlement Policy 1965–1985. Myths and Realities*, IWGIA Document, 57, Copenhagen: International Workgroup for Indigenous Affairs.
Partridge, W. L., Brown, A. B. and Nugent, S. J. B. (1982), 'The Papaloapan Dam and Resettlement Project: Human Ecology and Health Impacts', in A. Hansen and A. Oliver-Smith (eds), *Involuntary Migration and Resettlement: The Problem and Responses of Dislocated People*, Boulder, Col.: Westview Press, pp. 245–63.
PEEM (WHO/FAO/UNEP Panel of Experts on Environmental Management for Vector Control) Secretariat (1986), 'The Environmental Impacts on Population Resettlement and Its Effect on Vector-Borne Diseases', *Land Reform, Land Settlement and Cooperatives*, 1(2) pp. 73–91.
Pereira, C. (1981), 'Land Use Managemet on Tropical Watersheds', in R. Lal and E. W. Russell (eds), *Tropical Agricultural Hydrology: Watershed Management and Land Use*, Chichester: John Wiley, pp. 3–10.
Perfect, J. (1980), 'The Environmental Impacts of DDT in a Tropical Agro-Ecosystem', *Ambio*, 9(1) pp. 16–21.
Perry, J. A. and Dixon, R. K. (1986), 'An Inderdisciplinary Approach to Community Resource Management: Preliminary Field Test in Thailand', *Journal of Developing Areas*, 21(1) (October) pp. 31–48.
Petts, G. E. (1980), 'Long-Term Consequences of Upstream Impoundment', *Environmental Conservation*, 7(4) (Winter) pp. 325–32.
Petts, G. E. (1984), *Impounded Rivers: Perspectives for Ecological Management*, Chichester: John Wiley.

Posey, D. A. (1987), 'Etnobiologia e ciência de folk: sua importância para a Amazônia', in G. Kohlhepp and A. Schrader (eds), *Homem e Natureza na Amaẑonia/Hombre y Naturaleza en la Amazônia*, Tübingen: Geographisches Institut der Universität Tübingen, pp. 95–108.

Powles, H. (1986), 'Fencing Off Fish', *IDRC Reports*, 15(4) (October) pp. 18–19.

Prazmowski, A. (1987), 'The Fresher, the Better: Life in the Lake Victoria Fishery', *IDRC Reports*, 16(4) (October) pp. 12–13.

PRODIAT (Projeto de Desenvolvimento Integrado da Bacia do Araguaia–Tocantins) (1982), *Diagnóstico da Bacia do Araguaia–Tocantins*, Brasília, D. F.: Ministério do Interior.

Rachagan, S. S. (1983), 'Malaysia Abandons Tembeling', *International Water Power & Dam Construction*, 35(7) (July) pp. 43–7.

Ramberg, L. *et al.* (1987), 'Development and Biological Status of Lake Kariba – A Man-Made Tropical Lake', *Ambio* 16(6) pp. 314–21.

Reíning, C. S. (1982), 'Resettlement in the Zande Development Scheme', in A. Hansen and A. Oliver-Smith (eds), *Involuntary Migration and Resettlement*, Boulder, Col.: Westview Press, pp. 201–24.

Russell, E. W. (1981), 'Role of Watershed Management for Arable Land Use in the Tropics', in R. Lal and E. W. Russell (eds), *Tropical Agricultural Hydrology: Watershed Management and Land Use*, Chichester: John Wiley, pp. 11–16.

Salati, E. (1987), 'Amazônia: um ecossistema ameaçado', in G. Kohlepp and A. Schrader (eds), *Homem e Natureza na Amazônia/Hombre y Naturaleza en la Amazônia*, Tübingen: Geographisches Institut der Universität Tübingen, pp. 59–81.

Scudder, T. (1966), 'Manmade Lakes and Social Change', *Engineering and Science*, 24, pp. 18–20, 22.

Scudder, T. and Colson, E. (1972), 'The Kariba Dam Project: Resettlement and Local Initiative', in H. R. Bernard and P. Pelto (eds), in *Technology and Social Change*, New York: Macmilan, pp. 40–69.

Scudder, T. and Colson, E. (1982), 'From Welfare to Development: A Conceptual Framework for the Analysis of Dislocated People', in A. Hansen and A. Oliver-Smith (eds), *Involuntary Migration and Resettlement*, Boulder, Col.: Westview Press, pp. 267–87.

Sigaud, L., coord. (1984), 'Impactos sociais de projetos hidrelétricos', Rio de Janeiro: Museu Nacional (mimeo).

Siso, Q. G. (1983), 'Venezuela: el potencial hidroenergetico de un pais eminentemente petrolero', National report presented at XI International Course on Applied Geography, Quito: Pan American Center for Geographical Studies and Research (mimeo).

Spurgeon, D. (1987), 'ICRAF: A Decade of Agroforestry', *IDRC Reports*, 16(4) (October) p. 20.

Srivardhana, R. (1987), 'The Nam Pong Case Study: Some Lessons to Be Learned,' *Water Resources Management*, 3(4) (December) pp. 238–46.

Suryono (1987), 'Controlling Sediment in the Brantas River Basin', *International Water Power & Dam Construction*, 39(1) (January) pp. 25–9.

Szekely, F. (1982), 'Ecological Aspects of Large Hydroelectric Projects in Tropical Countries', *Water Supply and Management* 6(3) pp. 233–42.

Thomi, W. (1984), 'Man-Made Lakes as Human Environments: The Formation of New Socio-Economic Structures in the Region of the Volta Lake in Ghana, West Africa', *Applied Geography and Development*, 23, pp. 109–27.

Toomey, G. (1986), 'A Kenyan Community Fights Schistosomiasis', *IDRC Reports*, 15(3) (July) p. 8.

UNEP (United Nations Environment Program) (1978), 'Environmental Issues in River Basin Development', in UNEP (ed.), *Proceedings of the United Nations Water Conference on Water Management and Development*, vol. 1, part 3, New York: Pergamon Press, pp. 1163–72.

Weil, C. and Kvale, K. (1985), 'Current Research on Geographical Aspects of Schistosomiasis', *The Geographical Review* 75(2) (April) pp. 186–216.

Wicke, P. (1987), 'Prospects for Large and Small Hydro Development in Peru', *International Water Power & Dam Construction* 39(7) (July) pp. 47–50.

World Bank (1980), *Social Issues with Involuntary Resettlement in Bank Financial Projects*, OMS 2(33) pp. 1–7.

World Bank, Energy Department (1987), 'The World Bank and Hydroelectric Power', *International Water Power & Dam Construction Handbook 1987*, pp. 7–8.

Xiutao, W. (1986), 'Environmental Impact of the Sanmen Gorge Project', *International Water Power & Dam Construction*, 38(11) (November) pp. 23–4.

6 Development Planning and Mineral Mega-projects: Some Global Considerations

Fred T. Neto[1]

INTRODUCTION

The post-war period has witnessed several major government attempts to integrate Brazilian Amazonia geopolitically and socio-economically into the rest of the country. This effort was clearly intensified during the last decade, which turned out to be divided into two contrasting phases of Amazonian development planning: one marked by the active State role in a process of frontier colonisation oriented towards small-scale agriculture, and the other in which government action tended to be limited to the support of corporate activities, such as cattle ranching and mining.[2] This later phase was spear-headed by the regional strategy of the Second National Development Plan (PND II), and the Programme of Agricultural and Mineral Poles of Amazonia (POLAMAZONIA), both of which were formulated in 1974. Its emergence is usually explained by the failure of government colonisation schemes, and more importantly, the fierce opposition of powerful interest groups to a regional development strategy dominated by State intervention and small-scale activities.

This phase can, however, also be associated with global economic factors, such as the 1973 oil price rise, which created a serious trade imbalance (Brazil's import bill rose from US $6.2 billion in 1973 to US $12.6 billion in 1974), and contributed to a massive increase in foreign borrowing and the eventual escalation of the foreign debt. In an attempt to counteract these economic problems in the medium run, PND II proposed to intensify the process of import substitution industrialisation, develop native sources of energy, and increase agricultural, mineral and industrial exports. In line with this broader economic strategy, the new government policy for Amazonia emphasised the 'comparative advantages' of potential export commodities – such as beef, timber and minerals – in what were considered 'dynamic sectors of the international market' (Brasil, 1974, p. 66). At the same time, PND II gave particular emphasis to the development of the:

> Mining–Metallurgical Complex of Eastern Amazonia, comprising the Carajás–Itaqui integrated scheme (iron ore and steel-making), the (Trombetas–Belém) bauxite–alumina–aluminium complex, and innumerable

other undertakings associated with the exploitation of the hydroelectric potential of the Araguaia–Tocantins region.[3]

As both the balance of payments deficit and the foreign debt problem worsened towards the end of the 1970s, increasing attention was turned to the export potential of these mineral complexes – which were to dominate the following phase of regional planning during the 1980s.

As there is plenty of evidence to show that mining has been a major driving force of Amazonian development planning since the beginning of the PND II–POLAMAZONIA phase, it is surprising that very little of the extensive research on recent patterns of frontier expansion is focussed on the underlying dynamics of mineral development in the region. While this type of development is evidently interwoven with major government policies for Amazonia (which have their own independent national dynamic), it should also be seen as part of a globalisation process through which downstream mineral activities are located in less industrialised regions and countries. However, as most of the research effort is also centred on patterns of change *within* Amazonia – and occasionally within Brazil (Foweraker, 1981; Wood and Wilson, 1984) – it tends to overlook crucial *global* processes of economic change. Although lip service is often paid to international aspects of Amazonian mineral projects, little attempt is made to analyse their emergence as an integral part of global economic processes. It is to this task that this chapter is directed. The basic idea here is not to isolate a particular regional development pattern from its wider structural context, but to focus on major global processes in order to identify certain vital underlying forces behind that regional pattern.

THE BIRTH OF THE IRON-ORE COMPLEX

During the 1950s and early 1960s many Western industrial groups became increasingly concerned about their dependence on manganese supplies from a handful of countries – Brazil, India, Gabon and the Soviet Union – the last three of which were not considered totally reliable (see Park and Freeman, 1968, pp. 83–5; and Santos, 1986). Given that a large-scale manganese mine jointly-owned by the US multinational Bethlehem Steel and a private Brazilian company had been in operation in eastern Amazonia since the mid-1950s,[4] it was no accident that exploration subsidiaries of two other major American manganese consumers moved into the region in the mid-1960s. The discovery of a manganese deposit in south-eastern Pará by Union Carbide's Codim in 1966 quickly attracted US Steel's Meridional to the area. Following a series of 'chess moves' in which these two rival exploration companies tried to outwit each other, geologists of Meridional practically stumbled upon one of the world's largest iron-ore reserves, in the Carajás mountains, in 1967 (Almeida, 1981; Santos, 1986).

After three years of tough negotiations between US Steel and the Brazilian government, a joint venture was formed between the American multinational and the state-owned mineral group *Companhia Vale do Rio Doce* (CVRD) to develop an iron ore project. This project was never started because of US Steel's insistence that its viability would be compromised by the recessive repercussions of the 1973 oil price rise on the world iron and steel industry (Santos, 1986, p. 300; Soares, 1981, p. 12). Thus, only in the long run might there be a sufficient improvement in the world steel market to justify the implementation of such a huge mining project – whose output would otherwise compete with those of mines owned by the American multinational in Venezuela and Liberia. Contrary to US Steel's market predictions, CVRD not only expected world iron-ore demand to outstrip supply by the mid-1980s, but was also convinced that if the project was postponed, other iron-ore mines developed elsewhere would jeopardise its later implementation (CVRD, 1982, p. 94). As the world's largest iron-ore producer and exporter, it was vitally important for CVRD that the Carajás mine came on stream in the medium run, given that it would become increasingly more difficult to extract high-grade ore from its other mines in south-east Brazil.

This irreconcilable conflict between the global strategies of CVRD and US Steel led to a break-up of their joint venture in 1977, when the latter sold its stake to the former for US $50 million. Following US Steel's withdrawal, CVRD tried to obtain alternative funds for the implementation of the project: besides seeking foreign loans, the State enterprise began negotiations with major Japanese and European steel producers which had previously shown some interest to participate with equity capital in the Carajás venture. However, apart from a World Bank loan offer, those moves had little success, due to the world steel crisis of the late 1970s and the hefty price-tag of the project – which rose from less than US $1 billion in 1974 to US $3.5 billion by 1977. As a result, a new feasibility study undertaken by CVRD proposed to scale the project down from the original 50 million tonnes a year by 1986 to 35 million by 1987, at a revised capital cost of US $2.4 billion (at 1977 prices). The Carajás iron ore project (*Projeto Ferro Carajás*), as we know it today, was finally born (see CVRD, 1981b).

Although this scaled down project was officially approved by the federal government in 1978 – during which an 82-kilometre section of a proposed 900-kilometre railway linking the mine to the Atlantic coast began to be built – there is evidence to show that the Minister of Mines and Energy of the Geisel Administration was very sceptical about its viability, in view of the prolonged world steel crisis and the inability of CVRD to secure adequate outlets for its projected output (Soares, 1981, pp. 15–17; Pinto, 1982, pp. 77–8). Thus, while CVRD decided that it was capable of implementing Carajás on its own, the start of a small stretch of the railway was little more than a token gesture towards the completion of such a massive project. Given the reluctant attitude of the Geisel government towards this project, and the great deal of uncertainty about its sources of finance and market outlets, the very foundation of PND II–POLAMAZONIA's iron and steel complex was seriously weakened by the late 1970s.

THE TROMBETAS BAUXITE VENTURE

The first commercially viable bauxite deposits in Amazonia were also discovered in 1967 – in the Trombetas area of north-western Pará – by an exploration subsidiary of the Canadian aluminium multinational, Alcan. These exploration activities – started in the early 1960s – had been originally associated with a possible expansion of Alcan's aluminium plant in South-East Brazil as its main refinery in Quebec was adequately supplied by its mining subsidiary in British Guyana (where it also owned a small refinery), and by another foreign-owned mine in Dutch Surinam (Machado, 1985, pp. 132–4). This safe source of supply was, however, jeopardised by the independence of Guyana in 1966, and particularly by the declaration of a 'co-operative republic' in the early 1970s, whose first government decided to nationalise all mining enterprises in that country (Girvan, 1976, Chapter 4; Graham (1982, pp. 93–101). This helps to explain why the Canadian parent company became far more interested in the Trombetas finds towards the late 1960s.

Another Alcan subsidiary (*Mineração Rio do Norte*, MRN) was thus created to intensify mineral exploration in the area, and from 1969 onwards to develop a bauxite mine there. Although this project began to be implemented in 1971, it was mothballed only a year later, allegedly because of a slump in the world aluminium market. There is evidence to show, however, that this decision was based primarily on three main factors: the likelihood that Alcan's bauxite supplies from Guyana would not be threatened by the nationalisation of its mine there (in 1971), the swift implementation of the export-oriented Boke project in Guinea, and government pressures for Brazilian participation in the Trombetas project (Machado, 1985, pp. 156–63; Dantas, 1982a, p. 21). In 1974, following a long period of tough negotiations, MRN was transformed into a consortium led by CVRD and Alcan, with the participation of various other multinational mineral groups.[5] The Trombetas project was then revived at scaled up levels of output and investment, under pressure from both CVRD and the Brazilian government, which was particularly interested in its export potential.

Due to major disagreements between CVRD and its foreign associates, this US $400 million project came on stream two years behind schedule (in 1979), with an annual capacity to produce 3.3 million tonnes. In order to ensure the viability of the project, it was originally agreed that its whole output would be bought under long-term contracts by its foreign shareholders – paving the way for an export 'enclave' under foreign control (Sá, 1981, pp. 39–40; Dantas, 1982a, p. 22; Ibase 1983, pp. 67–8). This has greater significance if one notes that MRN's pricing policy tends to be biased in favour of its foreign end-users, which may also make use of the 75 per cent board majority clause to block any price changes contrary to their global interests (Dantas, 1982a, p. 22; Machado, 1985, pp. 167–73). Serious disagreements over prices are liable to occur in a consortium like MRN, between a state mining enterprise and aluminium multinationals. While the former is primarily concerned with the profitability of the mining operation, the latter prefer to keep ore prices as low as possible in order to allow greater profit margins in upstream operations. Given the export orientation of the Trombetas venture, the whole question of transfer pricing has become a problem not only to CVRD but also to Brazilian governments concerned with the generation of foreign exchange and tax revenues.

Although the government has the means to regulate export prices in general, it chose until very recently not to intervene in MRN bauxite prices so as not to undermine one of the main foundations of recent phases of Amazonian development planning: the vital role of foreign investments in mining projects. This was particularly important to major aluminium multinationals which had decided to diversify their sources of supply in response to the introduction of significant bauxite levies by leading producing countries, and the creation of a producers' cartel, the International Bauxite Association, in 1974 (Girvan, 1976; Graham, 1982). But while the Geisel government used the creation of this cartel to stress the importance of its 'free market' policies for foreign investors in the Amazonian bauxite sector, the integrated mining–metallurgical strategy of PND II–POLAMAZONIA also aimed to transform part of MRN's ore into higher value-added products (alumina and primary aluminium) in the region.

BACKGROUND TO THE ALUMINIUM COMPLEX

This complex can be traced back to an agreement signed in 1974 between CVRD and a consortium of Japanese primary aluminium producers (LMSA) to develop an export-oriented project in eastern Amazonia. In the following year, a feasibility study carried out by CVRD proposed a joint venture to produce 640,000 tonnes of aluminium and 1.3 million tonnes of alumina a year, which would have required a total capital investment of around US $3 billion at the time. This project was, however, heavily criticised by LMSA, which was particularly opposed to the financial burden of its support infrastructure and the prices of local inputs (bauxite and electricity). While these were the official reasons for the abandonment of the project in 1975, it is now clear that the Japanese had decided to take a very tough negotiating stance as part of their global strategy to secure cheap supplies of primary aluminium on a long-term basis.

As the 1973 oil price rise jeopardised the long-term viability of the (oil-fuelled) Japanese primary aluminium industry, its major producers decided to invest heavily in aluminium projects located in countries with abundant supplies of cheap energy and bauxite – such as Australia, Indonesia, Venezuela and Brazil.[6] The Japanese Ministry of International Trade and Industry (MITI) has played a major role in this strategic move by encouraging all representatives of the industry to join together and form *ad hoc* consortia (Tsurumi, 1976, pp. 46–9). As Murphy (1983, pp. 97–100) argues, equity participation of all Japanese companies in overseas projects has the effect of equalizing risks and costs for competitors after the same commodity – which increases their collective bargaining strength. In addition, by diversifying their supply sources over multiple projects, the Japanese have often played their participation in (or abandonment of) one project against another as a means to secure greater concessions from all host countries.

This helps to explain why during the long period of negotiations over the revival of the project, the Geisel government had to make significant concessions to accommodate Japanese demands (Machado, 1985, pp. 240–57). In the end it was agreed that the revived complex would be composed of two separate joint ventures between CVRD and the Nippon Amazon Aluminium Company (NAAC)[7] – *Alumínio do Brasil* (ALBRAS) and *Alumina do Norte* (ALUNORTE) – whose annual capacities were scaled down to 320,000 tonnes of aluminium and 800,000 tonnes of alumina, respectively. Although the lower capital cost of this revised project (US $1.3 billion at 1975 prices) was partly associated with its scaled down capacities, it was also a result of the Brazilian government decision to provide its support infrastructure (hydroelectric, port and urban schemes). By far the most complex and expensive of these was the Tucuruí hydroelectric project on the Tocantins River, to be financed and implemented by ELETRONORTE, the Amazonian subsidiary of the state holding company ELETROBRAS.[8]

As Machado (1985, p. 242) notes, this major Brazilian concession to NAAC was aimed to secure, in exchange, cheap Japanese loans for the Tucuruí scheme. However, as the Japanese remained firmly opposed to any financial involvement in what looked to be an eternal loss-making venture, ELETRONORTE had to secure funds elsewhere. During President Geisel's official visit to France in 1976, a pool of French financial institutions and capital goods producers agreed to lend US $232 million to ELETRONORTE, on condition that it bought a significant amount of French goods and services for the project. Although these funds allowed the construction of the dam works planned for 1976 and 1977, lack of additional funds slowed down the following phase of implementation during 1978 and 1979 (ELETRONORTE, c. 1982, pp. 8–9; Machado, 1985, pp. 200–2). By the late 1970s, as the Brazilian foreign debt crisis worsened, the shortage of public funds for the continuation of this project created a great deal of uncertainty about its completion – originally planned for the end of 1981 but eventually put back to 1984.

This was bound to have an adverse impact on the implementation of ALBRAS (and to a lesser extent on the Carajás project), which led Bourne (1978, pp. 138–9) to note 'how vulnerable any interrelated project was to delay in any connecting part'. In any case, although ALBRAS and ALUNORTE were both officially incorporated in 1978, preliminary works on the Barcarena site (near Belém) were started only a year later, and progressed at a very slow pace. Besides a certain Japanese reluctance to implement them according to schedule, CVRD had great difficulty securing government finance for this twin project. Apart from the general shortage of public funds, the state-owned National Economic Development Bank (BNDE) – which was expected to provide considerable loans to CVRD – had strong objections to the legal basis of the joint ventures as it allowed NAAC to have disproportionate control over major board decisions, pricing policy and technological matters (Dantas, 1982a, p. 29; Sa, 1981, pp. 46–50).

This led to complex renegotiations of the contract of association between CVRD and NAAC, which delayed the implementation of the project. To make matters worse, the estimated capital cost of ALBRAS–ALUNORTE rose from US $1.3 billion in 1975 to US $2.5 billion by the late 1970s (ALBRÁS, c. 1981). These legal and financial problems were exacerbated by the 1979 oil price rise, which raised serious doubts about the viability of a mega-project completely dependent on an unpredictable world aluminium market.

In sum, while the foundations of the PND II-POLAMAZONIA bauxite–aluminium complex were successfully established by the MRN venture, its other three connecting parts (ALBRAS, ALUNORTE and the Tucuruí hydroelectric scheme) were lying on rather shaky ground at the end of the 1970s.

THE GREATER CARAJÁS PROGRAMME

As far as mineral development is concerned, government policy for Amazonia was at a crossroads at the end of the PND II phase in 1979. On the one hand, increasing attention was focussed on the export potential of the mining–metallurgical complexes; on the other, the Geisel government and vital State agencies, such as BNDE, were hesitant to provide appropriate support for their implementation. This indecision may appear to have been broken by the establishment of General J. Figueiredo's government (1979–85). This new government is often seen as the main orchestrator of the most recent phase of Amazonian development planning – spearheaded by the Greater Carajás Programme (*Programa Grande Carajás*, PGC), and centred on export-oriented mineral projects. A closer examination of the facts will, however, show that some of the main tunes of this new phase have in fact been called by global corporate forces.

For a start, the completion of the PND II mineral complexes is not even mentioned in the following national development plan (*Plano Nacional de Desenvolvimento 1980/85*, PND III). This was drawn up only a few months before the PGC was officially instituted and its main stated economic goal for Amazonia was:

> the non-predatory exploitation of its natural resources with rigorous respect for its ecological balance and indigenous population (Brasil, 1980, p. 86).

This glaring omission of any new mineral development drive for the region was not only a reflection of continued government indecision towards the mining–metallurgical complexes, but also an early sign of a weaker government role in the current phase of eastern Amazonian development planning.

The origins of the PGC can be traced back to a study commissioned by CVRD and undertaken by Japanese consultants, whose report (IDCJ, 1980) evaluated mineral and agricultural investment opportunities along the 'Carajás export corridor'. A similar appraisal of potential investments over a much wider geographical area was presented in another pioneer document entitled 'Eastern Amazonia: A National Export Project' (CVRD, 1980), which argued that the national macro-economic strategy and Amazonian development planning should both give top priority to export-oriented mineral and agricultural projects in eastern Amazonia. Given that this CVRD document was also based on the Japanese study, and that it was used by the Brazilian government for the creation of the PGC, several analysts have criticised this disguised foreign influence over the embryonic stages of such a strategic national and regional development programme (IBASE, 1983, pp. 125–6, 1984, p. 8; Cota, 1984, p. 161). This has greater significance if one notes that the major underlying aim of the IDCJ study was to establish an alternative source of cheap agricultural and mineral commodities for the Japanese market (Dantas 1982b, pp. 19–22).

The foundations of such a programme were also determined by the global corporate strategy of CVRD, whose 'national export project' was aimed to ensure greater government support for the troublesome projects in which it was involved (Carajás, ALBRAS–ALUNORTE and MRN) or reliant upon (Tucuruí, Barcarena port and new town). These projects were thus skilfully inserted into a regional development programme that could simultaneously address major national economic problems:

> only a grand and systemic project of national dimensions, and with marked export characteristics, would be able to deal with the critical

period of the following years . . . [by] generating the foreign exchange we desperately need (CVRD, 1980, pp. 3, 9).

This national export project succeeded in catching the imagination of the Figueiredo government, which saw a historical opportunity to solve the foreign debt problem. Thus, when the Greater Carajás Programme was officially instituted by decree-law no. 1,813 (24 November 1980), the CVRD document was its basic point of reference.

Although this law outlined the programme only in very general terms, it established the fundamental tool of implementation of the PGC: a system of fiscal incentives for selected projects to be overseen by an interministerial council. The *ad hoc* programming region extends over an area of 895,000 square kilometres (10.6 per cent of the country) – covering parts of the states of Pará, Maranhão and Goiás – larger than the United Kingdom, Ireland and France combined (see Figure 6.1). Subsequent decrees authorised the interministerial council to grant exemptions of income tax, manufactured products tax (IPI) and import duties to projects established or expanded in this region until 1990 (SEPLAN, 1985). The most immediate priority of this programme was to ensure the completion of the PND II mineral complexes and their required infrastructure, in order to generate foreign exchange to service the foreign debt, and to correct the balance of payments deficit.

The implementation of these mega-projects – together with the generous fiscal incentives and infrastructural improvements – was also expected to attract considerable new investments into the region to accelerate its economic development. According to a more refined regional development plan also prepared by CVRD, the implementation of the PGC would require an estimated total investment of US $61.7 billion over the 1981–90 period, 63 per cent of which to be channelled into mineral and agricultural projects, and the rest into basic infrastructure (CVRD, 1981a, vol. 1, pp. 8–9). By far the most significant group of projects was related to mineral activities, which were expected to attract investments of almost US $35 billion. However, such ambitious estimates included many potential investments and should not hide the fact that 95 per cent of all mineral and agricultural investments approved by the interministerial council until 1985 was accounted by only four mega-projects which started to be implemented before the creation of the PGC: the Carajás iron ore project, ALBRAS–ALUNORTE, ALUMAR and the Tucuruí hydroelectric scheme (PGC, 1985b). A close examination of their implementation during the present decade will help to show the increasing influence of global corporate forces over the current phase of (Eastern) Amazonian development.

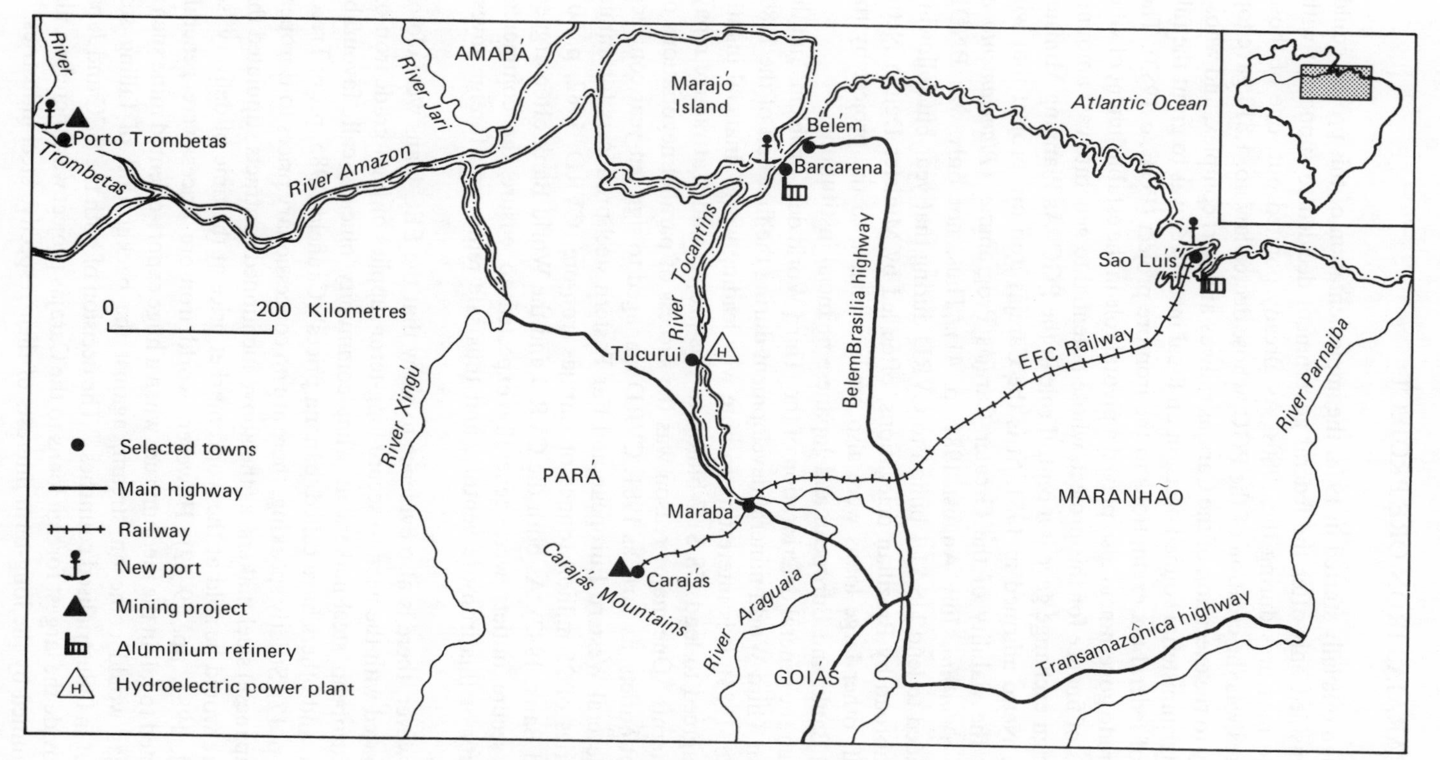

Figure 6.1 The Greater Carajás region
Source: *Jornal do Brasil*

THE CARAJAS IRON-ORE PROJECT

Although officially started in 1978, the implementation of this project could be intensified only after the federal government decided to support it with effective measures during the 1980s. As already pointed out, one of those measures was the creation of the PGC, whose decree-law no. 1,813 gave top priority to the execution of the Carajás railway and port complex, and whose first interministerial council decision on fiscal incentives was to grant the full range of federal tax exemptions to the iron ore project (PGC, c. 1983). The Figueiredo government also played a pivotal role in the establishment of vital sources of finance for this project, whose potential to generate vast amounts of foreign exchange gave it a central role in the PGC. As Planning Minister Delfim Netto affirmed in 1981, 'it is the Carajás iron ore project that will ensure the viability of the Greater Carajás Programme' (*Planejamento & Desenvolvimento*, July–August 1981, p. 41). Thus, not only was BNDE persuaded to lend US $1.1 billion to CVRD during that year but following visits abroad by Brazilian delegations, often led by Minister Delfim Netto himself, other large loans were also made available to the project by the World Bank, and European and Japanese financial institutions.

Given the general deterioration of the 'Third World debt problem' and the crisis in Third World mineral development during the first half of the 1980s (Walde, 1985), it is interesting to note why leading world financial institutions agreed to lend US $1.7 billion to this project under attractive repayment terms.[9] One major reason was the success of parallel negotiations on market outlets: as early as 1981, CVRD managed to sign ten-year contracts with several Western European and Far Eastern steel producers for annual deliveries of 25 million tonnes of Carajás iron-ore (CVRD, 1982, p. 107; World Bank, 1982). As both the CVRD and the World Bank often argued, these secure outlets were generally expected to ensure the commercial viability of this iron-ore venture, and thus safe returns on foreign investments.

However, there is also evidence to show that the EEC loans were closely associated with the need to secure long-term supplies of high-grade iron-ore for European steel-makers at what community officials call 'favourable prices' and others have called 'banana prices' (Caufield, 1985, p. 26; Treece, 1987, p. 17). Strictly speaking, there are no concessionary prices to European (or Japanese) steel-makers, as the above-mentioned contracts stipulated that the ore would be sold at the world market price at the time of delivery (see Brasil, 1981, pp. 79–82). However, world iron-ore prices were generally expected to fall in the medium run, when a huge export-oriented mine such as Carajás would come on stream, against the background of falling steel output in industrialised countries.[10] The decision of both the EEC and Japan to provide the largest foreign loans to the Carajás project was, therefore, also determined by the long-term interests of their respective steel industries.

But despite this access to massive foreign funds, CVRD was required to channel a significant amount of its own financial resources into the project. This led the state enterprise to place large amounts of debentures in local financial markets, and to enlarge its capital stock considerably between 1981 and 1984. Given that those debentures could eventually be converted into shares, the government stake in the company fell from almost 70 per cent in the early 1980s to just 56 per cent in 1985. While the proceeds of those debentures were vitally important for the implementation of the Carajás project, top CVRD officials often point out that this partial privatisation poses no threat to the state enterprise status of the company. In fact, despite having the majority of its capital stock controlled by the Brazilian Treasury and state-owned banks, CVRD enjoys a great deal of autonomy to operate without government interference (unlike most Brazilian state enterprises). This autonomy is closely associated with its active participation in competitive world markets and high profitability, which allows it 'to stay free of dependence on government subsidies and to invest and diversify according to criteria devised by company management' (Trebat, 1983, p. 102–5).

Having secured full government support, adequate funds and market outlets for the project, CVRD was then able to implement it swiftly during the first half of the 1980s. The highly-mechanised open-pit mine (code-named N4E) and complementary processing facilities came on stream in January 1986, with a capacity to produce 15 million tonnes of ore a year. This was raised to 35 million tonnes during 1987, when Carajás became one of the largest iron-ore mines in the world. The 890-kilometre railway (*Estrada de Ferro Carajás*, EFC) was officially inaugurated by President Figueiredo one year ahead of schedule, in February 1985. The deep-water port of Ponta da Madeira (280,000 dwt) was inaugurated by Figueiredo's civilian successor, President J. Sarney, in March 1986. The capital cost of the whole complex – including the railway and the port – was US $3.1 billion, which is expected to rise to US $4 billion after debt servicing and amortisation.[11] In 1986, the N4E mine produced 14 million tonnes of iron-ore (85 per cent of which was exported), whereas in 1987 this output was expected to rise to 22 million, most of which was destined for Japan, West Germany and Italy (*Brasil Mineral*, December 1986, pp. 16–17; *Brazil Report*, November 1986, p. 8).

THE ALBRAS–ALUNORTE ALUMINIUM COMPLEX

Although preliminary earthworks on the Barcarena site were started in 1979, the implementation of this complex was at a virtual standstill until 1981. In order to reinforce Japanese interest in this mega-project, a vital concession was indirectly made by the Brazilian government on the same day that the PGC was officially announced: the controversial contract signed between ALBRAS and ELETRONORTE, under which the latter would supply heavily subsidised energy to the former for a minimum period of twenty years. Thus, the creation of the PGC was once again a turning-point in the government's attitude towards the completion of another core mineral mega-project. As a result of indirect government pressure, BNDE was also persuaded to drop its opposition to the legal basis of the joint venture between CVRD and NAAC after they signed a supplementary contract of association in 1981 in the presence of the Brazilian Minister of Mines and Energy, and the Japanese Minister of International Trade and Industry. This contract paved the way for the provision of vital loans to the joint venture by both BNDE and a pool of 23 Japanese financial institutions led by Jeximbank.

A more direct form of government backing for this project was expressed by the actual provision of its support infrastructure. Thus, at the end of 1981, the state-owned PORTOBRAS began to build the port of Vila do Conde, adjacent to ALBRAS–ALUNORTE and included in the PGC. More importantly, the implementation of the Tucuruí project was also intensified during that year, when both banks of the troublesome River Tocantins were finally linked by the dam. This hydroproject was inaugurated by President Figueiredo at the end of 1984, with an initial generating capacity of 660 MW, which rose to 2,000 MW in 1986 and is expected to reach 4,000 MW by the end of the decade. Its capital cost is conservatively estimated at US $4.6 billion, two-thirds of which has been financed with foreign loans (PGC, 1985a).[13] Given that the Tucuruí power station will sell electricity at prices well below its production costs, one must question the government decision to allocate vast amounts of scarce resources to supply subsidised energy to export-oriented mega-projects.

Having secured appropriate funds, energy subsidies and the support infrastructure for ALBRAS–ALUNORTE, CVRD and NAAC began to implement this US $1.9 billion complex in 1982. The first phase of ALBRAS (160,000 tonnes a year) was officially inaugurated by President Sarney in October 1985, together with the port of Vila do Conde (40,000 dwt), and completed in the following year when its actual output was almost 100,000 tonnes (CVRD, 1986b). In his inauguration speech, President Sarney asserted that:

> my government will support the operation of the Albrás complex at full production in order to allow the transformation and industrialisation of Amazonian bauxite here in the region.[14]

Nevertheless, it emerged during the same occasion that NAAC was blocking the start of its second phase, because of the low aluminium prices in the international market and the lack of sufficient electric energy.[15] While the aim of this energy argument was to transfer the cost of building a second transmission line (from Tucuruí) on to the Brazilian government, the low world aluminium prices were in fact beneficial to NAAC, given that it was entitled to buy the whole output of the first phase for transformation into higher-value added products in Japan. This helps to explain both its reluctant attitude towards the second phase, and the more aggressive line taken by CVRD on the disposal of the first phase output.

This new line led to an agreement with NAAC in mid-1986, under which each partner would be entitled to a quota of production equivalent to their equity share in the project (51 and 49 per cent, respectively), despite fierce Japanese protests (*Brasil Mineral*, June 1986, pp. 8–9). In exchange, NAAC was allowed to buy ALBRAS aluminium at a concessionary price of US $25 a tonne below the world market price 'so as to make it more competitive with Indonesian and Australian exports to Japan' (Brasil Mineral, July 1986, p. 5). But when this agreement expired at the end of 1986, CVRD and NAAC could not agree on a new pricing formula for shipments to Japan, which the Japanese consortium seems to have linked to the implementation of the second phase of ALBRAS (LACR, 15 January 1987). In any case, this phase can only be started after the question mark over the second transmission line is cleared: although ALBRAS will be its exclusive user, the joint venture has decided not to contribute one single dollar towards its US $62 million cost. As neither CVRD nor ELETRONORTE are prepared to finance this line, both have agreed to pass the bill to the government, whose decision has become so vital for the start of the second phase (*Brasil Mineral*, December 1986, p. 10). Given the complex round of negotiations, it is very unlikely that ALBRAS will be in full production (320,000 tonnes a year) by the end of the decade, as was officially expected in late 1986 (CVRD, 1986b).

But if there are problems with the completion of ALBRAS, the implementation of ALUNORTE has been little short of a disaster. This project was almost abandoned in 1982 due to a serious world alumina crisis, and a suspicious offer by the Aluminium Company of America (Alcoa) to supply alumina to ALBRAS at a price less than half the expected cost of production of ALUNORTE. Although these developments threw serious doubts on the viability and continuation of ALUNORTE, CVRD was firmly opposed to its abandonment, and even postponement as late as the end of 1982 (*Minerios*, December 1982, p. 39), for its bauxite–aluminium complex in eastern Amazonia would be undermined by dependence on the supply of imported alumina from the leading aluminium multinational. Although NAAC was also concerned about such a dependent link, given that its sole interest in ALUNORTE was to secure the supply of alumina for the production of aluminium to be exported to Japan, the Japanese consortium became even more sceptical about the timetable of the project.

As a result, it was agreed in early 1983 that the inauguration of ALUNORTE should be postponed from 1985 to 1989, so as to coincide with the completion of the second phase of ALBRAS. During that period ALBRAS would import its alumina from the Surinam Aluminium Company (Suralco), jointly-owned by Alcoa and Royal Dutch Shell's mining–metallurgical subsidiary, Billiton Metals. However, NAAC's increasing reluctance to support this project after the inauguration of ALBRAS, brought its implementation to a virtual standstill. In early 1987 the Japanese finally decided that they would not invest any further in ALUNORTE (LACR, 12 February 1987, p. 3), which is now certain to be postponed at best or at worst abandoned. This is a clear example of how one of the foundations of a core PGC mega-project can be undermined by its dependence on global corporate forces and unstable world commodity markets. These forces have, in effect, blocked the regional development strategy to create an integrated mining–metallurgical complex – partly controlled by Brazilian capital – as MRN's bauxite cannot be transformed into primary aluminium by ALBRÁS without the ALUNORTE refinery.

The postponement of ALUNORTE has also helped to maintain the dependent relationship between MRN and those dominant global forces. Although MRN's bauxite output rose from 62,000 tonnes in 1979 to a peak of 5 million tonnes in 1984, over 80 per cent of the latter had to be exported, mainly to Alcan and its foreign consortium associates, whereas most of the rest was sold to Alcoa's aluminium complex in São Luís (ALUMAR). Following several years of significant financial losses, this was the first year that MRN made a sufficient profit (US $62 million) to allow the payment of dividends to its shareholders (MRN, 1985). This profit was mainly a result of over-production as this was also the first year that actual output (5 million tonnes) exceeded the nominal capacity of the project (3.3 million tonnes a year). This was not only beneficial to CVRD and the government as it generated US $70 million of net foreign exchange, but also to major foreign consumers able to purchase greater amounts of ore at lower real prices.[16]

The world aluminium crisis of the mid-1980s contributed to maintain this downward pressure on prices, which was not only partly responsible for a significant drop in output (to 4.1 million tonnes) in 1985, but also provoked a serious dispute between CVRD and its MRN foreign associates. While the former has sought to sustain prices that would continue to ensure the profitability of the venture, the latter have made concerted moves to lower MRN bauxite prices in order to allow greater profit margins in upstream operations. This irreconcilable conflict of interests led the Ministry of Mines and Energy to fix a minimum export price and to halt MRN bauxite exports in 1986 (*Brasil Mineral*, June 1986, p. 5; LACR, 6 June 1986, pp. 2–3). Export deliveries were suspended until the end of 1987, when the International Chamber of Commerce in Paris was expected to arbitrate the dispute. This dispute only serves to reinforce the argument that the national bauxite–alumina–aluminium complex of eastern Amazonia – which is so central to the PGC – has been seriously undermined by global corporate forces.

THE ALUMAR ALUMINIUM COMPLEX

The fourth core PGC mega-project, the *Alumínio do Maranhão* (ALUMAR) alumina–aluminium complex near São Luís, is characterised by three unique features: it is wholly-owned by foreign multinationals, it was announced only in the 1980s and it took just four years to come on stream. ALUMAR is jointly-owned by Alcoa's Brazilian subsidiary (60 per cent) and Shell's Billiton Metals (40 per cent) – whose US $1.5 billion investment makes it the largest privately funded project ever undertaken in Brazil. This project was originally put forward by Alcoa in January 1980, as part of its global strategy to relocate an increasing part of its production capacity to regions able to produce cheap bauxite and electric energy.[17] Alcoa's decision to invest in Brazilian Amazonia was also reinforced by the country's cheap labour, lax pollution controls and considerable government incentives.

There is plenty of evidence to show that during the earliest stages of its implementation, the project counted on the exceptional support of the federal government, which tended to emphasise its ability to generate even more foreign exchange than ALBRAS (MME, 1981, p. 26; PGC, c. 1982, p. 24). This support was expressed by the swift approval of Alcoa's controversial 'Saõ Luís project' (as it was originally called) by the influential Iron, Steel and Nonferrous Council (CONSIDER) of the Ministry of Industry and Commerce, in 1980. In that same year, ELETRONORTE agreed to supply the project with virtually unlimited amounts of subsidised energy from Tucuruí, in a deal comparable to that of ALBRAS: an additional 10 per cent discount on the heavily subsidised A 1 tariff for a minimum period of 20 years, during which electricity costs cannot exceed 20 per cent of the world market price of aluminium. As a result of additional support from the Maranhão state government, Alcoa was given disproportionate access to scarce local resources, such as 6 million cubic metres of water a year and up to 10 square kilometres of land, not to mention official connivance at the disturbing socio-economic and ecological impacts of the project on local communities (English, 1984; Galvão, 1984).

This virtually unrestricted national and local government support removed the most difficult political obstacles often associated with foreign mega-projects, and reduced the period of negotiations between Alcoa and the Brazilian government considerably. As a result, Alcoa was able to begin the implementation of its aluminium complex as early as August 1980, and to obtain a huge US $750 million loan from a Citibank-led pool of 28 foreign banks later that year. These funds allowed the immediate acceleration of the project, which was supported by Alcoa's extensive experience in the field and its reliance upon the established urban–industrial infrastructure of São Luís (unlike ALBRAS which is relatively isolated from Belém and thus subject to greater logistic problems). With the creation of the PGC, the São Luís project was further supported by (a) considerable fiscal incentives (exemptions of income tax, IPI and ICM), (b) its central role in such a major integrated regional development programme, and (c) the intensified execution of the Tucuruí hydroelectric project and its transmission line to São Luís.

That central role was formalised at the beginning of 1981 when the Brazilian government went so far as to declare Alcoa's project 'relevant to the national interest', in view of its great potential to generate foreign exchange and to stimulate regional development (PGC, c. 1982, p. 24; ALUMAR, c. 1984). Such a high status also allowed Alcoa to expand its influence in the downstream stages of the Amazonian aluminium industry by means of a controversial deal approved by the government in mid-1981, under which the US multinational bought the considerable bauxite reserves owned by the American shipping magnate D. Ludwig in the Trombetas area. Although these reserves should have been legally brought back under State control, there is evidence to show that the government yielded to intense pressures exerted by Alcoa, which linked the viability and implementation of its Saõ Luís project to the ownership of these bauxite reserves (Santos, 1982, p. 14; Machado, 1985, pp. 137–9). The formal association with Shell–Billiton in September 1981, when the project became officially known as the ALUMAR consortium, served only to reinforce that influence over the Amazonian bauxite industry in view of Billiton's stake in MRN, and to strengthen the execution of Alcoa's alumina–aluminium complex, whose massive costs and risks could now be shared with another leading multinational.

But the swift implementation of ALUMAR was, above all, a result of Alcoa's determination to establish a strong presence in Amazonia as fast as possible. This was not only associated with the 'cold war' waged on ALBRAS, as Alcoa's attempt to freeze ALUNORTE showed, but also with the difficulties faced by its other aluminium mega-project in Australia, which raised the strategic importance of ALUMAR to the American multinational.[18] In contrast with the erratic implementations of ALBRAS–ALUNORTE, ALUMAR was officially inaugurated in mid-1984, exactly according to its original schedule. This complex was composed of a refinery with an annual capacity to produce half a million tonnes of alumina, a smelter able to produce 110,000 tonnes of aluminium a year, and an exclusive bulk load terminal capable of handling 50,000 dwt vessels. According to a ten-year contract signed with MRN at the time, the entire bauxite requirements of its refinery was to be supplied by the Trombetas consortium, although this contract has been jeopardised by the MRN price dispute and by ALUMAR'S willingness to import cheaper bauxite.

Despite Billiton's announcement before the inauguration of ALUMAR that it was not prepared to invest in the expansion of the smelter because of the unpredictable world aluminium market, Alcoa decided to press ahead with the implementation of the second phase of the project. This decision was closely associated with the sale of a 35 per cent stake in Alcoa's Brazilian subsidiary to a local construction group (Camargo Correa) for a sum of US $240 million. The proceeds of this sale were thus used to finance the US $200 million expansion of ALUMAR. But as these funds were derived from tax exemptions granted to Camargo Correa during the construction of the Tucuruí dam, the subsequent approval of such a controversial deal by the interministerial council of the PGC meant that the second phase of ALUMAR was to be entirely financed by the programme's generous fiscal incentives.

That phase was officially inaugurated by President Sarney in the beginning of 1986, during which almost 200,000 tonnes of metal were produced. ALUMAR is now the largest aluminium smelter in Brazil, with an installed capacity of 245,000 tonnes a year, which may still rise to 380,000 as Billiton has recently proposed investing a further US $150 million in the implementation of a third phase, provided that ALUMAR receives firm government guarantees of further supplies of subsidised energy (*Brasil Mineral*, April 1987, pp. 8, 33). The evidence from this PGC mega-project amply reveals how the present eastern Amazonian development strategy has been influenced by global corporate forces.

CONCLUSION

By laying emphasis on mining–metallurgical activities with high financial risk and upfront capital requirements, the PND II–POLAMAZONIA regional strategy strengthened the role of multinational corporations in Amazonian development, given that they are normally better prepared to provide capital, technical expertise, and to absorb project output by marketing it worldwide. In any case, there were exogenous reasons for foreign mineral groups to move into countries such as Brazil, which were seen as politically stable, rich in untapped natural resources and supportive of big business. As it has been shown, these reasons range from insecurity of supplies and the creation of producer cartels to energy constraints and a general preoccupation to increase world mineral supply so as to maintain low ore prices. At the same time, CVRD's move into Amazonia has not only created a national basis from which to bargain with the multinationals, but also added a national corporate dimension to mineral development.

If there was a certain corporate emphasis in the previous phase of eastern Amazonian development, the current phase has been strongly influenced by global corporate strategies (including those of an autonomous state enterprise like CVRD) and, to a lesser extent, by national economic priorities, although the foreign debt crisis should not be seen in isolation from the global context of Brazilian dependent development. As we have observed, the PGC has its roots in a study carried out by a Japanese government agency oriented towards the agricultural and mineral resource needs of that country, and upon which CVRD formulated tentative development plans for eastern Amazonia. These plans emerged at a time when the future of its mineral mega-projects was in the balance, but also when the Brazilian government seemed to have run out of ideas as far as eastern Amazonian development planning was concerned. The government thus jumped on the Carajás train of CVRD, which after all had devised a regional development strategy that would 'solve' the Brazilian balance of payments and foreign debt problem. This helps to explain why the PGC was primarily based on the export-oriented mega-projects, which were simply expected to attract further mineral and agricultural investments into the region.

This weaker role of government planning in regional development is closely associated with the increasing influence of global corporate strategies, which puts CVRD in the forefront of the national struggle against the controversial activities of powerful foreign mineral groups in eastern Amazonia. Although CVRD has sometimes made significant efforts to defend national interests, foreign groups tends to have greater bargaining powers irrespective of their type of relationship with the Brazilian mineral group. As has been shown, even the Carajás iron-ore project has been subjected to external influence by means of interrelated financial and sale agreements. In the cases of MRN and ALBRAS–ALUNORTE, this influence has been reinforced by the direct participation of powerful multinationals, such as Alcan or foreign consortia such as NAAC, which are thus able to exert more direct forms of control over those projects. However, their global strategies have sometimes clashed with that of CVRD, often with adverse impacts on the implementation and operation of their joint ventures. These problems tend to be nonexistent or insignificant in joint ventures wholly-owned by foreign multinationals – such as ALUMAR – which can also have more effective control over the project.

The mineral strategies of these foreign groups are closely associated with the process of globalisation of productive activities which has been particularly evident in the aluminium industry, one of the best examples of a truly global industry with worldwide sourcing and markets.[19] This industry has been subjected to major structural changes over the past fifteen years, which have encouraged the powerful aluminium multinationals (the so-called 'six sisters') and the main Japanese producers to locate energy-intensive smelters in countries with abundant supplies of cheap energy. The 'comparative advantage' of Brazilian Amazonia thus lies in the subsidised hydroelectricity of Tucuruí and, to a lesser extent, the bauxite of Trombetas, cheap labour, fiscal incentives and lax pollution controls. The creation of the PGC was not, therefore, the underlying cause behind the emergence of huge aluminium complexes and mining projects in eastern Amazonia, although the programme provided valuable support for those projects through tax exemptions, energy subsidies and basic infrastructure.

But, in addition to this transfer of scarce public resources to export-oriented mega-projects dominated by global corporate forces, the current eastern Amazonian mineral development drive is likely to have serious socio-economic implications for the region, in view of the way it has become integrated into the global economic system. First of all, the Greater Carajás development model is clearly undermined by its heavy dependence on adverse world commodity markets, whose potential (if not propensity) to create recessive multiplier effects on primary export regions should not be underestimated (the disastrous socio-economic repercussions of the collapse of the Amazonian rubber economy earlier this century is a good example to have in mind). Secondly, contrary to official claims that the PGC mineral mega-projects were also designed to benefit the regional population, it is evident that vast natural resources and scarce capital are being primarily used to serve extra-regional interests, either in the Centre–South or abroad. Last, but not least, while the regional population is receiving few economic benefits, it is bearing the tremendous social and environmental costs (well-documented in the recent literature and elsewhere in this volume) of a process of global restructuring carried out by powerful corporate forces.

This calls for a radical re-think of Amazonian mineral policy, in the context of the recent democratisation of Brazilian society, and along the lines of a regionally-oriented natural resource development model. One such model is succinctly put forward by Weaver (1981, pp. 97–8), on the basis of three general premises: that 'the proportion of resources allocated to earning foreign exchange should be critically scrutinized and regulated, . . . [that] the bulk of resources should be used in meeting regional production needs, . . . [and that] restraint and conservation should be the by-words'. The current political and economic instability in Brazil points to a sharp reduction of further investments in Amazonian mineral projects, not only by cautious foreign investors but also by the heavily-indebted CVRD, which registered a massive US $344 million loss in 1987 (the first since its foundation in 1942). Although eight medium-sized pig-iron projects are now set to come on stream before the end of the 1980s – and are certain to accelerate the rate of deforestation in the region in view of the lack of strict government controls on the exploitation of the native rainforest for the production of charcoal – there is evidence (Neto, 1988) that not even half of the 2 million tonnes a year target set by official regional plans (MME, 1981, pp. 22–4; CVRD, 1981a, vol. 1, pp. 33–5) for the 1981–90 period is likely to be achieved.

This slowdown of the mineral development drive in the region offers a handy opportunity for corrective action, whose effectiveness will depend, to a great extent, on the ability and willingness of the Brazilian government to confront global corporate forces (the government intervention in the MRN price dispute shows that this task is by no means impossible). Broadly speaking, an immediate priority is to direct a more significant part of the wealth being generated by the mega-projects towards the promotion of a development strategy genuinely aimed to benefit the regional population.

Unless the government is prepared to take decisive steps towards this aim, which appears unlikely, Brazilian Amazonia is doomed to remain merely a source of cheap primary commodities to serve corporate interests and the expansion of economic centres beyond its boundaries.

Notes

1. The author would like to thank the Brazilian National Council for Scientific and Technological Development (CNPq) for the financial support which made this research possible. Thanks are also due to Anthony Hall and David Goodman for their constructive comments on an earlier draft. The responsibility for all opinions expressed in this chapter rests solely with the author.
2. Although these two distinct phases can be broadly identified with the governments of Generals E. Médici (1970–4) and E. Geisel (1974–9), it is worth noting that State planning and private initiative were not actually opposed to each other, nor exclusive to each phase. Thus, while the value of tax rebates granted to private cattle ranches rose from £26 million to £40 million between 1969 and 1974 (Branford and Glock, 1985, p. 46), a policy of 'semi-directed' colonisation by small farmers continued to be implemented in the north-western frontier during the second half of the 1970s (Valverde et al., 1979; World Bank, 1981).
3. See Brasil (1974, p. 66). Note that all translations from Portuguese into English are my own and therefore approximate.
4. This was the first large-scale mining project to be established in Brazilian Amazonia (it is situated in the federal territory of Amapá). According to a vice-president of Caemi, the Brazilian partner of Bethlehem Steel, this joint venture (named ICOMI) was able to be set up in the 1950s only because the Soviet Union was threatening to cut off manganese supplies to the West during the Cold War, and because the closure of the Suez Canal in 1956 disrupted supplies shipped from India (see Bourne, 1978, pp. 142–3).
5. The present equity distribution of MRN is as follows: CVRD (46 per cent), Alcan (24 per cent), Royal Dutch Shell–Billiton (10 per cent), Companhia Brasileira de Alumínio (CBA), Brazil's only private aluminium producer (10 per cent), Reynolds (5 per cent) and Norsk Hydro (5 per cent). Despite the Brazilian majority, major board decisions have to be approved by at least three-quarters of shareholders – which gives the foreign associates disproportionate control over the project.
6. See Tsurumi (1976), and Pepper *et al.* (1985, pp. 274–9). As a result of this long-term strategy, together with the 1979 oil price rise, Japanese primary aluminium production fell from a peak of 1.1 million tonnes in 1977 to 770,000 in 1981, and only 140,000 in 1986 (*Financial Times*, Aluminium Survey, 28 October 1987).
7. This is a powerful consortium of 33 leading representatives of the Japanese aluminium industry: the five LMSA producers, ten trading companies, sixteen industrial consumers, a private bank and the state-owned Overseas Economic Co-operation Fund, which is the leading shareholder.
8. As the original ALBRAS scheme would consume 60 per cent of Tucuruí's energy capacity, the CVRD–LMSA joint venture was expected to take an equivalent equity stake in this hydro project (originally estimated to cost US $1.1 billion). Thus, the ALBRAS project could be revived only after the

Japanese had managed to transfer this heavy burden to the Brazilian government; unlike the rival Asahan aluminium venture in Sumatra – a joint venture between a Japanese consortium similar to NAAC and the Indonesian government – which comprises two hydroelectric power stations.

9. These loans were made available by the European Coal and Steel Community (US $600 million), a pool of Japanese financial institutions (US $500 million), the World Bank (US $300 million), a West German government fund and American banks (*World Mining*, January 1983, pp. 52). Over 90 per cent of these loans carried a fixed interest rate of 8 per cent a year, as opposed to Brazil's troublesome foreign debt, which is subject to flexible interest rates.
10. Average world iron ore prices gradually fell from US $20 a tonne in 1981 to US $17 in 1985 – when the Carajás project was about to come on stream – only to be followed by sharp cuts imposed by Japanese (and some Western European) steel-makers in 1986 and early 1987. See Franz *et al.* (1986), and the commodities section of the *Financial Times* on 31 October 1986, 19 February 1987 and 17 March 1987.
11. Interview with B. A. Santos, director of CVRD's geology and mining subsidiary, at the company headquarters in Rio de Janeiro on 19 December 1985. This iron-ore venture was then expected to have a profit margin of as little as 1 dollar a tonne (after transport and financial costs), based on the late 1985 fob sale price of US $16 a tonne.
12. ALBRAS is entitled to an extra 15 per cent discount on the concessionary national tariff (A 1) for aluminium smelting, and to a ceiling on its energy costs (20 per cent of the world market price of aluminium). In addition, if the fob sale price of ALBRAS aluminium is below US $1,413 a tonne, it is exclusively allowed to pay the lowest electricity rate in the country (US $10.5 mills/kwh). According to the regional development superintendent of ALBRAS, this guarantee of subsidised energy on a long-term basis was one of the most important factors behind the implementation of the project (interview with C. Bentes, at the company headquarters in Belém on 13 November 1985).
13. Another estimate made by a respected Amazonian journalist puts this figure at US $5.5 billion, which will rise to US $7 billion if one includes debt servicing and the amortisation of costly foreign loans (L. F. Pinto, personal communication, November, 1985). In my Ph.D. thesis (Neto, 1988), there is evidence to show that part of this excessive financial cost was associated with the urgency to implement the project.
14. The whole speech of President Sarney was taped by myself during the official inauguration of ALBRAS, in Barcarena on 24 October 1985.
15. Collective interview given by the president of ALBRAS–ALUNORTE (R. N. Teixeira) on 24 October 1985.
16. According to MRN annual reports, its fob average sale prices actually fell from almost US $31 a tonne in 1982 to US $29 in 1983 and 1984. A company manager told the author in November 1985 that this downward pressure on prices was essentially the result of 'the crisis of the aluminium industry at the international level' (interview with M. Xavier, at MRN headquarters in Belém).
17. Interview with T. Mosci, a top ALUMAR manager, at the company headquarters in São Luís on 6 December 1985. He also told the author that although two countries were singled out by this global corporate strategy – Australia and Brazil – the latter had the slight advantage of being closer to the American market and having a large internal market.
18. In the same year that its São Luís project was started, Alcoa began to

implement another aluminium mega-project in the south-eastern Australian town of Portland. One of the reasons for starting the two projects at the same time was to play one off against the other so as to secure concessions (particularly on energy tariffs) from both host countries. But while the Brazilian Minister of Mines and Energy (C. Cals) decided to play Alcoa's game in earnest (Brasil, 1981, pp. 137–41), the Victoria state power company firmly resisted Alcoa's demand of lower tariffs, which led to the abandonment of the original Portland project.

19. While this specific globalisation process can be associated with a more general process of international economic restructuring (Muegge, Stohr *et al.*, 1987), other recent studies (Henderson and Castells, 1987) have cautioned against the use of general models of internationalisation, such as the new international division of labour thesis of Frobel *et al.* (1980), on the basis that those processes should be historically and empirically specified in relation to particular industries and regions.

References

ALBRAS–ALUNORTE (n.d., c. 1981), *Alunorte e Albrás n.p.: ALBRAS.*

Almeida, E. B. de (1981), 'A Descoberta do Ferro de Carajás', *Ciências da Terra*, 1, pp. 22–4, Salvador: SBG.

ALUMAR (n.d., c.1984), *O Brasil Ingressa no Restrito Clube dos Grandes Produtores Mundiais de Alumínio*, n.p.: ALUMAR.

Bourne, R. (1978), *Assault on the Amazon*, London: Gollancz.

Branford, S. and Glock, O. (1985), *The Last Frontier: Fighting Over Land in the Amazon*, London: Zed Books.

Brasil, República Federativa do (1974), *II Plano Nacional de Desenvolvimento (1975–79)*, Rio de Janeiro: IBGE.

Brasil (1980), *III Plano Nacional de Desenvolvimento (1980/85)*, Rio de Janeiro: IBGE.

Brasil, (1981), *Simpósio Alternativas para Carajás*, Brasília: Senado Federal.

Brasil Mineral (1986, 1987), various issues, São Paulo: Signus.

Brazil Report (1986, 1987), various issues, London: Latin American Newsletters.

Caufield, C. (1985), *In the Rainforest*, London: Heinemann: Picador (Pan Books).

Cota, R. G. (1984), *Carajás: A Invasão Desarmada*, Petrópolis: Vozes.

CVRD, Companhia Vale Rio Doce (1980, *Amazônia Oriental: Um Projeto Nacional de Exportação*, n.p.: CVRD.

CVRD (1981a), *Amazônia Oriental: Plano Preliminar de Desenvolvimento*, 2 vols, n.p.: CVRD.

CVRD (1981b), *Projeto Ferro Carajás*, n.p. CVRD.

CVRD (1982), *Companhia Vale do Rio Doce: 40 anos*, n.p.: Nova Fronteira.

CVRD (1986a), 'Carajás: Mina de Ferro N4E', *Revista CVRD*, 7(25) Rio de Janeiro: CVRD.

CVRD (1986b), 'A CVRD e o Alumínio da Amazônia', *Revista CVRD*, 7(26) Rio de Janeiro: CVRD.

Dantas, M. (1982a), 'A Questão do Alumínio', in O. Varverde *et al.* (eds), *A Amazônia Brasileira em Foco*, 14 Rio de Janeiro, CNDDA, pp. 7–49.

Dantas, M. (1982b), 'Tudo Que Voce Pergunta (e o Governo Nao Responde) Sobre o Projeto Carajás', in O. Klein *et al.* (eds), *Salvar Carajás*, Porto Alegre, L&PM, pp. 9–34.

ELETRONORTE (n.d., c. 1982), *Usina Hidreléctrica Tucuruí 8000 MW*, n.p.
English, B. A. (1984), *Alcoa na Ilha*, São Luís: Caritas.
Financial Times (1986, 1987), various issues, London.
Foweraker, J. (1981), *The Struggle for Land*, Cambridge: Cambridge University Press.
Franz, J. *et al.* (1986), *Iron Ore: Global Prospects for the Industry 1985–95*, Washington, D.C.: World Bank.
Frobel, F., Heinrichs, J. and Kreye, O. (1980), *The New International Division of Labour*, Cambridge: Cambridge University Press.
Galvão, R. X. (1984), 'Alcoa, a "besta-fera" nas terras de São Luís', *Pau Brasil*, (November–December) pp. 22–6, São Paulo: DAEE.
Girvan, N. (1976), *Corporate Imperialism: Transnational Corporations and Economic Nationalism in the Third World*, London: Monthly Review Press.
Graham, R. (1982), *The Aluminium Industry and the Third World*, London: Zed Books.
Henderson, J. and Castells, M. (eds), (1987), *Global Restructuring and Territorial Development*, London: Sage.
IBASE, Instituto Brasileiro de Análises Sociais e Econômicas (1983), *Carajás – O Brasil Hipotêca Seu Futuro*, Rio de Janeiro: Achiamé.
IBASE (1984), *O Capital Japonês no Brasil*, Rio de Janeiro: IBASE.
IDCJ, International Development Centre of Japan (1980), *A Preliminary Study on Regional Development of the Carajás Corridor in Brazil*, Tokyo: IDCJ.
LACR, *Latin American Commodities Report* (1986, 1987), various issues, London: American Newsletters.
Machado, R. C. (1985), *Apontamentos da História do Alumínio Primário no Brasil*, Ouro Preto: Gorceix.
Minérios, Extração & Processamento (1982) (December issue) São Paulo: EMEP.
MME, Ministry of Mines and Energy (1981), *Grande Carajás Program*, n.p.: MME.
MRN, Mineração Rio do Norte (1985), *Relatório Anual 1984*, n.p.: MRN.
Meugge, H., Stohr, W. B. *et al.* (eds) (1987), *International Economic Restructuring and the Regional Community*, Aldershot: Avebury.
Murphy, K. J. (1983), *Macroproject Development in the Third World*, Boulder, Col.: Westview.
Neto, F. T. (1988), 'National and Global Dimensions of Regional Development Planning: A Case-Study of Brazilian Amazonia', *Ph.D thesis*, University of London.
Park Jr, C. F. and Freeman, M. C. (1968), *Affluence in Jeopardy – Minerals and the Political Economy*, San Francisco: Freeman, Cooper and Co.
Pepper, T. *et al.* (1985), *The Competition – Dealing with Japan*, New York: Praeger.
PGC, Programa Grande Carajás (n.d., c. 1982), *Grande Carajás Program – A Challenge for all segments of Brazilian society*, n.p.: PGC/Banco Itaú SA.
PGC (n.d., c. 1983), *Atos Declaratórios do Conselho Interministerial*, Brasília: PGC.
PGC (1985a), *Programa Grande Carajás: Histórico, Objetivos, Atividades e Legislação*, Brasília: PGC.
PGC (1985b), *Projetos Aprovados (Posição em 31 Julho 1985)*, Brasília: PGC.
Pinto, L. F. (1982), *Carajás, O Ataque ao Coração da Amazônia*, Rio de Janeiro: Marco Zero.
Planejamento & Desenvolvimento (1981) (July–August issue) Brasília: SEPLAN.
Sá, P. C. de (1981), 'A CVRD e a Indústria do Alumínio', n.p.: CNPq (mimeo).
Santos, B. A. (1982), 'Carajás e o desenvolvimento regional', *Revista Brasileira de Tecnologia*, 13, (5) pp. 9–18, Brasília: CNPq.

Santos, B. A. (1986), 'Recursos Minerais', in J. M. Almeida Jr (ed.), *Carajás – Desafio Político, Ecologia e Desenvolvimento*, pp. 296–361, São Paulo: CNPq/Brasiliense.

SEPLAN, Secretaria de Planejamento da Presidência da República (1985), *Programa Grande Carajás – Legislação e Normas*, Brasília: SEPLAN.

Soares, M. C. (1981), 'Projeto Carajás: Origens e Desenvolvimento – Uma Visão Institucional', n.p.: CNPq (mimeo).

Trebat, T. J. (1983), *Brazil's State-Owned Enterprises*, Cambridge: Cambridge University Press.

Treece, D. (1987), *Bound in Misery and Iron: the impact of the Grande Carajás Programme on the Indians of Brazil*, London: Survival International.

Tsurumi, Y. (1976), *The Japanese Are Coming*, Cambridge, Mass.: Ballinger.

Valverde, O. *et al.* (eds), (1979), *A Organização do Espaco na Faixa da Transamazônica*, vol. 1: Sudoeste Amazônico, Rondônia e Regiões Vizinhas, Rio de Janeiro: IBGTE.

Walde, T. (1985), 'Third World Mineral Development in Crisis', *Journal of World Trade Law* (January–February) pp. 3–33, Geneva: Jacques Werner.

Weaver, C. (1981), 'Development Theory and the Regional Question: A Critique of Spatial Planning and its Detractors', in W. B. Stohr and D. R. Taylor (eds), *Development from Above or Below?*, Chichester: John Wiley.

Wood, C. H. and Wilson, J. (1984), 'The Magnitude of Migration to the Brazilian Frontier', in M. Schmink and C. H. Wood (eds), *Frontier Expansion in Amazonia*, Gainesville: University of Florida Press, pp. 142–52.

World Bank (1981), *Brazil: Integrated Development of the Northwest Frontier*, Washington, D.C.: IBRD.

World Bank (1982), *Staff Appraisal Report: Carajás Iron Ore Project* (6 July) Washington, D.C.: World Bank.

World Mining (1983) (January issue) San Francisco, Miller Freeman.

7 Frontier Security and The New Indigenism: Nature and Origins of the Calha Norte Project[1]

João Pacheco de Oliveira Filho

There are many government projects and programmes for Amazonia – the Plan for National Integration (PIN), POLAMAZONIA, PROBOR, POLONOROESTE, and the Greater Carajás Programme (PGC), amongst others – whose roots go back to the 1970s at a different moment in Brazil's economic and political history, but whose consequences are felt up to the present time. They derive from a diversity of circumstances, of funding sources and executive institutions, generally expressing the interests both of those who control the Brazilian economy and of their overseas partners. The assumptions and directives of modernisation theory are advanced in the hope that the principles of a market economy will gradually come to characterise settlement of the interior. Thus, agricultural and agroindustrial activities are emphasised, along with a concern for indicators of economic efficiency and the balancing of migratory flows, in the hope of reconciling these economic goals with social programmes of directed assistance to small farmers.

The technical corps of these projects is composed of various disciplines (economists, sociologists, agronomists, engineers, specialists in public administration, etc.) brought together in bodies and commissions under the Ministry of the Interior. The formalisation of this type of planning under the New Republic has been the responsibility of the Development Plan for Amazonia (PDA), prepared by the Superintendency for the Development of Amazonia (SUDAM) as well as other sectors of the Interior Ministry such as BASA and SUFRAMA, and technicians from Amazonia's states and territories.

The Calha Norte Project,[2] (PCN) however, because of its intrinsic features, institutional affiliations and historical roots, is unlike these programmes. It derives from a different set of objectives, and has its origins in another sector of the Executive. Its official commencement came with Statement of Intent (*Exposição de Motivos*) no. 018/85, prepared by General Bayma Denis, General Secretary of the National Security Council, and Head of the Military Household of the Presidency of the Republic. In this document, the President is requested to approve the formation of an interministerial working group whose job it would be to work out terms of reference and guidelines for government intervention in 'the region north of the rivers Solimões and Amazonas', with a view to 'overcoming the great obstacles posed by the environment to developing the area', bringing about 'its effective integration into the national context'. According to EM 018 the onus on this group would be to 'consolidate and present a development plan for the region', including among its members representatives from the Planning Secretariat (SEPLAN), the Ministry of Foreign Affairs (MRE) and the Ministry of the Interior (MINTER), apart from the National Security Council (CSN) itself.

In its final report, dated December 1985, the working group presented its findings in a 37-page document which contained a plan for operations and budget for the five-year period 1986–90. The work is entitled, 'Development and Security in the Region to the North of the Rivers Solimões and Amazonas', subtitled 'Calha Norte Project', the name by which it was later to become known in official circles and by which, much later, the press got to hear of it. Although its formal structure remains unchanged (these four ministries signed EM no. 770, asking the President to approve the conclusions of the working group), it is possible to see a reorganisation of decision-making, thus: (a) members of the group met in the General Secretariat of the CSN, using its facilities and technical assistance (b) other federal bodies working in Amazonia, such as the military ministries and Ministry of Finance (concerned with taxation and excise issues), made a 'valuable contribution' to the work (PCN, p. 7) and enjoyed an 'effective participation' in the formulation of this plan (EM 770, p. 2), (c) even though SEPLAN has formal responsibility for coordinating the working group, given the nature of the problems and concerns stated as well as the few resources allocated to it, its role is limited to dealing with financial matters and passing on funds to the executive organisations, (d) there is no mention whatsoever of collaboration from state government representatives in the region,[3] although this provision is clearly made in EM 018 (p. 4).

This document, approved by the President on 19 December 1985, is the most complete available account for the purpose of analysing the Calha Norte project. As far as involvement of other public bodies is concerned, this project is implemented through annual budgets, published in the Official Journal (*Diário Oficial*), and announced by the Planning Secretariat. The rationale for EM 018 of 31 August, 1985, became known only in October 1987 following its arrival at the National Congress that month to complement material due to be examined by a Parliamentary Commission of Enquiry which was investigating allegations made by the newspaper *O Estado de São Paulo* against the Indigenist Missionary Council (CIMI).

A MILITARY PROJECT?

The most notable feature of the Calha Norte project is its secretive nature, having been prepared by a small group of people, in isolation from other government departments. The National Congress was not asked to give an opinion nor kept informed on the subject, which it only got to know about officially in October 1987, two years after the project was born. The public became aware of what was happening only through the denunciations and criticisms made by CIMI in October 1986, 15 months after EM 018 appeared and almost a year after the project was approved.

Replying to journalists, spokespersons from SEPLAN and other civilian ministries minimised their participation in the PCN and stressed that further clarification should come from the CSN. General Bayma Denis, hotly pursued by journalists, rejected criticisms of the project as unwarranted and fanciful, referring to the precarious nature of Brazilian settlement in that frontier region and suggesting that the scheme was designed to solve existing problems there (such as border security, smuggling, drug trafficking and proximity to areas active with guerrillas). From then on discussion of the subject became polarised into two extreme camps; one side accused the PCN of causing severe harm or even the extinction of native groups, while the other, in rare expressions of opinion by qualified observers, called such criticisms entirely unfounded and rhetorical, emphasising that the project was simply in defence of national interests.

How could the PCN have remained unknown to other official bodies and to public opinion itself at a time of such intense political activity in Brazil, during a period of democratic transition, when the Constituent Assembly was being set up and when the government enjoyed widespread support for its economic policy (the so-called 'Cruzado Plan')? The answer to this question is to be found in Decree 79.099/77 of the Geisel government, which preceded the process of 'political opening' and which sets out norms and directives for the circulation of restricted documents.

In contrast to EM 018 (normally Statements of Intent have a limited circulation and are rarely published in the Official Journal), the working group's report is stamped 'confidential' at the top and bottom of each page. According to Decree 79.099/77 secret subjects are divided into 'top secret', 'secret', 'confidential' or 'reserved', access to each type being granted only to those with appropriate credentials and who can demonstrate a need to become acquainted with the subject. Even more restricted than 'reserved', documents classified as 'confidential' are those, 'knowledge of which by unauthorised persons would harm the national interest', or which 'are likely to cause severe official embarrassment'.

The practice of secret government also includes a whole repertoire of administrative action devoid of any obligation to inform the public, as is the case with secret decrees (only the numbers of which are published in the

Official Journal, without revealing their content), secret programmes (such as the parallel nuclear programme), plans considered to be 'in the national interest' and even secret bank accounts which are related to issues of 'national security'. According to a recent study by a journalist from Brasília, although these practices are long-standing (ten years old) and were the product of a different political–institutional context, they have continued in full force under the New Republic right up to the present day (Simonetti, 1987).

Furthermore, it should be noted that the PCN does not have the formal, systematic structure of a programme or a plan, nor does it constitute a group or unit which operates in a permanent fashion. What we are talking about is a coherent set of guidelines and objectives with which all government initiatives in the region must be compatible. In contrast to the model of other official interventions, the PCN does not possess a clearly defined decision-making forum (such as the broad council, composed of representatives from various ministries and institutions, which passes judgement on key issues in the Greater Carajás Programme). Nor does it constitute a technical body which implements its programme in specific situations (such as GETAT and GEBAM, which are basically executive units). Together with its secretiveness, this makes the PCN difficult to visualise in the sense that there is no-one who is clearly accountable, who could act as a spokesperson for social groups affected by its decisions. In the final analysis it is the President of the Republic who, with all the weight of his authority, directly and exclusively, decides and stamps with his seal of approval all action to be taken.

In an effort to become invisible and immune from outside scrutiny, the PCN has become subdivided into special projects under the wing of specific executive bodies to which SEPLAN, charged merely with overall financial management, channels funds which are thus allocated. The goals and budget were established by the interministerial working group proposed by EM 018. In principle any adjustments and specific details became the responsibility of ministries in charge of each special project. But how is it possible to talk about annual adjustments, reprogramming and continuous reassessment of goals and values if there is no clearly defined subject? Nothing is rigidly stipulated but there seems to be no administrative confusion since monitoring and evaluation is also carried out by another solid and powerful body, the Secretariat-General of the CSN. This group, as directly advisory to the President, has no powers to give opinions or make decisions independently, but always acts on behalf of the country's most powerful executive.

By analysing the funds allocated to different executive agencies the fundamental priorities of Calha Norte become transparently clear (see Figure 7.1). Most resources go to the Ministry of the Army (46 per cent of the total) and are intended for activities such as: (a) building, enlarging and improving barracks (48.4 per cent of this special project), (b) acquisition of military equipment (27.4 per cent), (c) purchase of vessels for river transport

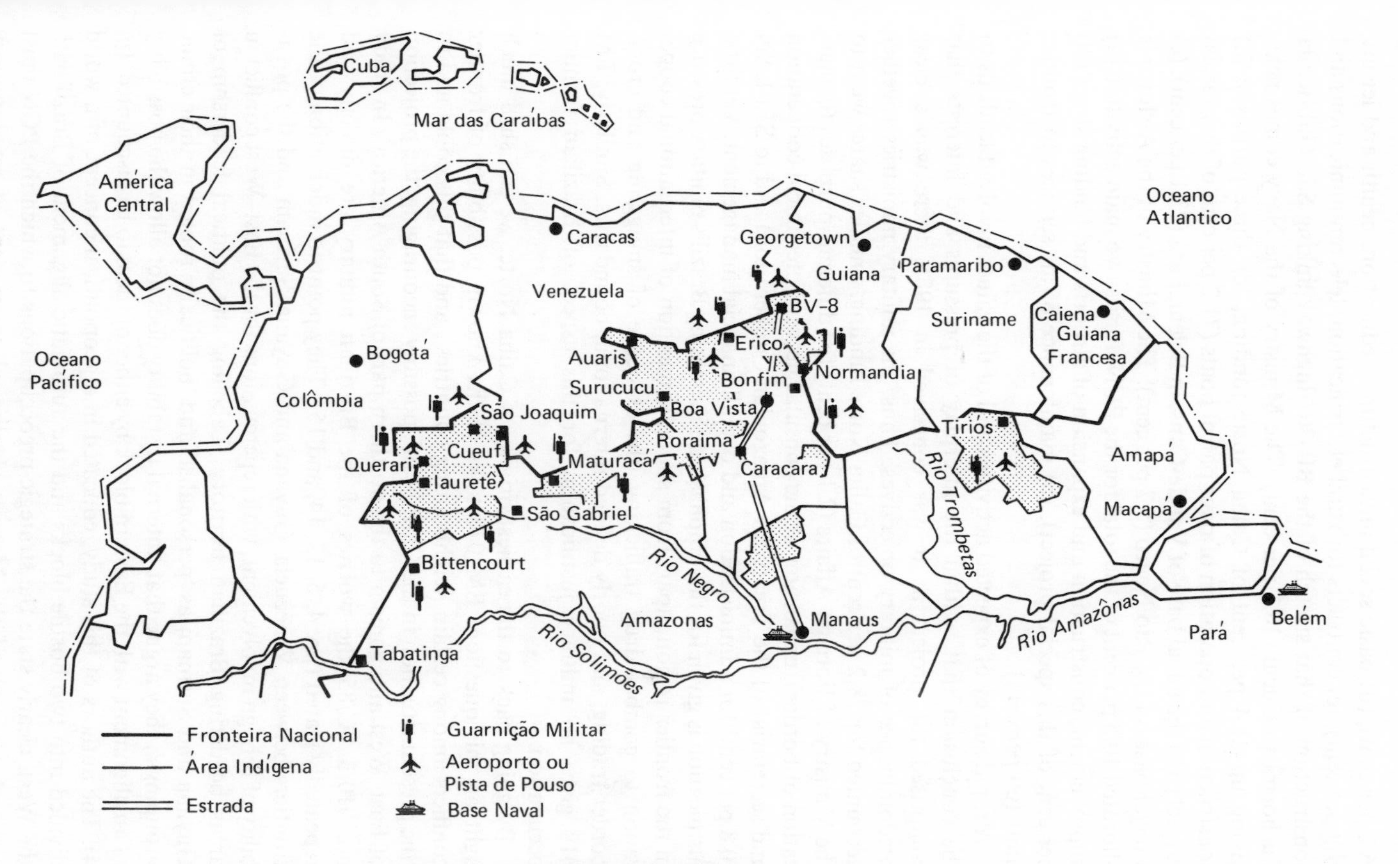

Figure 7.1 The *Calha Norte* project

(6.2 per cent), (d) basic social infrastructure – education, health and leisure (7.9 per cent), (e) contracts for rural electrification, telecommunications and construction of that stretch of the BR-307 highway linking São Gabriel da Cachoeira to Cucuí (10.1 per cent). The Ministry of the Navy comes next, taking up 21.4 per cent of Calha Norte funding, for three purposes: (a) construction and operation of river patrol boats (71.2 per cent of this special project), (b) the naval base of Val-de-Cans, in Belém, Pará (23.1 per cent), (c) a naval base on the Rio Negro (5.7 per cent). The Ministry of the Airforce is allocated 10.5 per cent of the total for the PCN, for its two main activities: (a) improvement, construction and expansion of airports and landing strips (40 per cent of this special project), (b) maintenance of air services to frontier units (60 per cent).

A rapid survey of expected activities and of the values involved leads us to the conclusion that, within the interplay of pressures and interests that comprised the project as it was conceived in 1985, there was a clear predominance of military objectives. Thus, the military ministries together accounted for 78.2 per cent of Calha Norte funding. Smaller shares went to the Ministry of Foreign Affairs (2.1 per cent) for different purposes (demarcation of borders, expansion of the consular network, technical cooperation and secretariat of the Treaty for Amazonian Cooperation) and to SEPLAN (0.8 per cent) for administration and overall financial management. Yet little justification is given for this emphasis on increasing Brazil's military presence in the frontier region, apart from sporadic mention of international cooperation to combat drug trafficking, the existence of smuggling and cross-border trading, and poorly defined international boundaries. Similarly, EM 018 gives no military justification for this aspect of Brazilian frontier occupation.

By going back to the earliest origins of Calha Norte, we can shed much light on this question. EM 018 talks of the known 'possibility of frontier conflicts among certain neighbouring countries', and that this could add to 'the present situation in the Caribbean', possibly encouraging 'the projection of East–West antagonism to the northern part of South America'. In study no. 010/3 SC/85 the worries of the Brazilian military are directly and repeatedly stated (pp. 4, 5, 12, 14, and 15). They point to indefinition of the frontiers between Venezuela–Guyana and Guyana–Surinam and the possibility of foreign involvement, with repercussions for the East–West conflict in areas bordering Brazilian territory. Labelling the political leadership of Guyana and Surinam as 'personalist' and 'politically rooted in their ethnic groupings', they argue that 'internal instability does not allow the possibility of an alignment with the Eastern block by either of them to be discounted' (p. 4). The authors of this study, educated in a geopolitical concept of a world divided into two hostile blocks and the automatic alignment of Brazil with the West, clearly state the strategic preoccupations to which the PCN must respond: 'in view of the Marxist ideological component of revolutionary movements in the Caribbean, the vulnerability of Guyana and Surinam to Cuban influence, awaiting the possibility of an "invited intervention" in these countries, there is a great need for the rearticulation and reorganisation of military forces' (p. 16).

A third feature is that Calha Norte is an impact project, that is, one in which investment is concentrated during an initial phase, followed by a huge reduction in spending, which is then limited to simple maintenance. Of the total estimated funding for the PCN, namely 628,892 million cruzados at October 1985 prices, 35.1 per cent was allocated for 1986, increasing to 40.2 per cent in 1987. Thereafter the proportions decrease significantly, falling to 12.45 per cent in 1988, 6 per cent in 1989 and 6.3 per cent in 1990. 75.3 per cent of the total is allocated for the first two years of the PCN towards basic infrastructure (construction and extension of barracks, airports and naval bases) and the acquisition of military equipment and boats for river transport. This phase of the project, in which budgeting was decidedly for military ends, coincided with a period in which, due to its confidential nature and invisibility, the PCN was almost entirely ignored or discussed on the basis of very limited information.

THE INDIAN PROBLEM WITHIN THE PCN

Among the basic goals of this project – increasing military presence in the area, demarcating frontiers and improving bilateral relations – that of 'defining an appropriate indian policy for the region' figures prominently (PCN, p. 4; EM 770, p. 3). Such an emphasis is also apparent in the budget, with MINTER/FUNAI absorbing 18.9 per cent of total project funding. In that section dedicated to indians, Calha Norte has the same general characteristics already mentioned. Indigenist activities are viewed as top secret, justified on the grounds of their being a subject of 'extreme political sensitivity' which could 'give rise to heightened domestic expectations' (PCN, p. 2). Further on it elaborates, 'indian affairs are among the most important areas of government policy-making and, in the recent past, they have been exploited at the expense of this country's positive image' (PCN, p. 4). Thus, FUNAI's activities in the frontier zone enter the world of confidentiality, being communicated to newspapers only through official statements, through visits by accompanied officials or spokespersons from the protective agency. According to this view of the world which, it is hoped, will be imposed on other public bodies, the critics of indian policy or those who work with native groups in the region but who do not work for FUNAI are labelled enemies of Brazil, because their behaviour is against the so-called 'national interest'.

The common direction of proposed indigenist activities is reflected in concrete actions. For the first year of the project top priority was given to

Roraima, which on its own would receive 49 per cent of total funding allocated to FUNAI for the various regions within Calha Norte. In 1987 the bulk of funds would go to the Upper Rio Negro (36.4 per cent) and the Amapá/Tumucumaque region (44.4 per cent). The balance would be spent proportionately to cover maintenance costs. Within each area the pace of activities also reflects this impact strategy. Of the total funding set aside for Roraima 56.5 per cent is for the first year, and 10.9 per cent for each of the following four years. In the Upper Rio Negro most resources are concentrated in 1987 at 41.1 per cent, falling to 12.4 per cent in subsequent years. The Upper Solimões would receive the bulk of its funding in 1986, or 22.9 per cent, dropping to 19.3 per cent in the four subsequent years for maintenance of infrastructure.

Perceptive readers cannot fail to have noticed the discrepancy in the PCN between its declared new indian policy and the absence of any discussion of the principles upon which it is based. Even as regards the type of structure expected for the indian agency, the PCN goes no further than a few scattered references, a feature which also applies to those concrete measures for which funds have been allocated. As far as the first aspect is concerned, namely the new principles upon which indian policy is based, this will be discussed below since it has implications for criteria used to demarcate indian lands; it was absent from the 1985 plan and is related to a more complex, later debate over government policy for the occupation of Amazonia.

The PCN is far more explicit about the second issue, FUNAI's new structure, to the extent that it is compatible with the military objectives of the project. This views the expansion of indigenist strategy as a form of 'increasing Brazilian presence' and 'strengthening the expression of national power' in frontier regions (EM 770, pp. 3–4). The document stresses the precariousness of FUNAI's infrastructure in the region, omitting any reference to qualitative change and highlighting as a goal the 'broadening and strengthening of FUNAI's activities' (EM 770, p. 3). Only a few indications are given as to the possible reorganisation of indigenist activities, affirming that it should be the product of 'intensified fieldwork' (PCN, pp. 5 and 23). It emphasised the need for a 'permanent presence in the area', and that certain situations and reserves 'required special attention' (PCN, p. 23). Further on there is talk of 'the basic principle of strengthening indian posts', pointing out the need to create decentralised administrative units (p. 24). The distribution of funds for indigenist activities reinforces the impression of continuity between the proposed structure and that which it is supposed to replace. Expenditure on conventional units such as indian posts (*postos indígenas*), points of contact (*frentes de atração*), regional administrations (*administrações regionais*) and handicraft shops (*casas do índio*), including their construction, repair and maintenance (27.2 per cent), as well as staff expansion (28.7 per cent), construction of airstrips (19.3 per cent) and acquisition of equipment (1.6 per cent), accounts for as much as 77.3 per cent of the total FUNAI budget under PCN.

In Roraima, it is proposed to build and repair six indian posts, to construct five landing strips, and to maintain ten posts and a contact point over five years. In the Upper Rio Negro, the project anticipates building nine indian posts, three airstrips, the repair of one and maintenance of ten indian posts and one handicraft shop. In the Upper Solimões, the repair, re-equipment and maintenance of seven indian posts. In the Amapá–Tumucumaque region, the repair of four posts and three landing strips, the construction of one *casa do índio* and an administrative centre, the maintenance of seven posts, one handicraft shop and a regional administration.

In terms of describing concrete activities to be implemented by the PCN there is nothing substantially new when compared with existing FUNAI practice. Indian posts, *casas do índio* and regional units have long formed a basic part of FUNAI's welfarist infrastructure, being the most obvious expression of the failure of this indigenist model, which neither benefits the indians nor exerts a positive influence on their lives. The problem with such a framework of action arises not just from the areas not reached by FUNAI, but mainly from its authoritarian and bureaucratic slant, its ignorance of indians' needs and aspirations, its poor capacity for living with (or even allowing) initiatives and organisational forms originating from within the population which it manages. This structure is anachronistic and deficient not simply in terms of its regional distribution, but also because of its inefficiency in bringing tangible benefits to its target population (the so-called 'indigenous communities') and failure to create alternative means for allowing the socio-cultural continuity and development of these groups in a way which is compatible with their traditions and current aspirations.

It is important to note that all the points mentioned in the PCN are included in the so-called restructuring of FUNAI, made known to the public in February 1986 when an interministerial working group (MINTER, SG/CSN, MJ, MIRAD, MA, MEC, MS, MT) was formed to prepare a proposal for 'modernising and strengthening FUNAI'.[4] Preliminary studies were carried out on the initiative of the Ministry of the Interior during the second half of 1985, the first decentralised FUNAI unit being created, for the south-west, in December. The criterion which guided this restructuring was the desire to improve its efficiency and administrative rationality, sustained by an abstract analysis of systems and organisations. Both the studies and first measures were undertaken simultaneously with the drafting of Calha Norte, which mentions 'the need to reorganise the structure of FUNAI' (p. 23) and even refers to the decentralised units by the same name used in FUNAI's new statutes, that is, 'regional administrations' (p. 24).

Within its brief of intensifying Brazilian presence along the frontier zone, the PCN has led to a restructuring of FUNAI activities based on the idea of military organisation as the model for indigenist activity. The document justifies the channelling of new funds to FUNAI on the grounds that it has a degree of technical competence and qualified personnel to deal with indian problems, comparable with those of the military for tackling questions of national defence. It so happens, however, that apart from its blunders and evident loss of public credibility in recent years, this approximation between FUNAI and the armed forces is unsatisfactory due to the very role of the indian agency. An indian post can serve as a frontier observation point and bastion of the national interest, but its primary function is to assist the indians and preserve their habitat.

In a complete reversal of this principle, what the PCN appears to be doing is establishing a national presence, omitting to protect the land occupied by indians; it is increasing the degree of control over native populations, not improving standards of assistance. The network of 34 indian posts, 11 airstrips and four regional offices alone is sufficient to reproduce a military logic of occupation and defence of colonial territory. Based on the principle of hierarchy and the setting up of communications networks, it reconciles the decentralisation of certain administrative areas with the constitution of a command unit which reaches regional offices and local scattered representatives located near the villages.

Only 12.2 per cent of the funds allocated to FUNAI under the Calha Norte project will be used for studies to identify indian territories, to pay for demarcation and even land-titling, which includes the transfer of non-indian squatters, payment of compensation and relocation. In other cases the project has adopted an impact strategy, since funding for land issues is distributed evenly throughout the five years.

Two mechanisms are envisaged for controlling indian groups. On one hand FUNAI is contracting almost 200 new employees; 30 for Roraima, 84 for the Upper Rio Negro, 42 for the Upper Solimões and 40 for the Amapá–Tumucumaque area. As civil servants, or rather 'servants of FUNAI', such people must make contact with indians, gather and transmit information, and obey orders from their superiors. In their unequal dealings with indians they build up a network of relationships, facilitating acceptance of their (official) role through local clients. They are always duty-bound to obey their superiors in the hierarchy and to keep quiet over subjects or events which could be considered embarrassing to the administration, especially those classified as secret. On the other hand FUNAI also runs community development projects, for which funding to the tune of a constant 6.3 per cent of the agency's total regional budget is channelled. In theory, these projects are designed to improve the living conditions of communities, with no expectation by the government that the funds will be repaid. In practice, they are set up in a most precarious manner by people who are ignorant of the local reality or who have no planning capacity, operating in the most irrational way imaginable through the use of individual grants intended to create dependency, bribe and corrupt indian leaders, neutralise critics and fragment groups. In conclusion, it is clear that this restructuring of FUNAI, intended to fit in with the peculiarities of remote Amazonian frontier zones, has not resulted in any improvement in the operational standards of this indian agency. It is, in fact, an attempt to expand Brazilian presence there, as well as that of federal institutions, to guard the borders and maintain direct control over indian communities.

DIAGNOSIS FOR THE REGION

The PCN prioritises four basic issues: (a) increasing military presence in the area, (b) improving bilateral relations, (c) demarcating borders, and (d) appropriate indian policies for the region. This was a deliberate choice made by planners within a particular political context, also conditioned by the interrelationships and preoccupations of the various ministries which made up the working group. It was easier for CSN representatives to focus on a discussion of strategic goals, requiring the equal participation of military ministries, than to discuss with technicians from SEPLAN and MINTER the concept of economic planning and strategies for the development of Amazonia. Calha Norte was designed as an essentially military project, within the limits and sacrosanct principles of national security. The reduced participation of other government bodies and the absence of state representatives are a direct product of this tactic, which would only change during 1987.

But military or strategic interests form only a part of the CSN's original manifest objectives, as outlined by the PCN document and by EM 770. There are other 'fundamental and pressing needs' for the Calha Norte region which were highlighted by earlier CSN studies and which appear in EM 018 with the same clarity as the four points mentioned above. They are: the expansion of transport facilities, accelerated production of hydroelectricity the establishing of economic development poles and the improvement of basic social infrastructure (p. 3). These are more clearly defined processes which involve many other federal bodies (MT, DNER, MME, ELETRONORTE, MA, COBAL, EMATER, EMBRAPA, BASA, DENTEL, SESP, SUCAM, FUNDACAO EDUCAR, MEC, etc.) which depend for their implementation on integration into an economic master plan (SEPLAN/IPEA) and on regional planning (SUCAM/MINTER).

In the internal document already referred to above, the CSN presents a diagnosis of economic and psycho-social problems associated with the current settlement process in Amazonia. The first feature to be highlighted is the 'vast demographic emptiness' of the region, which covers 14 per cent of Brazil but has only 1,620,000 people (1.2 per cent of the total population), 60 per cent of whom are concentrated in urban nucleii. The overall population density of the region is very low (1.2 people per square kilometre), situated mainly by rivers, leaving the interior virtually uninhabited. The document suggests that although the population is growing, which is positive, this is merely the quantitative aspect; its 'poor quality', a negative factor, is considered much more important (study no. 010/3 SC 85, p. 9). It is worth noting the stereotypical view taken by this analysis (p. 8) in relation both to the indians (whose habitat is labelled as a dangerous demographic emptiness, and a situation which should be countered in the frontier zone) as well as to the local Amazonian population along the valleys and to those migrants from the North–East (these so-called 'souls in search of new opportunities, adding few improvements to the already precarious living standards of the majority of the native population').

Another crucial aspect of this diagnosis is its assessment of perspectives for agricultural development on the Amazonian frontier. It is observed that, once forest cover is lost, the region's soils are subject to rapid erosion and declining fertility (p. 2). In contrast to the proponents of agricultural occupation in Amazonia, from Tavares Bastos to Vianna Moog, CSN analysts do not consider agriculture a priority for regional development. The PCN document clearly states the rationale behind this view: 'experience in Peru and Bolivia suggest that agriculture will always be less profitable than the cultivation of *coca*. In other words, any small farmer resettlement project for Calha Norte could result in an increase in such production (p. 11).

In contrast to primary activities, supposedly characterised by 'the rudimentary nature of its productive forms' in extractivism and agriculture, CSN analysts point to the region's mineral potential. 'Geological research by the RADAM project confirms the existence of large mineral reserves at various locations within the area under study' (study no. 010/3 SC/85, p. 7). They highlight initiatives such as the National Copper Reserve, within GEBAM's area of jurisdiction, the extraction of manganese in the Serra do Navio, the cassiterite reserves of the Upper Rio Negro, of kaolin near the rivers Jarí and Capim (AP/PA), gold deposits, and diamonds and cassiterite in Roraima, in the Surucusús mountains and the headwaters of the rivers Uraricaa, Uraricoera, Uaicas and Araçaca.

One limiting condition, the diagnosis adds, is the lack of an infrastructure which would facilitate positive steps towards economic development. Electricity supplies are inadequate for the region, coming from hydroelectric schemes at Coaracy Nunes (AP) and Balbina (AM), the latter still under construction. Transport infrastructure is equally poor, especially overland, which is at present centred on two major axes; that between Jarí (AP)–Oiapoque of 812 kilometres, and that from Manaus–Caracaraí–Boa Vista–Marco BV8, on the Brazil–Venezuela border of 800 kilometres, the BR-210 or Perimetral Norte, of which less than 500 kilometres are transitable (p. 7). Reference is also made to the inadequacy of social services, which embrace education and public health, but only within the larger urban centres (p. 9).

A NEW INDIGENIST POLICY

It is within the framework outlined above that the PCN study discusses the indian question. The first difficulty it acknowledges is 'the location of indigenous reserves in areas which are coincidentally rich in minerals' (p. 13). 'The big problem with mineral exploration in Roraima, as in other areas to the north of the rivers Solimões and Amazonas, is that the richest deposits are situated in indian territory, or in that which is presumably indian territory, in particular the area inhabited by the Yanomami indians' (p. 7). The crux of the matter is this paradox: the existence of rich mineral reserves and the lack of any legal means of exploiting them. Or rather, as the same document points out further on, 'the limits imposed upon the exploitation of mineral resources in indian areas by laws relating to indigenous affairs' (p. 14). It is for this reason that the PCN requires the revision of legal and administrative provisions which determine indian policies, whatever the mandate of FUNAI.

Thus, apart from specific proposals for localised activities the CSN, together with FUNAI, mounted an intensive lobby in the Constituent Assembly on the question of mining within indigenous areas. During the preliminary drafting phase of the new Constitution, in the sub-commission and Commission on Social Order, suggestions from the CSN including amendments presented by various deputies (José Lourenço, leader of the PFL, Nilson Gibson and others), centred upon the idea of legalising mining within indian areas. Conversely, congressmen briefed by scientific organisations and those defending indigenous rights attempted to outline and have passed measures which would limit the negative consequences of such mining activities. This pressure on the legislature was articulated with moves in the executive, and resulted in Order (*Portaria*) no. 01/87 from FUNAI/DNPM, signed on 18 May 1987, which allowed State enterprises and, under exceptional circumstances, private companies also, to mine within indigenous areas. This gives an entrepreneurial impetus to FUNAI's administration, which reaps benefits from royalties derived from the extraction of non-renewable resources from the lands of its client population.

The CSN studies then go on to criticise the size of indigenous territory which is considered excessive. It is argued that in the Calha Norte region the indian population possesses 'an enormous stock of land' but accounts for only 2.5 per cent of the population or 40,500 individuals (study no. 010/3 SC/85, p. 9). Further on the same document contends that, just within Roraima, indian lands amount to 26 million hectares (p. 10), and that 'in general there is strong pressure to legalise indigenous areas located within the area under study, in order to minimise encroachment by the surrounding society and producing focii of serious social tension' (p. 11).[5]

The proposed solutions to this question do not derive equally from initiatives designed exclusively for the Calha Norte region, and responses have involved both the legislative and the executive branches of government. In a press interview (*O Estado de Saõ Paulo*, 24 October 1987) and in a circular to members of the Constituent Assembly dated 25 October 1987, the president of FUNAI, Romero Jucá Filho, compares land and population ratios in indigenous areas with the situation in other states and foreign countries, in order to criticise article 198 of the existing Constitution, which guarantees recognition of indian rights and their habitat. In clear contradiction with his role as official protector, the president of FUNAI affirms quite explicitly that indigenous lands, 'in absolute and relative terms', are 'positively exaggerated'. Exercising the authority and legitimacy which, one supposes, his position confers, he opposed the continuation of article 198 and endorsed a replacement put forward by Deputy Bernardo Cabral (PMDB/AM), prepared in August 1987, which reversed much important work previously undertaken by various commissions and sub-commissions.

One journalist stated, certainly with an accuracy that she could hardly have imagined, that the figures presented came from a study that had landed on the desk of FUNAI's president (*Correio Brasiliense*, 27 October 1987). In fact, it is true that the data used by Jucá Filho can be found in study no. 007/3 SC/86 (pp. 4, 6, 11 and annexes) of the SG/CSN. In its conclusion this document makes it clear that what its authors consider to be a more appropriate indigenist policy, is in fact a new position regarding the limits of indian land and the use of its natural resources:

> In order to avoid adverse consequences for future generations arising from the break-up of national territory, a historic decision must be taken to block the process, especially in the frontier zone, which is backed by the Constitution, of conceding and demarcating indian territory. A proper policy towards indigenous lands must be implemented quickly and firmly, which would integrate the forest dwellers progressively and harmoniously into the national community (p. 17).

By means of two decrees signed by the President of the Republic on 23 October 1987, two important modifications were introduced to the laws which guide the demarcation of indian lands. Decree 94.945 changed the composition of the interministerial working group which considers the boundaries of indigenous areas. Responsibility for its coordination was moved up the official hierarchy to the Ministry of the Interior, while FUNAI became a mere member. The CSN became a permanent member of the group. The state land institutes gained seats and acquired voting rights, similarly to MIRAD, every time the issues under discussion refer to their respective states. Under the same decree the participation of state land agencies is not limited to final decision-making but also takes place during the initial phases, including that of identification (which embraces the questions of both ethnicity and territorial definition). The second decree creates a new law which can be applied to lands which have been in the possession of indigenous groups since time immemorial, thus attempting to regulate article 17 of Law 6.001/73 (the Indian Statute). This concerns the concept of the 'indigenous colony', which is to be applied to lands belonging to 'acculturated indians', as opposed to 'indigenous areas' which involve the territory of 'unacculturated indians', or those in the initial phase of acculturation (see Decree 94.946/87).

The two measures are complementary both in terms of their preoccupations and the interests which they favour. The first, which was guided by studies that preceded the SG/CSN and which even details the recommended changes (study no. 007/3 SC/86, pp. 17–18), introduces into the process of delineating indian areas two provisions capable of rectifying those restrictions considered to be 'anomalous'. First, considerations of a strategic nature, in which the presence of the CSN must be taken into account and, secondly, the objections of local landowners must be heard, so that indigenous lands will not be defined without first being subjected to agreements and compromises with the regional power structure. In other words, central government, through its specialist advisors (SG/CSN), does not wish that indian policy should be rigidly applied and clash directly with regional economic or political interests, creating conflicts which could have negative repercussions on political–electoral alliances or party structures.

The second decree appears to tackle another question, that of reconciling the drawing up of indigenous area boundaries with exploitation of their natural wealth (of the soil and subsoil) within regional development projects. It is a direct continuation and extension of the FUNAI/DNPM Joint Order mentioned above. As EM 018, which accompanies this decree, systematically argues, there are two possible positions: 'Those who believe that the forest dweller can live in isolation from the national community suggest that the simple demarcation of land will guarantee their survival. Those who desire the gradual and peaceful integration of the forest dweller maintain that, apart from demarcation, there is a need to provide social services'. At the moment FUNAI, having followed the CSN's recommendations in other studies, believes firmly in the second option. The classification of indian lands as either colonies or areas, with the consequent labelling of their inhabitants as agriculturalists or not, is no more than an administrative tactic for assigning distinctive roles to FUNAI and other government bodies in their common objective of ensuring full economic use of the basic resources of the indians' habitat, such as timber, minerals and others.

This legislation opens the way for justifying economic actions designed to extract profits from indigenous lands, based on suggested indicators of acculturation,[6] based on the assumption that indians are aware of the repercussions of their actions on Brazilian society and, therefore, are capable of reaching legal agreements and signing contracts with other public or private institutions. Furthermore, if the natural resources within indigenous areas can be economically exploited, assuring the owners of unexpected gains and new forms of payment, tribal territories will not need to be as large as was predicted earlier, when natives were considered to be simply gatherers or farmers.

THE NEW INDIGENIST FIELD

There is a third issue in CSN documents relating to the presence of indians in the Calha Norte region which should be mentioned. This manifests itself in a tremendous preoccupation over the possibility that external interests could determine the drawing up of reserve boundaries in the frontier zone, thus creating nationalities which might then wish to seek independence from Brazil. In study no. 010/3 SC/85 mention is made of moves to create a Yanomami National Park as an initiative which, by affecting 500 km of the Brazil–Venezuela border, would 'seriously compromise territorial integrity', apart from representing a 'threat to national sovereignty', resulting from the 'efforts of foreign interests to establish boundaries for an area designed to preserve an indigenous nation' (p. 11). In the same document reference is also made to 'the hidden danger of indigenous communities being transformed into independent states' (p. 15).

Identical concerns are present in the actual PCN document, where it is stated that: 'for a long time there have been visible pressures, from both national and international sources, seeking to create a Yanomami State at the expense of Brazilian territory' (p. 5). In another document, in a clause within the Statement of Intent which accompanies study no. 029/3 SC/86, these accusations are broader and more vicious: 'suspicions of the existence of manipulation of indian lands are confirmed by observing the high frequency with which these areas expand and converge in the direction of regions where discoveries have been made of rich mineral deposits, and in the direction of border areas where groups of the same ethnic type exist within the neighbouring country'.

The existence of such preoccupations came to the public's notice in a batch of measures during 1987, starting with a campaign against CIMI undertaken largely by the newspaper *O Estado de São Paulo* (which also published several SG/CSN studies on the subject with a view to making their allegations more credible). The campaign was continued with the installation of a Parliamentary Commission of Inquiry charged with ascertaining the truth behind these accusations and which became immediately polarised between conservatives (who tended to criticise the social actions of the Church) and liberals (who straight away perceived the fraudulent nature of the evidence). Representatives from the newspaper spoke at the Inquiry, alleging that they were unable to prove their allegations, on the basis of which a final report was prepared. In spite of this, the conservatives continued to hold further sessions of the Inquiry, using it as a means of mobilising political forces opposed to the group of deputies and senators who criticised the recent actions of FUNAI and the incursions of the CSN into the realm of indian affairs. By 1988 new measures were introducted to withdraw missionaries and researchers from indigenous areas, to initiate campaigns designed to discredit opponents and to encourage police intimidation against local organisations (as happened with the Centre for Indigenist Work, the CTI, with the Kaiova–Nandeva project and the MAGUTA, or Centre for Documentation and Research in the Upper Solimões).

All these measures are described and proclaimed in documents and studies prepared by the SG/CSN. Thus, just as an example, we can cite the conclusions of study no.029/3 SC/86, which refer initially to the 'mobilisation of different political currents which support the government in opposition to national and foreign pressure groups which support indian demands' (p. 29). This is followed by reference to the unleashing of 'campaigns for enlightening public opinion regarding the serious and growing risks to the Brazilian nation arising from the poisonous manipulation of the indian problem by foreign and national pressure groups' (p. 30). It goes on to suggest the creation of mechanisms for controlling the activities of these organisations in Brazil, even recommending 'prosecutions and penal action' against those held responsible for causing offence to the authorities, incitement to passive resistance or open rebellion, or the encouragement of more agressive demands' (pp. 30–1).

Thus, following developments during 1987, it is possible to see which are the components of this new indian policy which appear in the PCN's document. That is, reformulation of the means used to establish boundaries for indigenous lands and legitimation of the use of indians' natural resource endowment for the purposes of economic development and funding of the protective agency. These points are compatible with the already-mentioned process of FUNAI's restructuring, expansion of its infrastructure in the region, increasing control over indian communities, as well as attacks upon and prohibiting of the presence of missionaries or researchers, stimulating the replacement of FUNAI officials whose approach is based on a more traditional indigenism. With this operation complete and new guidelines drawn up by the CSN, it remains only for FUNAI's current management to take care of the public image, publicising the arguments and denunciations listed in the CSN studies.

CONCLUSIONS

At the start of this chapter it was stated that the PCN did not share the same origins as other programmes either for Amazonia as a whole or for specific sub-regions. It was then demonstrated how, at the moment of its birth, the PCN was diverted in the direction of military interests, avoiding publicity and minimising wider participation, and approved as a secret project concerned with national security. This position was maintained for one or two years, meeting its basic targets, which were quickly reached through impact measures. It mentioned the need for a new indigenist policy, but in practice limited itself to encouraging the expansion (and a certain rationalisation) of FUNAI infrastructure, harnessing it as a means of strengthening the expression of national power in the region.

From mid-1987 onwards, with the advent of a new political situation in which the Democratic Alliance rapidly collapsed and the power of parties (especially the PMDB) was diluted by the emergence of a multiparty bloc in support of the President and of conservative principles, Calha Norte became something very distinctive. It now proposed modifying indigenist policies in accordance with the interests of public institutions and private companies in the exploitation of Amazonia, avoiding any conflict between FUNAI and regional power-holders. It started to act publicly, establishing links with state and municipal governments, looking for support from political leaders and taking part in the debate with businessmen and planners about appropriate types of development for the region. During this phase the CSN's role was transformed from that of invisible mentor to that of formally coordinating the PCN, with SEPLAN losing even its financial management function. In 1988 responsibility for releasing PCN funds, previously with the Secretary-General of SEPLAN, was transferred to a lower level in the hierarchy, the Special Secretariat for Economic-Social Coordination.

However, to be able to follow the genesis and transformations of Calha Norte does not signify being able to define precisely what are its economic and social roots. This is what will be attempted in conclusion. I believe that Calha Norte leads us in the direction of the Executive Groups which operate(d) under the direct supervision of the CSN, such as GETAT and GEBAM, along the lines discussed by Almeida (1980). Using his analysis, the PCN can be seen to follow on from other similar bodies and was constituted to deal with strategic interests. In other words, how to reconcile a sparse regional population, large indigenous reserves, rich mineral deposits, economic development and frontier security. As an advisory body to the President of the Republic, the CSN collects information and provides analyses of the problems in question.

Once the general principles are approved, such groups acquire the power to arbitrate in conflicts within a certain sphere, in this case defined by the convergence of the issues mentioned above. Yet while sharing some jurisdictional space with other institutions of public administration, they impose themselves on other organisations, labelling them as inefficient or out of touch with reality. Despite this connection, however, they cannot be viewed as identical to the CSN. The function of the latter, as an advisory body, is merely to collect information and prepare studies on subjects of interest to the federal government. The Executive Groups, on the other hand, have a specific role to settle conflicts but are denied close access to the real centre of power. Their function includes not only defining military and police activities, but also making contacts and negotiating with the various parties involved. Their presence is not so much military as political.

As with other Executive Groups, the reference point for the PCN is primarily a geographical area. Like all focii of power not subordinated to the direct control of other powers within the republic, and able to ignore the scattered criticisms of society, the PCN is not circumscribed by defined spatial limits; sometimes it legitimises specific policies (such as that of indian affairs), while at others it widens its scope to the frontier zone or to Amazonia as a whole. As they are not instituted on the basis of their own inherent characteristics or qualities, nor do they constitute structurally defined units, such groups have a tendency progressively to extend their sphere of influence until they meet strong opposition. They never establish general or impersonal criteria, nor do they operate along predetermined lines. They remain highly flexible, tackling each case in response to a particular diagnosis and set of alliances.

Seen from this perspective, Calha Norte is a window through which we may observe certain adjustments which have taken place in the central nucleus of power. These have resulted from the accumulated experience of military personnel within the CSN in dealing with problems of land and regional development, the product of strong conflicts of interest and a legal *impasse*, producing a situation of social tension which is seen as a threat to public order and national security. This is a model of official action which evolved during the final years of military rule, as a means by which central government could combat rural violence, establish direct links with regional power-holders, obtain political support for the so-called 'process of democratisation', and restructure local party organisations to provide electoral support for the government. With the aim of subduing focii of tension, they also set up a dialogue with social movements, providing temporary solutions which fragmented and manipulated their leaders. This is the geneology of the PCN, and of several of its component parts already referred to.

With the onset of the New Republic, this interest group became neither victor nor vanquished, and acquired a marginal position within a civilian government conducted on the basis of party directives and with a very different posture towards public administration. In the first instance it was possible only to slow down the deactivation of GETAT (instead of the summary extinction requested by party and union leaders) and the military command of Amazonia, to which General Otávio Medeiros was transferred.

Calha Norte was a rehearsal for the gradual restrengthening of the CSN's influence upon the Presidency of the Republic, utilising a subject that appeared to be within military competence and which dispensed with the need to consult or mobilise political forces. While the democratic transition was facing considerable political obstacles, the CSN managed to recover its prestigious position as an advisory body linked directly to the President and enjoying his full confidence, a role which intellectuals and technicians connected with the parties were not given or did not wish to fulfil. This structure has been inherited from another Brazil and another political change (that of 'gradual and voluntary opening [*abertura*]', not of 'democratic transition') fed by the ideology of national security and by practitioners of social action (ranging from negotiation to the cooptation of leaders through to intimidation). This is what guides official policy for the frontier region, while at the same time controlling the actions of FUNAI and moulding the shape of Brazilian indigenist policy.

Notes

1. First presented at the Symposium on Large Projects and Their Impacts on Indian and Peasant Communities in Brazilian Amazonia, at the 46th. International Congress of Americanists, Amsterdam (July 1988). This chapter was written as part of the 'Study Project of Indigenous Lands: Invasions, Soil Use and Natural Resources', National Museum, Federal University of Rio de Janeiro.
2. The full title is, 'Development and Security in the Region North of the Rivers Solimões and Amazonas – Calha Norte Project'.
3. Governors of the federal territories of Roraima and Amapá are nominated by the Executive, and their administrations are linked directly to the Ministry of the Interior.
4. In fact this group never met collectively and the expectation of FUNAI–MINTER–CSN was that other bodies would provide advice on specific points, approve studies already completed and measures in the process of being implemented. There were two purposes to this restructuring: (a) to prepare new statutes for FUNAI, making it legally possible to carry out an administrative decentralisation, and (b) to facilitate a total staffing review, sacking and transferring old employees and appointing new ones without the ideological slant of *sertanismo* or *rondonismo*. Once this was complete, it can be said that FUNAI changed not only its organisational structure but, principally and fundamentally its operational ideology, its customs and mental attitudes, as well as the internal and external social relations practised by *sertanistas* and 'authentic indigenists'.
5. Of note is the pragmatic position taken by this analysis in the face of constitutional and legal moves; if the law is against the interests of existing power relations, and if they are not considered a threat to the State, then the law is changed.
6. According to Order no. 520 of 4 May 1988 FUNAI established certain criteria for determining the degree of acculturation of indigenous groups; knowledge of the Portuguese language, the ability to accumulate within an exchange economy, dependence for goods and services upon the national economy, professional skills and productive activities, and the ability to receive help of the same type given to non-indian groups in the region. Note that these take no account of cultural losses (an issue which was considered in FUNAI's official plans during 1981–2 with their so-called criteria of 'indianness'), but only of acquiring new standards. According to the criteria thus established, the vast majority of indians within Calha Norte are considered highly acculturated since they have been obliged, since their earliest

contacts with white men, to adjust and search for new practices and knowledge in order to adapt themselves to their economically subordinate and politically dependent position.

References

Almeida, A. W. B. de (1980), 'GETAT: Segurança Nacional e revigoramento do Poder Regional', in *Transformações Econômicas e Socais no Campo Maranhense*, São Paulo, 7, Edição da Comissão Pastoral da Terra/Maranhão.

Bastos, A. C. T. (1975), *O vale do Amazonas*, Brasília: Editores Nacional.

CIMI (1986), 'O projeto Calha Norte nas notícias da imprensa'.

Davis, S. (1987), *Vítimas do Milagre*, Rio de Janeiro: Zahar Editores.

Imbiriba, M. de N. O. and Lopes, Fabio Sepuleda (1900), 'Experîencias Nacionais em Desenvolvimento Amazônico'.

Jucá Filho, R. (1987), Entrevista ao jornal *O Estado de São Paulo*, 24 September 1987. Carta aos Constituíntes, com anexos estatísticos, Brasíllia (25 September 1987).

Pinto, L. F. (1986), 'O 1° plano de desenvolvimento da Amazônia da Nova República vem ai', *Aconteceu*, 361, CEDI.

Pinto, Lucio Flavio (1986), 'Calha Norte: O projeto especial para ocupação das fronteiras, CECI (April) (mimeo).

Simonetti, E. (1987), 'O lado secreto do Governo Federal', *Gazeta Mercantil*, 30 May and 2 June 1987.

Velho, O. G. (1975), *Capitalismo Autoritário e Campesinato*, Rio de Janeiro: Zahar Editores.

Vianna Moog, C. (1936), *O ciclo do Ouro Negro. Impressões da Amazônia*, Porto Alegre: Livraria Globo.

Documents Cited

Lei 6001 (19 December 1973), Dispõe sobre o Estatuto do Índio.

Decreto 79.009/77 (1985), *1° Plano de Desenvolvimento da Amazônia da Nova República*, MINTER.

Estudo n°010/3ªSeção (31 May 1988), Secretaria-Geral do Conselho de Segurança Nacional, *Desenvolvimento e Segurança na Região ao Norte dos Rios Solimões e Amazônas – Projeto Calha Norte* (December 1985), Relatório do grupo de trabalho criado pela EM 018/85, Brasília.

Exposição de Motivos n°770/85 (19 December 1985), Aprova o Projeto Calha Norte.

Estudo n°007/3ªSC/86 (1986), Da Secretaria-geral do Conselho de Segurança Nacional.

Estudo n°029/3ªSC/86 (1986), Da Secretaria-Geral do Conselho de Segurança Nacional.

Portaria n°43/SG/SEPLAN, de 1 September 1986.

Exposição de Motivos n°18/SEPLAN/MINTER/CSN/MRE (1985).

Exposição de Motivos n°054/87 (1987), Acompanha o Decreto 94.946/87.

Portaria n°020/SG/SEPLAN (13 March 1987).

Portaria n°099, de 31 de Março de 1987, do Ministério do Interior, Dá novo estatuto à Fundação Nacional do Índio.

Portaria nº01/FUNAI/DNPM (18 May 1987), Normatiza a Pesquisa e Exploração Mineral em Áreas Indígenas.

Decreto nº94.945 (23 September 1987), Modifica a composição do Grupo de Trabalho Interministerial, que decide sobre a Delimitação das Áreas Indigenas.

Decreto nº94.946 (23 September 1987), Cria a figura jurídica de 'Colônia Indígena'.

Portaria nº01 (17 March 1988), Secretaria Especial de Coordenação Econômico Social, SEPLAN.

Portaria nº520/FUNAI (4 May 1988), Estabelece critérios para avaliação do grau de Aculturação dos Grupos Indígenas.

Part II
Environmental Destruction, Social Conflict and Popular Resistance

8 Environmental Destruction in the Brazilian Amazon[1]

Philip M. Fearnside

EXTENT AND RATE OF ENVIRONMENTAL DESTRUCTION

Environmental destruction in Amazonia takes many forms, such as deforestation, loss of animal and plant populations from the remaining forest, disturbance and pollution from mining, flooding by hydroelectric dams, and elimination of tribal peoples and their cultures. The various types of destruction are all linked to the advance of deforestation. The vast extent and explosive rate of deforestation hastens the demise of natural ecosystems, closes the door to the most promising human uses of the region and provokes impacts that are regional, and in some cases global, in scope.

The extent of deforestation remains poorly known despite the existence of remote sensing tools capable of monitoring land surface changes rapidly were sufficient funds and effort allocated to the task. The last LANDSAT analysis covering all of the Brazilian Amazon is for imagery from 1978 (Tardin *et al.*, 1980; Fearnside, 1982). Of the region's nine states and territories, 1980 LANDSAT data are available for only six: Acre, Goiás, Maranhão, Mato Grosso, Pará and Rondônia (Brazil, IBDF, 1983; see Fearnside, 1984a, 1986a). For Rondônia and half of Mato Grosso 1983 LANDSAT data are available (Brazil, IBDF, 1985; Fearnside and Salati, 1985). Data for 1985 on the NOAA weather satellite are available for Rondônia and the eastern part of Acre (Malingreau and Tucker, 1987, 1988; see Leopoldo and Salati, 1987); 1987 data are available only for Rondônia (J. P. Malingreau, personal communication, 1988).

In addition to many of these data being out of date, other deficiencies contributing to underestimates of deforestation include the sensors' inability to (1) distinguish secondary from primary forest, (2) register small clearings and (3) identify disturbances less than full clearcutting. Underestimation also results from the practice adopted by the agency responsible for monitoring deforestation (the Brazilian Institute of Forestry Development, IBDF) as a means of countering the problem of cloud cover: using some images from years previous to the nominal year for the estimate. Both the absolute area cleared and the rate of increase are underestimated by the decrease in the resolution of the sensors used for the most recent data: AVHRR's coarser resolution as compared to the earlier LANDSAT data increases the area of unregistered small clearings. Despite these limitations one must draw the best conclusions possible.

The satellite data indicate that deforestation has been following an exponential trend in many parts of the region. As will be discussed later, this reflects some of the underlying forces driving the deforestation process. The exponential nature of deforestation means that the cleared area could rapidly expand to encompass the entire region if the trend were to continue unopposed by some restraining force. Notwithstanding a number of constraints on the speed with which the forest can be dispatched, the process will continue to its endpoint of total destruction unless positive steps are taken to restrain deforestation and to redirect development to less destructive alternatives.

Each year a larger area of forest is cut than in the year before. If one makes the conservative assumption of a linear trend since the last two data points available for each state or territory, the area cleared annually as of 1988 is over 25,000 square kilometres – a tract about the size of Belgium. The limitations of satellite data mentioned earlier also make this an underestimate of deforestation. Assuming a linear trend up to 1988 within each state or territory implies that the total classified as 'cleared' is 322,000 square kilometres or 6.5 per cent of the area of 4,975,567 square kilometres, which is designated as Legal Amazonia. Assuming exponential trends to 1988 since the last available data would imply about 17 per cent of the region cleared, which is probably too high. The actual amount probably lies between these bounds; the linear assumption value of about 6–7 per cent represents a safely conservative estimate.

Deforestation is highly concentrated in certain parts of the region, especially those nearer to the source of population flows from Brazil's Central–South and North–East regions. When the Legal Amazon is viewed as a whole, however, the amount cleared is still small relative to the total area of the region. This fact frequently lulls policy makers and others into dismissing concerns about deforestation as 'alarmist'. Environmental destruction is seen as too far removed in space and time to warrant concern when more immediate economic and social pressures are clamouring for attention. Such a stance is sadly mistaken. The explosive rate of deforestation is far more important than the absolute area cleared so far. The exponential form of the curve is highly deceptive to most people, even to those who have lived their lives in daily contact with such exponential trends as inflation. An exponential trend can cause a seemingly insignificant area of clearing to expand to the size of Amazonia within a few years.

Making projections into the future can be useful as an illustration of the logical consequences of the trends. However, *projections* should not be confused with *predictions*, which are prophesies about how the future will actually unfold. The system of forces that drives deforestation is too complex for long-term predictions using simple linear or exponential equations. Nevertheless, simple projections make it obvious that speedy and firm government actions are needed to strike at the forces behind deforestation's present exponential trend.

In 1982, I published a projection of deforestation that has been widely quoted, and more widely misquoted. Based on data through 1978, continuation of an unrestrained exponential trend throughout Amazonia would lead to complete deforestation of the region by 1991 (Fearnside, 1982). More recent data are now available indicating that deforestation has proceeded at a somewhat slower pace than that implied by the trend observed through 1978. A number of factors explain the downward shift of the curve, including the limitations Brazil's economic crisis has imposed on the funds available for clearing as fast as Amazonian landholders might like. Many such restraints are temporary. The reasoning behind the 1982 projection remains as valid as ever: policy-makers would be well advised to consider carefully the power of exponential trends such as those produced by positive feedback processes underlying deforestation (Fearnside, 1987a). Deforestation is not a problem that will 'take care of itself' without conscious decisions by national leaders (Fearnside, 1985a). The trends are the result of identifiable forces that are subject to government control in many ways. Deforestation is not a foreordained process that plays itself out like a Greek tragedy: it is subject to human will.

IMPACTS OF DEFORESTATION

Deforestation provokes many serious consequences that rob Amazonia's residents of a potentially sustainable future. The ecological functions and potential economic uses of the forest are being traded for rapidly-degrading cattle pastures. Some of the potential impacts of this massive transformation can be expected to extend far beyond Amazonia. While planners often mistakenly associate deforestation with such positive terms as 'development' or 'progress', an examination of deforestation's impacts provides ample basis to justify bearing the financial and political price of containing forest clearance.

Impacts on Soil

Soil erosion results in loss of nutrient capital and agricultural productive potential that is permanent on the scale of human planning. Although erosion losses can be compensated through use of fertilisers, Amazonia's vastness and lack of deposits of key elements, such as phosphates, makes this impractical on any significant scale. Although erosion may seem to many planners to be too slow a process to worry about, it can cause substantial damage within a few years.

When the forest is removed, the soil becomes compacted upon simple exposure to sun and rain (Cunningham, 1963). Increased soil temperature shifts the equilibrium between oxidation and formation of organic matter such that less of this critical material is present to maintain soil structure (Greenland and Nye, 1959). In the case of the pasture that dominates land use in deforested areas in Brazilian Amazonia, the trampling of cattle further speeds the process of soil compaction. Infiltration of rainwater into the soil is decreased by an order of magnitude in Amazonian pastures as compared to adjacent forest (Dantas, 1979; Schubart *et al.*, 1976). The result is that rainwater runs off over the surface rather than sinking into the soil. The runoff causes both sheet erosion and gulleying.

Annual crops such as rice and maize are well known to be subject to serious erosion, especially in Amazonia, where much of the rain falls in torrential downpours rather than in slow drizzles that can sink into the soil. Measurements of the lowering of the soil surface on the Transamazon Highway indicate that the level commonly falls at a rate of about 1 centimetre per year; most of this surface lowering is the result of erosion rather than compaction since the rate is significantly correlated with slope (Fearnside, 1980a; 1986b). Recent direct measurements of erosion under various land uses confirm the rapid loss of soil when either bare or under annual crops (in preparation).

Perennial crops and tree plantations are often seen as a solution to erosion. Although most such land uses are far better than annual crops, the protection they provide can be less than one might think. Some perennials with clear areas between the plants, such as black pepper, have erosion rates similar to annual crops (Fearnside, 1980a). Much ground is maintained bare under coffee, which has been expanding in Rondônia and northern Mato Grosso. The spectre of exhausted coffee lands abandoned earlier in this century in the state of São Paulo provides a reminder of the power of erosion.

In silvicultural plantations that require replanting every few years, severe erosion often occurs when the ground is exposed between crops and when the planted trees are still young. This is evident in the more steeply-sloping portion of the Jarí estate (Fearnside and Rankin, 1982a). Rubber and oil palm in Malaysia (Brünig, 1977, p. 189) and teak in Trinidad (Bell, 1973) provide examples of erosion under tree plantations. Nevertheless, trees are undoubtedly much to be preferred over annual crops or cattle pastures.

Cattle pasture is the most important factor in soil degradation in Amazonia because of the wide extent and relatively long time that this land use remains in place. A mythology has developed in Brazil concerning the powers of cattle pasture to protect and improve the soil – a mistaken view that had significant effects on incentives programmes and official planning. In 1974, the government agricultural research agency (EMBRAPA) announced that pasture improved the soil, thus making it a 'rational means to occupy and increase the value of these extensive areas' (Falesi, 1974).

Unfortunately, available phosphorus (P_2O_5) declines under pasture following the initial peak caused by the ash deposited from burning the forest: after 10 years P_2O_5 it insufficient to maintain pasture growth (Fearnside, 1980b; Hecht, 1981, 1983). Reasonably high levels of pH or soil cations cannot compensate for lack of phosphorus (Fearnside, 1980b), which is the limiting factor for pasture growth in much of Amazonia (Koster *et al.*, 1977).

Pasture has also been mistakenly viewed as protecting the soil from erosion. 'Protection from erosion' is pointed out in EMBRAPA recommendations for pasture use on poor soils (e.g., Brazil, EMBRAPA–IPEAN, 1974, p. 34). The RADAMBRASIL land capability mapping, which is widely used for land use planning in Amazonia, classifies land as suitable for pasture if it is 'susceptible to erosion' and 'inappropriate for use of agricultural machinery' – that is, too steep for tractors (Brazil, Projeto RADAMBRASIL, 1978, vol. 16, p. 383).

Recent direct measurements of erosion under pasture at Manaus (Amazonas) and Ouro Prêto do Oeste (Rondônia) indicate much higher erosion under pasture than under forest.

Pasture yields decline steadily as the combined result of soil nutrient depletion, compaction and invasion by inedible weed species. Measurements of dry matter production in Ouro Prêto do Oeste (Rondônia) indicate that 12-year-old pasture produces only half as much as 3-year-old pasture. The strong seasonal cycle of grass production severely limits the fraction of this production that can be effectively converted into beef. Reduced dry matter productivity leads to reduction of beef yields to virtually zero within about 8 years (Fearnside, 1979b).

In addition to the immobilisation of phosphorus in forms that plants are unable to utilise, as occurs in pasture, other forms of degradation include the removal of cations, such as calcium and magnesium ions, through leaching when the soil is exposed to the region's heavy rains. In addition, the availability of sites for holding cations in the soil is reduced by depletion of soil organic matter and by the migration of clay particles to lower layers in the soil profile. Cycling of nutrients through the cattle concentrates them in the unevenly-dispersed dung, thus reducing their availability for pasture grass growth. Soil nutrients are also removed in the beef exported from the system.

The stock of nutrients in the system represents an equilibrium between inflows and outflows. The increases in outflows described above shift nutrient equilibria to lower levels. Nutrient inflows are largely through contributions dissolved in rainwater and from atmospheric particulates. In the case of phosphorus, a significant source is believed to be Saharan dust transported across the Atlantic Ocean by wind (see Talbot *et al.*, 1986). These inputs may have increased slightly over historical levels because of increased aeolean erosion in Africa provoked by human impact on the vegetation there. Further inputs of nutrients come from smoke from burning within the

Amazon Region, a phenomenon that has increased dramatically in recent years. A small input may also come from industrial pollution. Although these added inflows act to partially compensate for losses, the greatly increased outflows in pasture as compared with forest lead to a steady degradation of the nutrient capital.

Impacts on Rivers

Deforestation increases runoff as the combined result of soil compaction causing decreased infiltration of rainwater and reduced leaf area causing decreased evapotranspiration. An order of magnitude increase in runoff under pasture as compared to forest has been observed in Manuas (Amazonas) and Ouro Prêto do Oeste (Rondônia) for measurements over a one year period (in preparation). As deforestation in Amazonia proceeds, rivers in the region can be expected to have reduced water flows in the low water period and higher and more irregular floods in the high water period. The changes in the flood cycle will be particularly damaging to agriculture in the *várzea* (floodplain), where farming depends on precise timing of agricultural activities in accord with the river's annual cycle.

Hydroelectric schemes in Amazonia might appear to be benefited by the increased runoff, but the schemes would suffer negative impacts that more than outweigh the gain from increased stream flow. The flow increases would be concentrated at the height of the flood season, when most dams woud be obliged to pass the bulk of the flow over their spillways anyway. The runoff from deforested areas would also contain silt from soil erosion, greatly speeding the sedimentation of the reservoirs.

Impacts on Climate

Rainfall in Amazonia is closely tied to the presence of forest. As in other areas, the forest plays a role in inducing the water vapour present in the air to fall as precipitation. Another link to rainfall has perhaps greater importance in Amazonia than in other areas because of the region's vastness. This is the input from evapotranspiration to the stock of water vapour in the atmosphere over Amazonia and neighbouring regions. Several lines of evidence indicate the importance of water recycled through the forest. One is the simple comparison of the water flow from the Amazon River with the amount of rain falling in the catchment basin. The flow at Óbidos is only about half (46 per cent) of the rainfall in the basin above that point, indicating that the other half is returned to the atmosphere as evapotranspiration (Villa Nova *et al.*, 1976).

The impact of widespread deforestation on the water cycle is a serious concern because of its potential negative effects on forest survival in Amazonia and on agriculture both in the region and in the neighbouring

Central–West region, where rich farmlands produce much of the country's crops (Eagleson, 1986; Salati and Vose, 1984). A feedback to forest survival is expected, where increased drying provoked by deforestation would kill the more sensitive trees in dry years, thereby opening up the forest canopy and further drying the microclimate within the forest, leading to still more tree deaths (Fearnside, 1985b). Eventually rainforest species would be replaced by more drought-tolerant trees, such as those characterising the *cerrado* vegetation of Brazil's central plateau.

The length and severity of the dry season varies tremendously from one year to the next in Amazonia even without the impact of large-scale deforestation (Fearnside, 1984b). Since trees in mature Amazonian forests are believed to live 200 years or more, a very severe drought once every 50 years or so could have a tremendous impact on the forest. Radioisotope ratioing of atmospheric water vapour samples indicates that rainfall is most dependent on water recycled through the forest precisely in the dry season (Salati *et al.*, 1979). Increased probability of a very long dry season could be disastrous even if mean annual rainfall were to remain unaffected. Radioisotope ratioing indicates that half of the rainfall between Manaus and Belém derives from the forest, and that the importance of the forest increases with distance from the Atlantic Ocean (Salati *et al.*, 1978, 1979). More than 50 per cent of the rainfall would therefore be expected to come from the forest in the western Amazonian states of Rondônia and Acre, where cleared areas are now exploding exponentially.

The patterns of ecological succession following forest removal in Amazonia could change in the future in a way that increases the climatic impact and decreases the human use potential of the deforested areas. At present, clearings in Amazonia are colonised by woody secondary forest trees, such as *Cecropia* and *Vismia*. This pattern is not fixed by some divine decree – it could give way to a grassy dysclimax as occurs in South-East Asia. In Asia, the notoriously aggressive grass *Imperata cylindrica* prevents return of woody vegetation over wide areas, a barrier to forest recovery that is made virtually impassable by frequent burning of the grass. In South America, *Imperata cylindrica* does not now occur, although its congeneric *Imperata brasiliensis* does. In Peru's Gran Pajonal, for example, this species serves to impede colonisation by woody second growth (Scott, 1978). In highly degraded pastures in Brazilian Amazonia grasses such as 'rabo de cavalo' (*Andropogon* spp.) sometimes play a similar role. Even where some woody species are present, biomass accumulation can be minimal for more than a decade after abandonment of highly degraded pasture (Uhl, 1988). The danger of deflecting succession to a non-woody dysclimax is undoubtedly increased by such abuses of the soil as the now increasingly common practice in Northern Mato Grosso of bulldozing secondary vegetation in degraded pasture.

Serious consequences can be expected should large areas of forest be replaced by grasses, whether they be productive pasture or inedible weeds. The difference between forest biomass and the much lower biomass of grass (or of stunted woody growth) is proportional to the amount of carbon that is released to the atmosphere by the conversion. Carbon, which makes up about half of the dry weight of wood, is converted to carbon dioxide (CO_2) either by burning or by the decomposition of unburned wood. A further release of CO_2 results from oxidation of part of the organic matter stock in the soil – a consequence of higher soil temperature under pasture than under forest. Carbon dioxide is the principal cause of the 'greenhouse effect' – the increase in global temperature, especially near the poles, caused by the trapping of heat by an atmospheric blanket of gases that impede the passage of infrared radiation to space. Expected consequences include both a rise in sea levels and disruption of the present pattern of agriculture by moving climatic zones toward the poles. Most CO_2 is released by burning fossil fuels, but Amazonian deforestation could be a significant contributor to this wider problem in the coming decades. Conversion of Brazil's Legal Amazon from its original vegetation to pasture would release approximately 50 billion metric tons (gigatons = Gtons) of carbon. If this were to occur over a span of 50 years (an optimistic assumption), one gigaton would be released annually, or 20 per cent of the present global total from all sources (Fearnside, 1985c, 1986c, 1987b).

Carbon dioxide is not the only contributor to the greenhouse effect. Trace gases such as methane (CH_4) and nitrous oxide (N_2O), although present in much lower concentrations, have impacts that are now recognised as potentially rivalling those of CO_2 (Dickinson and Cicerone, 1986; Ramanathan *et al.*, 1985). Methane and nitrous oxide are both produced by burning forest and pasture (Crutzen *et al.*, 1979). Both gases have been increasing in concentration in the atmosphere over the past decades because of emissions from industrial and other sources: CH_4 by 1.1 per cent a year and N_2O by 0.2 per cent a year (Weiss, 1981; Mooney *et al.*, 1987). Unlike carbon dioxide, which is partially reabsorbed by the biosphere through photosynthesis, these trace gases remain in the atmosphere for long periods. Methane is removed very slowly, mainly by reaction with OH in the troposphere although a small amount is consumed by forest soils. Nitrous oxide is degraded only through photolysis in the stratosphere. Although the cycle of reburning pasture or secondary forest makes no net contribution to atmospheric carbon dioxide (except to the extent that average biomass decreases with each succeeding cycle), each reburning makes an addition to the stocks of CH_4 and N_2O with no associated contribution to the removal process.

Methane release increases from conversion to pasture both from the initial burning and because the soil changes from a consumer to a producer of methane, at least in the dry season (Goreau and Mello, 1987). The rumens of the cattle that occupy the pasture are one of the major sources of methane

(Ehhalt, 1985, p. 11). Termites are another potential source of increased methane emission from pasture. Termites are known to produce methane but the amount is a matter of controversy: a factor of 10 difference separates high estimates (Zimmerman *et al.*, 1982, 1984) from low estimates (Rasmussen and Khalil, 1983; Collins and Wood, 1984). Pasture, especially degraded pasture, is known to harbour many termites; unfortunately good comparative data on forest termite abundance are lacking.

Nitrous oxide is released by burning, but after this pulse the soil under pasture appears to produce less N_2O than the soil under forest, thereby counteracting some of the initial release (Goreau and Mello, 1987). In the years after the pulse from initial clearing, reburning of pasture or of shifting cultivation fallows continues to release some N_2O through combustion. Deforestation's long-term impact on N_2O is uncertain (Mooney *et al.*, 1987, p. 928). Over the next few decades, however, the net effect of conversion of Amazonian forest to pasture is expected to be an increase in the flux of this gas to the atmosphere (M. McElroy, personal communication, 1987).

Impacts on Forest Species

Widespread deforestation presents a serious threat to many species of plants and animals. The diversity of life in Amazonia is legend (Anderson and Benson, 1980; Prance, 1978; Prance *et al.*, 1976). What most exposes this diversity to destruction by deforestation is the highly localised distribution of many species. Because of this endemism, species can be eliminated without deforesting a very large area.

Another characteristic of the forest that magnifies the impact on species of a relatively small amount of deforestation is the requirement of large areas of continuous forest for many species to maintain reproductively viable populations, together with the required sources of food, pollinators, dispersal agents and other ecosystem components. A study being undertaken by the World Wildlife Fund–US (WWF–US) and the National Institute for Research in the Amazon (INPA) near Manaus is investigating some of these interrelationships as a range of sizes of forest fragments degrade following isolation in cattle pasture (see Lovejoy *et al.*, 1984). The need for large areas is already apparent. Climatic changes could make the areas needed to ensure survival even greater. Since climatic zones could shift by hundreds of kilometres, the reserves that would need to be created to buffer against the forced migration of sensitive species may well already be too large to expect in practice.

The question of species extinctions from environmental destruction in Amazonia appears to carry little weight with the decision-makers whose actions most directly affect Brazilian rainforests. Gilberto Mestrinho, then governor of the state of Amazonas, justified a plan to export the skins of jaguars and other wilds species by saying: 'the conservationists shouldn't worry. It is not man who decimates species on a wide scale, but rather nature itself, which closes the cycle of life of animals' (*A Notícia*, 25 June 1983, p. 5). He went on to cite now-extinct life forms that had lived in Amazonia millions of years ago. While extinction is, indeed, the eventual fate of all species, the evolutionary and environmental consequences could be severe from the unprecedented deluge of extinctions expected if present deforestation trends continue (Eckholm, 1978; Ehrlich and Mooney, 1983; Lewin, 1983; Wolf, 1987).

Impacts on Options for Forest Use

Wood and Latex

Deforestation destroys many of the most socially and environmentally attractive options for development in Amazonia. Loss of natural ecosystems directly eliminates economically-valuable species such as trees for hardwood timber, trees now producing about a score of extractive products, including rubber and Brazilnuts, and the many medicinal plants whose economic exploitation is presently miniscule.

Pharmaceutical Products

The pharmacological potential of Amazonian forest has scarcely begun to be tapped. It is humbling to realise that almost all of the drugs used in modern medicine were first discovered as products of naturally occurring organisms – from the penicillin mould to the Madagascar periwinkle now used to treat child leukemia (Caufield, 1985, pp. 220–1). Even aspirin was originally derived from willow leaves. Only after the medicinal effectiveness of a compound is recognised is the effort expended to synthesise it without the help of the organisms that produce it naturally. Loss of Amazonian forest is considered a serious potential setback to efforts to find cures for human cancer (see Myers, 1976, 1979, 1984).

Pathogens are continually evolving resistance to drugs used in treatment, making it necessary to have a constant flow of new drugs just to remain in the same place in the battle against disease. The sudden surge to prominence of chloroquine-resistant strains of malaria-causing *Plasmodium* during the 1960s precipitated a rush to tap long-neglected sources of natural quinine in South America (Oldfield, 1981).

In addition to new forms of old diseases, entirely different diseases also continually appear. The recent arrival of Acquired Immune Deficiency Syndrome (AIDS) should provide ample justification for not burning our stocks of potential pharmacological compounds. The notion that the shining achievements of modern medicine permit us to dispense with a major portion of these stocks represents a potentially fatal form of hubris.

Genetic Material

The potential for obtaining valuable genetic material from the forest is another opportunity that is sacrificed by deforestation. Like medicinal plants, genetic resources are irreplaceable – they cannot be bought back with the money earned through deforestation. Germplasm can be valuable both in supplying new crops to agriculture and in providing a store of varieties of already-cultivated species.

A strong popular tendency exists to view agriculture as a technology that has been fundamentally fixed in its basic configuration of crop plants since neolithic times, or at least since the early days of intercontinental travel. In fact, many of agriculture's most fundamental problems remain unsolved, and rely on very temporary technological 'fixes' to continue high levels of productivity. One problem is soil erosion under the annual crops that are the mainstay of farming in most of the world. An obvious solution would be to use more perennial crops, thereby reducing both the bare space between crop plants and the fraction of the cropping cycle when the soil is left bare between plantings. *Zea diploperennis*, a perennial relative of maize became known to the scientific community in 1978 (see Iltis *et al.*, 1979) and was subsequently saved from extinction in Mexico in one of the last vestiges of its threatened habitat (Iltis, 1983, p. 57).

Another basic limitation on agriculture at present is the inability of almost all major crop species to fix nitrogen. In the tropics as a whole, nitrogen is the element that most commonly limits crop production (Webster and Wilson, 1980, p. 220). Manufacturing and supplying nitrogen fertilisers to farmers depends heavily on petroleum – a non-renewable resource with rapidly-approaching limits. Green manures and interplanted legumes are a means of alleviating the nitrogen demands of crops, but the dream remains unfulfilled of having an array of crops capable of using directly the vast quantities of nitrogen present in the air. Surveying Amazonian plants and soil bacteria for nitrogen-fixing ability has barely begun (see Sylvester-Bradley *et al.*, 1980).

Another fundamental limit to agriculture is the inability of most plants to solubilise phosphorus when bound in 'unavailable' compounds of iron and aluminium in the soil. Were crop plants to gain this ability – either by themselves or through appropriate mycorrhizal symbionts – another impending limit to fertilizer-based agriculture would be less threatening.

The ability of Amazonian plants to make efficient use of scarce nutrients is a feature markedly lacking in the crops favoured by present-day agriculture. The dwindling of nutrient and fossil fuel stocks in the world will make this ability more and more valuable as time passes. Great potential value exists in incorporating new capabilities into the repertoire of crop plants, either by adopting new species, breeding wild relatives with present crops, or by genetic engineering techniques.

Pest resistance is another area where germplasm from natural habitats can be indispensible. Most crop plants have been bred to remove the toxic secondary compounds that protect wild plants from devastation by herbivores (Janzen, 1973). The decreasing effectiveness of agrotoxins and the increasing environmental and public health problems they cause provide ample justification for trying to restore to crop plants some of their lost ability to synthesise their own pesticides.

Geographical isolation provides the principal protection against diseases and pests for many crop plants. Rubber, for example, was taken from Brazil to South-East Asia at the end of the nineteenth century, thereby leaving behind such devastating diseases as the South American Leaf Blight caused by the fungus *Microcyclus ulei*. Cocoa, native to Central and South America, was taken to Africa and Asia where it grows free of witches' broom disease (*Crinipellis perednisciosa*). Coffee was brought from Arabia and the horn of Africa to the new world, freeing it of coffee rust (*Hemileia vastatrix*). The protection afforded by geographical isolation is only temporary, and genetic resistance becomes essential when the pests and diseases finally catch up with their far flung host plants. The need to protect the natural sources of resistance was dramatised by the case of coffee when the coffee rust arrived in South America in 1970 and subsequently spread through Central America. In 1964, the last remnants of forest in Ethiopia had provided invaluable genetic material for developing resistant strains; the opportunity might well have disappeared had the germplasm collection expedition been delayed by only a few years (Oldfield, 1981, p. 311).

Maintaining disease resistance in cultivated plants requires continual changes in the plant population's genetic material in order to keep pace with the evolution of pathogens. Because the life cycle of disease-causing organisms is much shorter than that of the plants – especially perennials – the disease-causing micro-organisms have an inherent advantage in the race. Genetic material conferring resistance to crop diseases is best obtained from wild populations that have been coexisting with the diseases for millennia.

Forestry Management

Destruction of the forest sacrifices the opportunity for sustainable management of this resource. Once the forest ecosystem has been traded for a vast tract of pastureland, re-establishing any kind of forest is costly and difficult, and regaining the original ecosystem can be considered impossible (see Gómez-Pompa *et al.*, 1972). A variety of systems of managing the forest is under testing; the greatest barriers to their use lies in the political decisions necessary to make them economically more attractive than such present money-making activities as planting pasture for land speculation (see Fearnside, n.d.a.).

Impacts on Tribal Peoples

For Amazonia's indigenous tribes, forest destruction means either death or loss of cultural identity when acculturation transforms the survivors into the lowest stratum of the dominant society. The process of removing tribes from the forest areas they still occupy has accelerated as new areas are targeted for mining, hydroelectric dams and military bases. The most potent enemy of the indigenous peoples in their ongoing struggle to maintain their lands and cultures is the tendency of the dominant society to consider disappearance of tribal peoples as either inevitable (and therefore not worth the effort to reverse) or as something that has already occurred. Many people think of the decimation of tribal peoples as a part of history rather than as a process that is still going on today and is, above all, by no means complete.

The litany of affronts to indigenous peoples in the Brazilian Amazon grows longer by the day. The Ecumenical Centre for Documentation and Information (CEDI) publishes a running catalogue of these events in the series *Aconteceu* (e.g. CEDI, 1986). Brazil's Legal Amazon has 368 indigenous areas in various stages of documentation, ranging from 'unidentified' areas where no measurements of areal extent have been made to 'regularised' areas where full legal protection applies (CEDI/Museu Nacional, 1987). Most of the land area and tribal population do not have legal protection beyond the little provided by the 'identification' stage that begins the long process of reserve creation but carries little guarantee that the land will not be subsequently usurped by other interest groups. Mining is a rapidly increasing threat that usually wins whenever conflicts of interest arise with indigenous peoples; at least 77 reserves are threatened (SBPC, 1986; see also CEDI/CONAGE, 1988). Lumbering and hydroelectric dams are also increasing, along with the relentless pressure of invasion by ranchers, speculators and small squatters. Invasions are closely tied to the construction of roads: once a road is built through a tribal area, the subsequent arrival of migrants takes place largely outside of government control. In some cases such invasions have even been informally encouraged by government officials. The expansion of highway networks is greatly speeded by large internationally-funded development projects such as POLONOROESTE and Carajás. In Rondônia, planned roads cut through six reserves (Fearnside and Ferreira, 1984), while in the Grande Carajás Programme area nine reserves are cut by highways, in addition to cuts by railway and electric transmission lines (Fearnside, 1986d).

Disappearance of the tribes and their cultures implies one cost that even the most narrowly pecuniary of economic planners should appreciate: the loss of knowledge of how to use the diverse forest species. The medicinal and other properties of the thousands of species present in the forest are prohibitively expensive to assess if done from random samples of the vegetation. Much more efficient is a programme of confirming the activity of species used by tribal peoples. Little of the knowledge of how to use forest species has been recorded (see Elizabetsky, 1987; Posey, 1983).

Recording and using the knowledge that is now the near exclusive domain of indigenous tribes and, to a lesser extent, of rubber-tappers and *caboclo* farmers should be done with all due haste because of the unique value of the knowledge and because it contributes a strong argument for maintaining intact significant tracts of the forests on which these groups depend for their survival. Some people fear that passing this knowledge to the dominant society would represent a 'last theft' from the tribes. The tribes' land and right to exist must be guaranteed independent of any economic value that the dominant society may see in preserving these cultures. Once all useful knowledge has been gathered from the tribes, they cannot be destroyed with impunity. At bottom, their right to existence is not a question of economic value but one of human rights.

CAUSES OF ENVIRONMENTAL DESTRUCTION

Efforts to control the process of deforestation will be ineffective unless they are founded upon a correct understanding of the forces that motivate forest destruction. The deforestation process varies greatly in different parts of the region (Fearnside, 1986a). Forest is converted to a variety of other uses, often for ulterior reasons rather than the direct products of the new undertaking.

Cattle Ranching

Cattle pasture dominates land use in deforested areas of Brazilian Amazonia, greatly magnifying the impact of a small human population on the forest (Fearnside, 1983a). The yield of beef is miniscule because of a steady decrease in pasture grass productivity caused by decline in available phosphorus in the soil, soil compaction, erosion, and invasion by inedible weeds (Fearnside, 1979b, 1980b; Hecht, 1981, 1983). The beef is almost all consumed within Brazil: the presence of hoof-and-mouth disease (afthosis) blocks exports of frozen beef to North America and Japan, thus sparing Amazonia the awesome force that international markets exert in Central America through the 'hamburger connection' (see Myers, 1981; Nations and Komer, 1983). Maintaining pasture productivity past the first decade or so requires inputs of phosphates (Serrão and Falesi, 1977, Serrão *et al.*, 1979). The level of inputs required could not be justified without massive subsidies and, on the vast scale of Amazonian pastures, are limited by the dimensions of this non-renewable resource (Fearnside, 1985d; see also Fearnside, 1987c). Amazonia has no known phosphate deposits, with the exception of a small deposit of phosphate-bearing bauxite on the coast of Maranhão (de Lima, 1976) and a hopeful but as yet unquantified find north of the Amazon River near Maicuru, Pará (Beisiegel and de Souza, 1986). Given the poor agronomic performance and unpromising long-term prospects of pasture, the reasons for this land-use dominating the landscape lie elsewhere.

One reason is the generous suite of financial incentives granted to large ranchers by the Brazilian government through the programmes administered by the Superintendency for the Development of Amazonia (SUDAM) and the Superintendency for the Manaus Free Trade Zone (SUFRAMA). These programmes not only grant exemption from income tax on the ranching operations themselves but also allow the enterprises to invest in the ranches the money that firms would otherwise pay as income tax on unrelated operations elsewhere in the country (Bunker, 1980; Hecht, 1985; Mahar, 1979). Special loans are granted with interest rates below the rate of Brazilian inflation (making the interest negative in real terms). The loan programmes create an additional motive to establish ranches as a front for receiving subsidised capital that, apparently, is sometimes in part diverted to more lucrative activities elsewhere (Mahar, 1979). Government subsidies account for up to 75 per cent of the investment in the ranches (Kohlhepp, 1980, p. 71).

Programmes for subsidising ranches grew rapidly in the 1970s, but have ceased to expand since. In 1979 SUDAM announced that no 'new' incentives would be granted in the 'high forest' area of the Legal Amazon, but maintained the programme of 'old' (already approved) incentives for the over 300 projects under way in the high forest region, plus the possibility of 'new' incentives for the wide area officially classified as transition forest along the southern fringe of the region. Most of the 'transition forest' area is, in fact, an interdigitation of high forest with scrubland (*cerrado*) vegetation, rather than an intermediate vegetation type. LANDSAT imagery of this region reveals that ranchers preferentially clear the higher biomass forest (Dicks, 1982).

Subsidised ranching is still an important factor in deforestation, but the country's economic crisis has reduced the amount of money available for this purpose. Because the strictures are mandated by lack of money rather than by basic policy decisions on the worth of pasture, the flow of funds to the ranchers can be expected to resume once Brazil's economy recovers. Brazil's president was recently quoted as saying that he didn't 'even want to hear' about the possibility of discontinuing the ranching subsidy programmes (*Isto É*, 15 July 1987, p. 65).

Much clearing by both large and small landholders is done without benefit of the subsidy programmes. Even in the heavily-subsidised ranching area on the Belém–Brasília Highway during the height of the SUDAM programme only about one-half of the clearing enjoyed fiscal incentives (Tardin *et al.*, 1978; see Fearnside, 1979c). The explanation for the bulk of the pasture is the key role of this land use in land speculation (Fearnside, 1979c; 1987a). The value of land in Amazonia has been steadily increasing at a rate higher than

Brazilian inflation, yielding handsome returns to anyone that can hold on to a claim and sell it to someone else. For example, during the 1970s, land values in Mato Grosso were increasing at an average annual rate of 38 per cent, *after* correction for inflation (Mahar, 1979: 124); pastureland on the Belém–Brasília Highway has similarly outstripped inflation (Hecht, 1985). Part of the reason for the land value increase is desire for investments in real property as a shelter from inflation – serving a role as a store of value (similar to gold bullion) rather than functioning as an input to production. Individual properties increase severalfold in value when they gain access to a road (a benefit provided by the Brazilian taxpayers and international banks that fund the highway construction programme). A similar jump in value occurs when a land claim is legitimised by a 'definitive title'. Replacing the forest with pasture is the cheapest way to occupy the area and protect it from takeover by squatters, neighbouring ranchers, or government agrarian reform programmes. Pasture also counts as an 'improvement' (*benfeitoria*) to justify the granting of a definitive title. Ironically, the investments in unproductive ranching enterprises are a significant factor in fuelling Brazil's inflation (Gall, 1980), thus forming a vicious circle leading to more and more pasture (Fearnside, 1987a).

Agribusiness

Agribusinesses account for a small portion of the cleared area relative to pasture, but one that could expand significantly. Large-scale plans exist for financing mechanised agriculture and associated industries in the Grande Carajás area (Brazil, Ministério da Agricultura, 1983; Fearnside, 1986e; Hall, 1987). Much of the agricultural portion of the programme is currently on hold awaiting funding; in contrast, the portions of the Grande Carajás scheme related to charcoal production have been rapidly expanding.

Forestry

The forestry plans in Carajás illustrate a common feature in Amazonian development: the 'phoenix from the ashes' phenomenon. The plan to use charcoal for processing iron ore was originally announced in 1980 by Nestor Jost, then head of the Grande Carajás Interministerial Programme (Fearnside and Rankin, 1982b). A 2.4 million-hectare *Eucalyptus* plantation scheme was announced, together with a plan to collect charcoal made from native forest by ranchers, farmers, and even indigenous tribes. The scheme was greatly reduced in the 1983 'Programa Grande Carajás Agrícola' plan (Brazil, Ministério da Agricultura, 1983; Fearnside, 1986e). Suddenly the charcoal plan reappeared on a huge scale, with a charcoal demand that would require over 700,000 hectares of *Eucalyptus* – almost ten times the area of Jarí's managed plantations (Fearnside, 1987d, 1988). Pig-iron production began in Açailândia, Maranhão on 8 January 1988 without fanfare (and without an environmental impact report).

The forestry plantations at the Jarí Project, used to produce pulp manufactured in the estate's own mill, were initiated by the North American shipping magnate, D. K. Ludwig, in 1968. Many features of the site, the project's founder and the concessions granted by the Brazilian government make it unlikely that similar undertakings will multiply in the region (Fearnside and Rankin, 1980, 1982a, 1985). Ludwig sold a controlling interest in the estate to a consortium of Brazilian firms in 1982; the price paid was a small fraction of the cost of establishing the enterprise. Jarí has suffered a number of biological problems, including poor growth of some of the first plantations that were located on inappropriate soil, much lower growth rates overall than originally anticipated and losses to a variety of pests and diseases (especially the fungus *Ceratocystis fimbriata* in Jarí's hallmark tree species: *Gmelina arborea*). The dramatic rise in pulp prices that Ludwig foresaw for the 1980s has not yet materialised. Although a profitable kaolin (China clay) mine in the estate has permitted the present operation to cover its operating expenses (but not its burden of debt service), the forestry sector has been losing money: in 1985 the loss was US$47 million (Fearnside, 1988). While some of Jarí's early problems can be attributed to uninformed decisions on the part of Ludwig himself, the continuing biological problems of the plantations in no way reflect poorly on the quality of management but rather indicate that large-scale forestry production in Amazonia is much more expensive and much more difficult than many planners might think. It is foolhardy to imagine that a plantation scheme in Carajás ten times larger than the one at Jarí can operate without major difficulties.

The likely result in Carajás is that charcoal production will be supplied from native forest for as long as accessible stands remain in existence. The initial decision to implant the pig-iron smelters, apparently taken without benefit of any analysis of the environmental impacts of supplying the charcoal, could lead to the entire economy of the affected area being pulled into feeding these enterprises with wood, much as a bird is drawn into feeding a cuckoo's chick in its nest (see Fearnside, 1987d).

When the first pig-iron smelter began production on 8 January 1988, the company (*Companhia Siderúgica Vale do Pindaré*) had drawn up a forestry management plan for producing wood for charcoal in the future. However, at the time that I visited the operation two weeks later, it had not yet purchased a tract of land for implanting the management scheme. Clearly, the management schemes are neither sufficiently detailed to require knowledge of a specific tract of land nor are the tracts pre-requisites for beginning operation. The Carajás pig-iron scheme is the latest in a long list of development misadventures in Amazonia where projects have been decreed before confirming their sustainability and level of impact (Fearnside, 1985a).

Alcohol
Alcohol is one product for which great potential has been proclaimed for development by agribusinesses (e.g., Abelson, 1975). The efforts to exploit this potential have so far met with mixed success. The Abraham Lincoln Sugar Cane Project (PACAL), begun in 1972 on the Transamazon Highway 90 kilometres west of Altamira, Pará has experienced a long series of problems. Originally intended for sugar, the mill now produces only alcohol. The site was located in an area that agricultural zoning had previously shown to be climatically inappropriate for sugar cane (Moraes and Bastos, 1972, Figure 8). The cane grown at the site has a low saccharose content, which has caused much of the production of surrounding farmers to be rejected by the mill, leading to severe social tensions. The social tension has been aggravated by grave administrative, technological and human-relations mistakes; for example, telling farmers to harvest their cane on a specified date whereupon the promised transportation is not delivered and the cut cane quickly loses its sugar content. On several occasions, the area's farmers have not been paid for many months after delivering their cane to the mill. A succession of firms running the operation have failed to establish a working relationship with the farmers, and have resorted to violence to keep the farmers in line. A larger cane alcohol project, with financing from the World Bank, is now being implanted by ALCOBRAS in Acre; the first 5,000-hectare plot of this 20,000-hectare scheme is nearing completion. Cane from the ALCOBRAS estate will be supplemented with purchases from local farmers. Social problems have already begun in the Acre scheme in the aftermath of expelling 80 families of rubber-tappers and small farmers from the area. A 5,000-hectare cane plantation and alcohol distillery will also begin production in late 1988 in Presidente Figueiredo, north of Manaus.

Manioc (cassava) alcohol in Amazonia, seen by Abelson (1975) as a potential solution to the coming end to fossil petroleum, has not proved the panacea that it was originally hoped to be. Producing alcohol from manioc is more expensive than producing it from sugar cane, in part because of the energy supplement the bagasse from the cane contributes to the process. At Sinop in northern Mato Grosso an agrochemical firm has produced manioc alcohol from tubers both grown on the company's own estate and purchased from surrounding farmers. Sweet potatoes and sorghum have also been used. In 1987, the firm discontinued using manioc because of the high cost and many headaches involved with the migrant labour force that harvested the tubers. The firm now uses sorghum grown in mechanised plantations on the estate for producing alcohol for use in beverages – a higher value product than the fuel alcohol obtained from manioc or sweet potatoes. The market limitations on expanding plantations of this kind are, however, much more severe than is the case for fuel crops.

Perennial Crops

Market limits severely restrict the areas to which many of the crops can expand that are favoured by agribusinesses. Because Amazonia is so large, any significant portion of the region planted to perennial crops would saturate world markets for these commodities. The prices of most products are already low from the farmers' point of view, with financial losses and changes in land-use resulting whenever the prices dip. Cocoa, for example, has been falling in price since its high in 1977, with the exception of a brief rise after the 1982–3 El Niño-provoked droughts in Africa and destroyed cocoa plantations there. A long-term fall in cocoa prices was foreseen by World Bank economists *before* the major cocoa planting effort in Rondônia was launched under the POLONOROESTE project (International Bank for Reconstruction and Development, 1981).

Plant diseases severely curtail the potential for conversion of large areas to perennials (Fearnside, 1980c, 1983b, 1985d). Cocoa and rubber are both native to Amazonia, and consequently have all the diseases to which they are heir waiting to attack them. Witches' broom disease (*Crinipellis perniscìosa*) in cocoa and South American Leaf Blight (*Microcyclis ulei*) in rubber already have a devastating effect on plantations. These diseases do not exist in Africa and South-East Asia, thus giving a competitive advantage to planters in those places. Other important perennials, such as coffee, black pepper and oil palm, suffer from diseases that have followed them from the continents in which these crops originated. Coffee is attacked by rust (*Helmileia vasatrix*), black pepper by the Margarita disease fungus (*Fusarium solani* f. *piperi*) and oil palm by a recently arrived shoot die-back. Disease has an unfortunate relationship with markets that reinforces the effect of either falling or rising prices. Because it costs money and effort to control disease, farmers are less motivated to make these outlays when the product price is low, thereby allowing the infestation to become worse and making it even more expensive to bring the disease under control.

Várzea *Development*

Várzea (floodplain) has several natural advantages over *terra firme* (upland) areas for supporting agriculture. The rivers adjacent to all *várzea* areas offer a permanent and cheap transportation route that upland areas lack. The soils, made up mostly of recent sediments deriving from the erosion of igneous rocks in the Andes, are inherently more fertile than the ancient highly-weathered soils of the uplands. More important for sustaining agriculture over the long term is the eternally renewable character of this fertility, with fresh deposits of silt laid down annually by the river's flood waters. This natural renewal is a key part of annual crop systems used by small farmers, but so far has not been so exploited by agribusiness.

The irrigated rice scheme at Jarí is a unique attempt to use the Amazonian *várzea* for agribusiness ventures. The silt from annual flooding, however, is excluded from the paddies by a polder (dyke); Jarí relies on fertilisers to maintain levels of soil nutrients. The plantation has 4,150 hectares of rice, with plans to expand to 12,700 hectares currently not being actively pursued (Fearnside, 1988; Fearnside and Rankin, 1980, 1982a, 1985). The expansion of irrigated rice to much wider areas in Amazonia, either by mechanised agribusiness, as at Jarí, or by small farmers, is technically possible but appears unlikely under present economic conditions (Fearnside, 1987c).

Water buffalo-raising for production of milk, cheese and meat has been expanded at Jarí to utilize 50,000 hectares. Large ranchers in other *várzea* areas of the lower Amazon, such as the Ilha de Marajó, have also adopted this method of exploiting the *várzea*. Water buffalo have been promoted by EMBRAPA in *várzeas* in the Amazon and Solimões (Upper Amazon) Rivers in the state of Amazonas, but have not yet reached the scale of lower Amazon developments. The *Estrada da Várzea*, under construction in the state of Amazonas, will bring settlement to infertile *terra firme* (upland) areas as a side effect of roadbuilding activity justified on the strength of the *várzea's* production potential, especially for water buffalo. Buffalo represent a means of using the *várzea* by large operators – competing with the subsistence and fibre crops of the small farmers that traditionally occupy this zone. Neither buffalo ranchers nor small farmers 'own' the *várzea*, since all land within 50 metres of a river's high-water mark belongs to the Brazilian Navy.

The annual deposition of silt by the flood waters makes prospects for sustainable agriculture good in the *várzea*. Long-term negative factors include loss of part of the *várzea* to rising sea levels in a greenhouse-warmed world, and the greater risk from higher and less regular floods provoked by watershed deforestation. The present constraints are more social than agronomic – most of the *várzea* is already occupied by untitled small farmers, which means that development by agribusiness is likely to imply their expulsion. A more desirable alternative might be to encourage the small farmers already present further to diversify their plantings, as is done by indigenous groups (Denevan *et al.*, 1984), and to make greater use of the rich fruit production of natural *várzea* forests as is profitably done near Iquitos, Peru (Peters, n.d.).

Lumbering

Lumbering is rapidly increasing in importance as a factor in Amazonian deforestation. Timber exploitation has, in the past, been much less prominent in Amazonia than in the tropical forests of Africa and South-East Asia because of the lower density of commercially-valuable trees in South America. The tropical forests of South-East Asia are dominated by a single family of trees: the Dipterocarpaceae. Despite a high diversity on the level of species, the wood of many of these is similar enough to be grouped into only six classes for the purpose of sawing and marketing – as though there were only six species rather than several hundred. Amazonian species, being less closely related to each other taxonomically, have a correspondingly more heterogeneous set of wood characteristics. Amazonian trees have so far defied efforts to group the species into a relatively small number of categories for processing and marketing purposes. Another disadvantage is the dark colour of the wood of most Amazonian trees, in contrast to the light colours that dominate in South-East Asian hardwoods. The light coloured woods are more easily substitutable for such temperate species as oak and maple in European and North American furniture manufacturing.

Decimation of the tropical forests of Africa is essentially complete from a commercial standpoint, while those of South-East Asia are rapidly nearing a similar end. Exports from Amazonia are therefore increasing. Timber removal from Amazonia has occurred through rapid proliferation of small sawmills, for example in Mato Grosso, Rondônia, Acre and Roraima. Many of these have moved from areas of Brazil where timber is already reaching its end, such as Espirito Santo and the Belém–Brasília Highway in Pará. A steady stream of trucks bearing either logs or rough-sawn lumber can be seen entering São Paulo from the neighbouring Amazonian regions.

Lumbering is becoming an important factor in incursions into indigenous areas in Rondônia, Acre and the western portion of Amazonas. Lumbering roads serve as entry routes for squatters who clear in the hope of securing land claims. Satellite imagery of Rondônia (AVHRR interpreted by C. J. Tucker at NASA, Greenbelt, Maryland, USA) shows that the burning of 1987 included areas in such Amerindian reserves as Pacaas Novas, Tubarões and Lajes. Several of these sites are known areas of logging penetration, such as the portions of the Pacaas Novas reserve supplying sawmills in Ouro Prêto do Oeste.

Lumbering in the uplands (*terra firme*) is rapidly destroying stocks of some of the most valuable species, including 'cerejeira' (*Amburana acreana*) and 'mogno' (*Sweitenia macrophylla*). In the flooded *várzea* forests – the first to be affected because of the ease of transporting logs by water; commercial species such as 'ucuúba' (*Virola* spp.) aré rapidly declining.

Some of the processing and logging is done by large firms such as Georgia Pacific, which has a series of approximately 60 properties near Portel, Pará (R. W. Bruce, personal communication, 1988) totalling about 500,000 hectares (Cardoso and Müller, 1978, p. 161). The company's veneer plant at Portel produces 150,000 cubic metres annually, and supplies approximately 25 per cent of the North American market for tropical hardwood veneer. So far most of the wood is purchased from private loggers outside of the company's estate (R. W. Bruce, personal communication, 1988). Most logging, however, is done by thousands of relatively small Brazilian operators rather than by large multinationals. In Amazonia as a whole, at least half of the logging activity is believed to take place in 'clandestine' operations outside of the control and tax-collection efforts of the Brazilian Institute of Forestry Development (IBDF).

The cutting of 'noble' hardwoods is spreading rapidly as road access improves to previously-remote areas and as market pressure increases. The less-noble woods are also increasingly finding markets, and it is this sector that has the greatest potential for expanding the impact of logging on deforestation. Contracts with less demanding markets, such as China and India, have been negotiated in some cases; for example, for wood from the Samuel Hydroelectric Project in Rondônia. Delegations from heavily deforested countries such as these have been visiting the region with increasing frequency in search of wood supply contracts. However, one contract to supply China with pig-iron (a product manufactured with charcoal) was recently cancelled (*A Crítica*, 8 August 1987).

Efforts continue to develop ways of using more of the forest's diverse species. The possibility that an entire forest can be simply ground up and shipped away for manufacture of chipboard or low-quality paper products is indicated by the use of this procedure in lowland Papua New Guinea. This is euphemistically called 'total harvest' by the Japanese firms that practice it there (Routley and Routley, 1977). So far, Amazonia has been spared the common sight in South-East Asia of mountains of wood chips being loaded on to ships for export. The dwindling of forest resources elsewhere, combined with continuing technological progress in using the available species, increase the likelihood of chipping becoming a factor in the destruction of Amazonian forests.

Chipping of selective native forest species is used as a supplement to plantation sources for pulpwood at Jarí. The number of species used for this purpose has decreased from 80 in 1983 to 40 in 1986 (Fearnside, 1988, p. 18). The reduction in species used maintains a more consistently high quality in the pulp; for lower quality paper or cardboard such standards need not apply.

The use of wood chips for fuelling thermoelectric plants is another possible contributor to deforestation. A series of wood-fuelled power plants is under construction in the states of Amazonas and Rondônia. Two (Manacapurú, Amazonas and Ariquemes, Rondônia) are already functioning. The expansion of this use depends heavily on the price of oil. High oil prices made the initial plants a priority in the early 1980s, but subsequent decline in oil prices has removed much of this incentive. For example, the Balbina hydroelectric scheme had a 7.5 MW wood-burning thermoelectric plant to supply the construction site. This was deactivated and replaced with oil generators in September 1987, slightly over a year before hydroelectric generation was to begin. Two 50 MW thermoelectric plants were to use wood from the area around the reservoir; the parts for these, which were already arriving at Balbina, were transferred to Manaus for conversion to an oil-fuelled supplementary plant there. The low price of oil is the key factor in the change of plans, not sudden awareness of the value of maintaining forest. Since the earth's stocks of petroleum are being rapidly depleted, oil prices are bound to rise in the future – thereby increasing the attractiveness of wood-fuelled thermoelectric plants.

Slash-and-burn Agriculture

'Shifting cultivation', with long fallow periods capable of regenerating the soil after a year or two of use under annual crops, must be distinguished from 'pioneer farming' practised by recent arrivals in the region. Both systems rely on 'slash-and-burn' to clear the forest, but the similarity between the systems largely disappears after this initial step. Pioneers coming to the region from other parts of the country fell and burn the forest in the same way as the first step in traditional shifting cultivation, but after the brief cropping period they either leave the fields fallow for a short time (insufficient to regenerate the productive capacity of the sites) or, more frequently, plant the area in pasture. Shifting cultivation as a sustainable practice requires a complex set of cultural traditions in the form of folk knowledge and respected customs such that farmers do not reduce the fallow period and set in motion the degradation process. Even though the system could potentially support a sparse population in a sustainable fashion, it is doomed to fail for pioneers because of population pressure, demand for cash generation, cultural bias against those who have secondary forests, and/or speculative motives for planting pasture instead. True shifting cultivation is minimal as a factor in deforestation in Brazil. Only indigenous peoples and some *caboclo* farmers use this traditional practice. Pioneer agriculture, however, is a major and growing force in Brazilian Amazonia.

Slash-and-burn pioneer agriculture has long been a major factor in Amazonian portions of Peru and Ecuador, but has been overshadowed in Brazil by the rapid increase of pasture on large ranches. The importance of slash-and-burn is increasing relative to large ranchers because of the shortage of funds for financing ranchers and because of the increasing expulsion of small farmers from southern Brazil. Slash-and-burn is increasing fastest in Rondônia, Acre and Roraima. The potential for spread of this kind of clearing by small farmers is much larger than what has been experienced so far, but the future course of its expansion depends on political decisions to which strong opposition exists. A far-reaching agrarian reform plan was announced by Brazil's President Sarney in 1985. The original plan called for the land for redistribution to be expropriated from large landholdings (Brazil, MIRAD, 1985, p. 30). If implemented in this way, the plan would help slow deforestation. However, landowners have exerted strong pressure to (1) stop the plan altogether and (2) have the plan interpreted to require first the distribution of government land. Since virtually all of the land still in the public domain is located in Amazonia, such an interpretation would make agrarian reform a mere euphemism for colonisation of the type that has given poor results on the Transamazon Highway, in Rondônia, and elsewhere. Colonists from southern states are already being resettled under the 'agrarian reform' plan on public land in such areas as Presidente Figueiredo in the state of Amazonas. Carried to its logical conclusion, using Amazonia as an escape valve for settling landless people spells disaster in both sacrificing the forest and implanting a non-sustainable form of agriculture on a massive scale. Brazil's Legal Amazonia has an area of five million square kilometres; if the entire region (including reserves and already-occupied land) were divided equally among the country's 10 million landless families, each would receive only 50 hectares (half the area of lots on the Transamazon Highway). The inability of Amazonia to solve the social problems of other parts of the country must be recognized by national policy-makers.

Hydroelectric Dams

The 2010 Plan

Reservoirs for hydroelectric power generation are claiming a greater and greater share of Amazonian forest. The potential for expansion of impacts from this sector is large: ELETROBRAS (the Brazilian government's power monopoly) has published a '2010 plan' outlining the possible construction of 68 dams by the year 2010, with the total rising to as many as 80 dams within a few decades (Brazil, ELETROBRAS, 1987). The 80 dams would flood roughly 2 per cent of Brazil's Legal Amazonia – a percentage that, while seemingly small, would provoke forest disturbance in much wider areas. Aquatic habitats would, of course, be drastically altered. Most of the sites that are favourable for hydroelectric development are located along the middle and upper reaches of the tributaries that begin in Brazil's central plateau and flow north to meet the Amazon River – the Xingú, Tocantins, Araguaia, Tapajós and others. This region has one of the highest concentrations of indigenous peoples in Amazonia.

The Tucuruí Dam

The Tucuruí Dam, which blocked the Tocantins River in 1984, flooded 2,430 square kilometres, including part of the Parakanã Indian Reserve. The dam was built before 23 January 1986 when Brazil's National Council of the Environment (CONAMA) established its Resolution no. 001 to operationalise Federal Law no. 6938 of 31 August 1981 by requiring environmental impact statements (RIMAs). Compilation of available environmental information (Goodland, 1978) was commissioned by ELETRONORTE, the branch of ELETROBRAS responsible for Amazonia. The World Bank refused to finance the dam construction because of environmental concerns. A more detailed series of reports was compiled by INPA (under commission for ELETRONORTE) during the period when the dam was under construction (Brazil, INPA/ELETRONORTE, 1982–1984). Problems include aquatic weeds, acid water provoking corrosion of the turbines, and sedimentation from the catchment basin that is experiencing rapid deforestation. The resettlement programme for residents in the submergence area has created social problems (Mougeot, 1987). Construction of the dam simultaneously with the environmental studies guaranteed that the maximum effect that the findings could have would be to suggest minor modifications in procedures once the dam was already a *fait accompli* (Fearnside, 1985a). Relegating research to a merely token role is an unfortunate tradition in Amazonian development planning (Fearnside, 1986f).

Despite recommendations that 85 per cent of the vegetation be removed from the area to be flooded, ELETRONORTE adopted a plan to clear only 30 per cent (*A Província do Pará*, 15 June 1982; Monosowski, 1986). Selective logging of valuable timber received higher priority, although this was carried out in only a small part of the area as a combined result of lower densities of valuable species than originally foreseen, the inexperience of the CAPEMI military pension fund that held the logging concession, and the short time available before filling the reservoir. A financial scandal led CAPEMI to bankruptcy in 1983 (*A Crítica*, 4 February 1983) after clearing only 0.5 per cent of the submergence area. An additional area adjacent to the dam was cleared by ELETRONORTE; assuming all of this 'critical" 100 square kilometre area was actually cut, the cleared total would be 5 per cent of the reservoir (see Monosowski, 1986). When vegetation left in reservoirs decomposes, the water becomes acid and anoxic (Garzon, 1984).

One of the most controversial features of the Tucuruí Dam is that the power generated does little to improve the lot of those who live in the area: a fact dramatised by the high-tension lines passing over hut after hut lit only by the flickering of kerosene *lamparinhas*. Most of the power from Tucuruí supplies subsidised energy for multinational aluminium plants in Barcarena, Pará (ALBRAS–ALUNORTE, of Nippon Amazon Aluminum Co. Ltd or NAAC, a consortium of 33 Japanese firms) and in São Luís, Maranhão (ALUMAR, of Alcoa). Companhia Vale do Rio Doce (CVRD) maintains 51 per cent and 61 per cent interests in ALBRAS and ALUNORTE respectively (*CVRD-revista*, 1983). The power is sold at roughly one-third of the rate charged to residential consumers throughout the country, and so is heavily subsidised by the Brazilian populace through their taxes and home power bills.

The role of research in planning, authorising and executing major engineering projects, such as hydroelectric dams, is a critical matter if decision-making procedures are to evolve that prevent the kinds of misadventures that now characterise so much of the development in Amazonia. The public relations focus of many of the environment-related activities, such as the highly-publicised effort to rescue drowning wildlife, is a matter of intense controversy. Research is used for similar purposes: for example, during a public demonstration in Belém against closing the Tucuruí Dam, leaflets were dropped by helicopter reassuring readers that INPA's research in the area guaranteed that there would be no environmental problems (Brazil, ELETRONORTE, n.d., 1984). No such endorsement had been given either by INPA or by the individual researchers involved in the study. Publication results by the researchers is subject to approval by ELETRONORTE, according to the terms of the funding contract. It is essential that both the studies themselves and their subsequent dissemination take place free of interference from any source.

The Balbina Dam

The Balbina Dam 146 kilometres from Manaus is the worst case of environmental destruction from hydroelectric development. When the water level in the reservoir reaches its normal full level of 50 metres above sea level, 2,360 square kilometres will be flooded. The reservoir will contain approximately 1,500 islands making the area of land affected much larger than that actually submerged. About one-third of the Waimiri–Atroari Indian Reserve will be flooded. Severe as these impacts are, the magnitude of the environmental and financial disaster at Balbina lies in the meagre benefits that the project will produce.

Balbina's nominal capacity is 250 MW: the sum of five generators of 50 MW capacity each. The amount of power that the dam will actually produce, however, is much less than this. An *average* output of 112.2 MW is expected (Brazil, ELETRONORTE/MONASA/ENGE-RIO, 1976, p. B-51). Of this, 64 MW represents 'firm power' at the maximum depletion of 4.4 m for which the turbines were designed (Brazil, ELETRONORTE/MONASA/ENGE-RIO, 1976, p. B-47). Losses in transmission reduce the firm power delivered to Manaus to only 62.4 MW (Brazil, ELETRONORTE/MONASA/ENGE-RIO, 1976, p. B-49). Although all dams generate less than their nominal capacity, at 26 per cent, Balbina's firm output is less than normal.

Balbina's 250 MW nominal capacity is itself miniscule for a reservoir of this size – about as large as the 2,430-square kilometre Tucuruí reservoir that will support a nominal capacity of 8,000 MW. Balbina sacrifices 31 times more forest per megawatt of generating capacity installed than does Tucuruí. Low output is a logical consequence of the area's flat terrain and of the Uatumã River's low streamflow – an inevitable limitation with such a small drainage basin (18,862 square kilometres: Brazil, ELETRONORTE, 1987). The drainage basin is only eight times larger than the reservoir itself – which must be something of a record in hydroelectric development.

Much of the reservoir is extremely shallow as a consequence of the flat terrain. The reservoir's 2,360 square kilometres, at the 50-metre level falls to

1,580 square kilometres at the 46-metre level, meaning that 780 square kilometres (33 per cent) is less than 4 metres deep. Average depth when full will be 7.4 metres (Brazil, ELETROBRAS, 1986, p. 6.12). The large shallow areas can be expected to support rooted aquatic vegetation, adding to the problem of floating weeds that could affect the entire reservoir. The combination of large surface area per volume of water in a shallow reservoir and high biomass of aquatic vegetation will lead to heavy loss of the stored water to evaporation and transpiration.

The Balbina reservoir will be a labyrinth of canals among the islands and tributary streams. The residence time in some of these backwaters will be many times more than the already extremely long average of 11.7 months (Brazil, ELETROBRAS, 1986, p. 6.12). Tucuruí, by contrast, has an average residence of 1.8 months, or 6.4 times less. Some parts of the reservoir may turn over only once in several years. The slow turnover means that the decomposing vegetation will produce acids that cause corrosion of the turbines. At the Curuá-Una Reservoir near Santarém, Pará, for example, power generation had to be halted temporarily in 1982, only five years after filling, to allow repairs to the corroded turbines at a cost of US$ 1.1 million (Brazil, ELETROBRAS/CEPEL, 1983, p. 34).The cumulative cost of maintenance in the first six years totalled US$2 million, or US$16,600 per installed megawatt per year – 70 times the cost for a comparable dam in the semi-arid North-Eastern part of Brazil (Brazil, ELETROBRAS/CEPEL, 1983, p. 44). Lost generating time is not included in the costs. Balbina's longer mean turnover time (355 days versus about 40 days at Curuá-Una) and its abundance of stagnant bays and channels, means that water quality and corrosion problems will be worse than at Curuá-Una. At the rate experienced at Curuá-Una, Balbina's maintenance can be expected to cost US$4.15 million per year, or 4.3 mils (US) per KW-hour of electricity delivered to Manaus (about 10 per cent of the tariff charged consumers). Repairs due to similar corrosion in the Brokompondo Dam in Surinam totalled US$4 million, or over 7 per cent of the construction cost, in the first 13 years of operation (Caufield, 1983: 62). Vegetation is being left to decompose in most of the Balbina submergence area: only a token 50 square kilometres (2 per cent) of the reservoir area was cleared before the dam was closed.

Balbina is particularly unfortunate because it is unnecessary. The dam is expected to produce firm power that could be counted on for only about one-third of the 218 MW 1987 level of power demand in Manaus; the average power delivered in Manaus (109.4 MW after 2 per cent transmission loss) would be half the 1987 demand. The dam will never supply this percentage of the Manaus demand because the calculations assume the 50-metre reservoir level – at first the dam will generate a substantially lower amount (a figure not yet disclosed by ELETRONORTE) because the reservoir level will be kept at 46 metres until water quality stabilises. Subsequent ELETRONORTE statements indicate that this concession is likely to be discarded and the reservoir filled directly to the 50 metre level as quickly as the availability of water permits. The percentage of power consumed in Manaus supplied by Balbina will shrink with each succeeding year as the city continues to grow: Balbina's average output (at the 50-metre level) delivered to Manaus corresponds to only 38 per cent of the 285 MW power demand ELETRONORTE projects for the city in 1996 when another dam, to be built 500 kilometres from Manaus at Cachoeira Porteira on the Trombetas River, is expected to make up the city's power deficit (Brazil, ELETRONORTE, 1987). Only one dam (Cachoeira Porteira) could have been built – with half the cost and half the impact – rather than building both dams. To make the futility of Balbina even more apparent, natural gas 500 kilometres from Manaus in the Juruá River basin could supply Manaus with power (Goldemberg, 1984). Recent discovery of oil and gas at Urucú, nearer Manaus, could also supply the city with power without Balbina.

The power from Balbina will largely benefit the international companies that have established factories in the Manaus Free Trade Superintendency Zone (SUFRAMA). That power will be subsidised for these firms at the expense of residential consumers throughout the country is an irritant to many Brazilians. SUFRAMA was established in Manaus in 1967 to compensate western Amazonia for the concentration of SUDAM's investments in eastern Amazonia (Mahar, 1976, p. 360). The financial and environmental costs are high when political decisions lead to the location of industrial centres in places where power generation is difficult. All of the consequences of supporting industries and population need to be considered before the initial decisions are made.

The Balbina Dam was closed on 1 October 1987. The dam was exempt from the environmental impact report (RIMA) because of its being under construction at the time when the report became mandatory, but was nevertheless required to obtain a Licence for Operation from the Amazonas state government's environmental organ, CODEAMA. CODEAMA's director was suddenly replaced only nine days before the dam was licensed (Melchiades Filho, 1987). The licence was granted on the *same day* that the last adufa (sluice base) was closed blocking off the Uatumã River. The precedent of making the environmental review process a mere token formality is perhaps the most far-reaching impact of this highly questionable project.

The momentum of the construction effort at Balbina not only succeeded in crushing the Brazilian environmental review process, but also managed to circumvent the environmental hurdles within the World Bank. The World Bank rejected Balbina on environmental grounds when presented as a separate project, but subsequently approved the first Brazilian Power Sector Loan – thus underwriting hydroelectric projects throughout the country. If, as World Bank officials say, no Bank money was spent directly at Balbina, then this was avoided by sheer luck. Whether or not the timing of ELETRO-

NORTE purchases means that Bank money was spent directly at Balbina, the relief these funds provided to ELETRONORTE's overstretched budget was undoubtedly critical to allowing the agency to bring the apparently low-priority Balbina project to completion.

Other Dams

Other dams planned or under construction in the region have many of the same problems as Balbina. The Samuel Dam, under construction on the Jamarí River in Rondônia, will flood a 656-square kilometre area for little power (216 MW installed) and a high cost (US$610 million). The Ji-Paraná Dam, now in the final planning stages on Rondônia's Ji-Paraná (Machado) River, will flood 107 square kilometres (6 per cent) of the Lourdes Indian Reservation of the Gavião and Arara tribes, plus 37.7 square kilometres (1.4 per cent) of the Jarú Biological Reserve (Brazil, ELETROBRAS, 1986, p. 6.23). The World Bank financed these reserves under the POLONOROESTE programme, but will also be financing their flooding under the loan for building the Ji-Paraná Dam. The migration encouraged by the BR-364 highway that was reconstructed under POLONOROESTE is, of course, what makes the power from these dams necessary. The explosive growth of population in the area is rapidly recreating the situation at Balbina, where the low capacity dams create impacts that only serve to postpone more definitive solutions – such as transmission lines from topographically favourable generating sites. The technology of long-distance power transmission has improved markedly since many of the hydroelectric projects were planned (Pires and Vaccari, 1986).

The Altamira Complex on Pará's Xingú River will flood a total of 7,365 square kilometres, of which 1,225 square kilometres will be for the Kararaô Dam (to be built first, downstream of the city of Altamira) and 6,140 square kilometres will be for the Babaquara Dam to be built upstream (Brazil, ELETRONORTE/CNEC, n.d., 1986). Kararaô and Babaquara are part of a chain of reservoirs in the Xingú River Basin Hydroelectric Project that would disrupt the lives of 4,000 indians (Survival International 1987a, p. 1). While some reports indicate a total of 21 dams planned in the Xingú Basin (SBPC, 1986), ELETRONORTE lists five (Brazil, ELETRONORTE, 1985). Despite public statements to the contrary, these dams are listed in the December 1987 version of the 2010 plan (Brazil, ELETROBRAS, 1987), including the Jarina Dam that would flood part of the Xingú Indian Park (CIMI, CEDI, IBASE and GhK, 1986).

The two reservoirs in the Altamira Complex itself (Kararaô and Babaquara) will flood portions of areas occupied by tribes from four different linguistic trunks – cultures as different as, for example, China and Europe (D. Posey, personal communication, 1987). The Kararaô Dam will have 4,675 MW of generating capacity installed and will cost US$5.52 billion (Brazil ELETRONORTE/CNEC, 1987). The entire complex is expected to cost US$10 billion, which would worsen Brazil's international debt crisis substantially (Environmental Policy Institute, 1987). The installed power capacity of the complex is a colossal 17,000 MW, making it the largest in the world. Almost all of this power would be transmitted to southern Brazil, rather than being used to create employment in the region (Brazil, ELETRONORTE/CNEC, n.d., c.1986). The first dam (Kararaô) will have substantially greater power output in relation to environmental and human impact than the other dams. This indicates the urgency of linking the environmental impact analysis of all of the dams, to avoid repetition of the problem caused by the Carajás mine and railway, where a high-value project causing relatively little direct disturbance was allowed to justify a series of extremely damaging subsequent projects that were not included in the evaluations of the initial scheme.

Military Bases

Calha Norte, meaning 'northern banks', is a programme to build or enlarge military bases and/or airstrips in 16 locations in a 150-kilometre strip along the portion of Brazil's border north of the Amazon River (Amazonas and Solimões Rivers). The plan was announced in 1986, and is already being funded and executed without benefit of any environmental review. Secrecy surrounding the plan has inhibited public discussion (Brazil, Universidade Federal do Pará, NAEA, GIPCT, 1987). Estimates of the indigenous population in the Calha Norte zone range from 50,000 (Matias, 1987) to 60,000 (Comité Interdisciplinar de Estudos sobre o Projeto Calha Norte, 1987, p. 5) out of Brazil's 220,000 Amerindians. The Calha Norte zone includes 84 indigenous areas, only 16 of which are demarcated (Comité Interdisciplinar de Estudos sobre o Projeto Calha Norte, 1987, p. 5).

One of the principal impacts of the programme is that it is impeding the demarcation of Amerindian reserves. Reserve demarcation is blocked in a 150-km wide strip not only in the Calha Norte zone but also along all international borders, affecting tribes in Acre, Rondônia and Mato Grosso (Survival International, 1987b, p. 2). The position of the tribes was further eroded in 1987 when the government abolished the concept of 'Indian Reserves' and replaced it with one of 'Indigenous Colonies' – thereby denying the tribes the special protections mandated by the country's constitution and legislation (thus allowing the land to be bought or otherwise taken away by non-Indians). Paving of the BR-364 from Rondônia to Acre has been financed by the Interamerican Development Bank on condition that 35 reserves in Acre be demarcated; disbursements have been suspended now that current policies block the demarcations, but the money for construction has not been returned (unlike the US$5 million loan received from the World Bank for the expense of actually demarcating the reserves themselves).

The impact of the bases being constructed under Calha Norte is potentially much wider than the immediate environs of the military installations. Although not contained in the current budget, the plan calls for building roads and promoting settlement in the area. The ministers of foreign relations, interior, planning and national security wrote in their exposition of motives to President Sarney: 'It is fundamental that the action of the government also contemplate expansion of highway infrastructure . . . and an increase in the colonisation of that frontier area' (Setubal *et al.*, 1986, p. 3). Once roads are built, settlers and speculators can be expected to enter and clear the forest regardless of official policies, as has occurred repeatedly elsewhere (Fearnside and Ferreira, 1984). Population flow from Rondônia is already rapidly pushing back the agricultural frontier in Roraima.

Military reasons are often convenient excuses for developments wanted for other reasons, as occurred in the case of the Transamazon Highway (Fearnside, 1984c; Kleinpenning, 1979). Deciding where to place roads and colonisation areas on the basis of geopolitical strategy rather than the agronomic potential of the soils is a sure formula for agricultural failure. None of the land in the Calha Norte area is shown by RADAM maps as suitable for agriculture (Brazil, Projeto RADAMBRASIL, 1974–1977, vols. 6, 8, 9, 11, 14).

The best example of the danger of allowing military reasons to determine the location of settlements is the Sidney Girão colonisation area, which was placed on Rondônia's border with Bolivia for strategic reasons (Muller, 1980); the area's poor soil resulted in such rapid abandonment of the lots that the government was unable to fill the project until long after all other areas in Rondônia were overflowing with landseekers. The project's failure has been officially recognised as due to poor soil (Valverde *et al.*, 1979).

Mining

Mining is another activity that is rapidly increasing as an agent of environmental destruction in Amazonia. Some of the impacts are direct, while others are indirect. Open pit mines obviously completely transform the environment in the specific localities affected, such as the iron mine at Carajás (Pará), manganese at Serra do Navio (Amapá), kaolin at Jarí (Amapá) bauxite (aluminium) at Trombetas (Pará) and cassiterite (tin) at various locations in Amazonas and Rondônia. The areas destroyed are small, although the destruction is total. Only the bauxite mine at Porto Trombetas has an active programme of revegetating the mine site (Knowles, 1988), although the Carajás iron project has plans for future revegetation of its pits (De Freitas and Smyrski-Shluger, n.d., 1983).

Waste from mining can be significant. The fines from the Trombetas bauxite mine form a 'red mud' that has completely filled the 200-hectare

Lago da Batata and suffocated trees along its margin and approaches; preparations are being made to transport future production of red mud back to the mine site itself. Devastated as the Lago da Batata is, it represents a tiny area in Amazonian terms. The Balbina reservoir, for example, will be over 1,000 times larger.

The silt from cassiterite mining is a large source of sediments in the drainage basins affected. In Rondônia, measurements in rivers with mining indicate much heavier silt loads than those in rivers without mining (Arnaldo Carneiro, personal communication, 1988). One negative effect of the increase could be more rapid sedimentation of reservoirs, including the Samuel Dam. One mining operation (Mineração Oriente Novo) released large amounts of sediment into the Rio Preto (a tributary in the Samuel catchment) until it was stopped in 1986 by a federal court order. Other operations in the Samuel catchment, such as the BRASCAN mines, store their fines behind small retaining dams. An undetermined amount of sediment would be released into the reservoir were these dams to break, as occurred in the state of Amazonas in 1987.

The incident in the state of Amazonas occurred at a cassiterite mine on the Pitinga River where Mineração Taboca (a subsidiary of Paranapanema) stores tailings in holding ponds for possible future use should the price of tin increase and justify more thorough extraction procedures. The dykes for four of these ponds broke in 1987, releasing its sediment into the adjacent Alalaú River, a tributary of the Rio Negro. The pollution affected fish in the Waimiri–Atroari Indian Reserve downstream. The reserve had already suffered many impacts from the mine, including being reduced in size in order to make room for the mining operation and having a road built through the area to connect the site with the Manaus-Caracaraí (BR-174) highway that also bisects the reserve.

Gold mining contributes greatly to the silt load of rivers. Much of the mining is done in river beds, either by dredging alluvium from the river bottom or by panning it from the banks. The river water is often a milky colour from the silt load far below the mining sites themselves. As with other minerals, roadbuilding spurred by gold strikes sets in motion the process of invasion and deforestation of the affected areas. The 'Rodovia do Ouro' through the Reserva Garimpeira de Tapajós in Pará was the first such road (*Veja*, 28 November 1984); similar highways in Roraima may follow soon.

Mercury pollution is rapidly becoming a public health crisis in Amazonia. Use of mercury to amalgamate the fine gold particles in the extraction process dumped an estimated 250 metric tons of highly toxic mercury into the rivers between 1984 and 1988 (J. Dubois, personal communication, 1988). The estimate of mercury discarded is derived from the weight of gold extracted and the 1.2 grams of mercury used per gram of gold; the amount of mercury could be much greater since much of the gold is smuggled out of the country illegally. Mercury concentrations in fish in the Madeira River (draining Rondônia) are as high as six times the levels permitted in food by the World Health Organisation (B. A. Forsberg, personal communication, 1988; Martinelli *et al.*, n.d.). In Itaituba, Pará – a gold mining centre on the Tapajós River – heavy incidence of human diarrhoea with blood is associated with mercury poisoning from eating contaminated fish (J. Dubois, personal communication, 1988; *O Liberal*, 1 February 1988). Fish supply a major part of the protein in the diet of Amazonian residents, including the indigenous peoples that inhabit some of the most active gold mining regions. The Madeira River is also a major supplier of fish to the cities of Manaus and Porto Velho.

Indirect effects of mining promise to be even greater than most direct effects. Roads built to the mining areas bring in population, with subsequent deforestation. The population of miners themselves add to this impact: what will become of the approximately 75,000 gold miners at the Serra Pelada 'anthill' in Pará is a major question if the often-postponed plans are put into effect to have a government-controlled mining firm mechanise the operation. A major impediment to such a move is the threat of the miners to invade the neighbouring Carajás iron project area in order to stake out prospecting and farming claims.

Amerindian areas suffer some of the most direct effects of gold prospecting. These include frequently bloody encounters with the gold miners, spread of disease, and the more subtle effect of providing motivation for not demarcating the tribes' land as reserves. Many delays and reductions in reserve demarcations are believed to result from the influence of large mining companies, the population of individual *garimpeiros* (prospectors) and the pilots, merchants and others that serve them. When demarcation is delayed, the areas are taken over by non-indians.

The presence of minerals can make possible agriculture and forestry projects that would otherwise be unviable. Examples include Jarí, where the forestry sector depends financially on the estate's kaolin mine (Fearnside, 1988). The AMCEL forestry operation in Amapá would likewise be improbable without the associated ICOMI manganese mine at Serra do Navio.

On a much larger scale, the entire Grande Carajás Programme is justified by the extraordinary mining potential of this region – where minerals such as iron, gold, copper and manganese were squeezed up from the earth's mantle at the point where the primordial continents of South America and Africa once joined. The Grande Carajás Programme includes a mammoth agricultural plan, the pig-iron smelting scheme with its associated forestry management and/or plantations for charcoal production, a railway and highway network, hydroelectric dams (including Tucuruí), power transmission lines and mineral processing facilities such as the Barcarena aluminium complex. The potential environmental impacts of these developments are unprecedented (Fearnside, 1986e).

POLICIES FOR SUSTAINABLE DEVELOPMENT

Overcoming Obstacles to Sustainable Development

Obstacles to sustainable development of Amazonia's natural resources include growth of the human population in this region to levels that exceed carrying capacity, severe inequalities in the distribution of resources (particularly land), policies leading to implanting unsustainable land uses (such as cattle pastures) and ultimately, the accelerating destruction of natural ecosystems. The rapid deforestation for cattle pasture that dominates land use in the Amazon Region of Brazil is in many ways a symptom of deeper causes that must be addressed if development is to be channelled to a wiser course. Simply outlawing deforestation is completely ineffective, as has been demonstrated in Brazil by the unenforced Forestry Code (Decree Law no. 4771 of 15 September 1965) limiting clearing to 50 per cent of any property and the 1986 law (Decree Law no. 7511 of 7 July 1986) prohibiting deforestation completely.

Many of the structural changes needed to strike at the root causes of deforestation will require years of effort. Nevertheless, a start must be made now. Population growth must be slowed both in Amazonia and in the regions from which migrants come. The effect of population growth is currently overshadowed by transformation of the agricultural systems in southern Brazil from those dominated by small farms producing labour-intensive crops to systems dominated by large agribusinesses growing mechanised crops, such as soybeans. The southern Brazilian states should make the social choices as to how to absorb the 'excess' population expelled by this transformation, so long as the solution is not the present one of simply transferring the problem to Amazonia. One means of absorbing more people in the rural areas is by redistributing large unproductive landholdings. Another is by favouring labour-intensive crops over mechanised ones in allocating agricultural credit. Ultimately, absorbing more people in the urban sector must be facilitated.

Adequate employment opportunities must be given to urban residents, including those who are attracted from the countryside: much more could be done to expand industry. For example, both the 12.6×10^3 MW hydroelectric dam at Itaipú and the 8.0×10^3 MW dam at Tucuruí have only a fraction of their generating capacities installed. More power could be had by simply mounting the remaining turbines and generators, without incurring any of the environmental and financial costs of building more dams and creating more reservoirs. Since both of these dams have transmission links to the cities in migrant source areas such as Paraná, the power could attract new factories that would employ some of the migrants that now leave for Amazonia, especially Rondônia.

It is unrealistic to think that Brazil can adopt agricultural patterns similar to those in North America and still keep over 30 per cent of its population in the rural zone. The rural population of the United States, for example, declined over the course of the twentieth century from a proportion similar to that of Brazil to less than 5 per cent today. If scarce capital resources are to create a vastly increased number of urban jobs in Brazil the location of cities must be planned more rationally than at present. Manaus, for example, grew from approximately 120,000 in 1967 to 1.3 million in 1987 because of population drawn to industries that have located themselves in a special duty-free zone. The city is now being provided with a hydroelectric dam: Balbina. Construction cost will total US$3,000 per KW of installed capacity. Similarly, the Samuel Dam in Rondônia which is being built to provide power to that new state whose population has been swollen by migration along the World Bank-financed BR-364 highway, will cost US$2,800 per KW installed because, like Balbina, it is on a small river in a flat region inappropriate for hydroelectric development. For comparison, when completed Tucuruí will cost US$675 per KW (4.6 times less than Balbina) and Itaipú US$1,206 per KW (2.6 times less than Balbina) (construction costs from *Veja*, 20 May 1987, p. 30). In other words, the same investment in a more topographically favourable site could produce several times more power, and generate proportionately more industrial employment. That employment could absorb many of the migrants now being forced to leave southern Brazil for Amazonia.

Brazil's policy of a 'unified' tariff for electricity means that industry and population can locate themselves where they choose, and the power authority is then obliged to take heroic measures to provide them with electricity. Power in unfavourable places like Manaus is subsidised by the consumers living nearer favourable sites like Itaipú. Were electricity sold at rates reflecting its cost of generation, industrial centres would relocate themselves and the total amount of urban employment would be significantly greater.

What Government Action Can Do

Measures needed to contain environmental destruction in Amazonia can be divided, somewhat arbitrarily, into short-term and long-term targets. It is important that action on long-term targets not be simply postponed in favour of concentrating on actions with immediate payoffs – such a course would be just as short-sighted as Brazil's rush to trade its rainforests for a few years of pasture production. Issues that must be confronted squarely now include population growth, resource distribution, and economic mechanisms to allow sustainable management of slow-growing biological resources like tropical forests.

At the same time, a number of important changes could be made literally at the stroke of a pen, provided that it is the right pen. Actions that require no consciousness-raising, long-term research programmes or similar slow activities include stopping highway construction in Amazonia, abolishing subsidies for cattle pastures and ending energy subsidies to the Amazon, such as the price unification policies for petroleum products and electric power. Land tenure policies could be changed to disallow pasture or annual crops as 'improvements' for establishing claims. Sustainable uses of standing forest should replace the present focus of agricultural research and, especially, of credit. The presently-favoured forms of agriculture can be sustained only if confined to a miniscule fraction of Amazonia – i.e. fertiliser-dependent agriculture and pasture.

Protection is needed in fact of natural forest reserves that are now only declared on paper. Governments cannot continue to renege on previous commitments to reserves whenever land is wanted for another purpose without condemning all of the remaining forest to the axe. The tendency to renege on commitments is a more fundamental menace than is the fact of explosive deforestation.

Despite the tremendous need for change, Brazil has made great advances in protecting examples of its natural ecosystems and incorporating environmental factors into development procedures. At the time of the Stockholm Conference on the Environment in 1972, Brazil was labelled the 'villain of Stockholm' for its role in leading the countries of the developing world in condemning any suggestion that these nations should protect their environments (Sanders, 1973). Today, Brazil has a Special Secretariat of the Environment (SEMA), a system of national parks, and a law requiring an Environmental Impact Report (RIMA) prior to approving any major development project. The legal and legislative advances in protecting the environment must be further fortified by building a corps of qualified people to carry them out and a tradition of serious consideration of the environment in development planning – especially in the early phases of project formulation before major developments become 'irreversible' *faits accomplis*.

What Foreigners Can Do

What can foreigners do to further sustainable development in Amazonia? Direct pressure on the government of Brazil or on the governments of other Amazonian countries can easily backfire to the detriment of whatever change might be desired by well-intentioned persons abroad. For example, in 1987 a petition signed by 45,000 Austrians in an effort to convince Brazil's Constitutional Convention to strengthen protection of indigenous peoples was instead used for a newspaper campaign against Brazilian indigenous rights groups and against the provisions of the draft constitution that (prior to their deletion) would have provided some protection for the tribes.

Taxpayers in countries that contribute to the budget of agencies such as the World Bank have every right to influence how their money is spent. These banks represent a great force, for good or for evil, in the developing countries where the money is applied. Expanding and improving the environmental sectors and procedures within these banks is a legitimate and effective focus for environmental concerns. The World Bank has already benefited from the efforts of environmental groups and national legislatures in the donor countries. The benefit for the recipient countries would be greatest if environmentalists, researchers and the direct and indirect employees of the Banks worked together.

One mechanism for applying funds from international sources to environmental problems in Amazonia is through agreements under which countries in the region are released from portions of their foreign debt on the condition that the money owed (or a specified fraction of it) be spent on establishing national parks or on other environmentally-beneficial activities. Such schemes both help relieve the burden of debt and achieve goals that the countries themselves ostensibly espouse. A major park in the Amazonian part of Bolivia has recently been created under an agreement of this type, but the device has not yet been used in Brazil.

The lending policies of banks, including commercial banks, should be tightened to reduce the temptation to run up heavy debts and to promote short-sighted land-use policies in order to pay them back. For example, the emphasis on producing soybeans in Southern Brazil may in part be an effort to generate export earnings to relieve the country's burden of debt; the mechanised cultivation of soybeans forces small farmers to migrate to Rondônia.

Direct investments from foreign countries also further the replacement of forests with unsustainable land uses. Examples include the Volkswagen and Armour-Swift–Brascan ranches in Pará and the Suiá-Missú Ranch (of the Italian multinational Liquigas) in Mato Grosso. French financing and equipment sales for the extensive and economically questionable hydroelectric projects at Balbina and Samuel is another example. Deforestation also results from the land speculation that investors employ as a means of turning Brazil's astronomical inflation to their advantage. The inflation is in part fuelled by ill-conceived projects from abroad that inject money into the economy without producing a corresponding flow of products for the consumers to buy with it. Examples include the inefficient dams and marginally-productive ranches in Amazonia, as well as the economically disastrous German-financed nuclear power plants near Rio de Janeiro.

Foreign countries also influence land-use choices in undesirable ways through purchases of the products generated by unsustainable activities. Demand for endangered animals and plants is one form of influence. Demand for beef, which is a major force in Central America through the notorious 'hamburger connection', is a much weaker influence in South

America because the presence of hoof-and-mouth disease (Afthosis) blocks export of beef in frozen form to North America and Japan. Plans for a 'sanitary pocket' in part of the Grande Carajás Programme in Brazil could unleash the full force of this connection in part of Amazonia.

In addition to refraining from damaging activities, foreigners can have a positive effect on land-use choices through direct inputs within limited fields of activity. One of these is scientific research, where both contributions of money (to such organisations as the World Wildlife Fund) and time are needed. Much work needs to be done in a broad range of fields. Particularly urgent is screening possible economically useful products that can be extracted from standing forest. Institutional mechanisms need to be developed that will facilitate this while guaranteeing that the economic rewards derived from any saleable products will remain in the South American countries where the forests are located. Field-based research in many areas of South America needs the collaboration of a substantial network of researchers elsewhere, especially taxonomists, pharmacologists and analytical chemists.

Training is another area where foreign scientists can make an important contribution to sustainable development. Although universities and research institutions in tropical areas of South America are rapidly growing stronger, they are still pitifully small when compared to the magnitude of the challenge of devising sustainable forms of development for Amazonia's vast, complex, and poorly-studied ecosystems. These institutions need to be strengthened quickly because of the explosive pace at which the options for sustainable use are being closed off by deforestation. At the same time, strict adherence to academic standards is essential, and the consequences of lapses are grave. The quality as well as the number of scientists in the region must be increased so that more of the intellectual activity of designing projects and interpreting results takes place in the Amazon, rather than at distant locations in southern Brazil or abroad. The negotiation of research and training programmes with countries such as Brazil is one of the most far-reaching and effective but difficult ways that international institutions can contribute to sustainable development in Amazonia.

In summary, environmental destruction in Amazonia is rampant and the forces behind the destruction are formidable. These facts, however, are not reason for despair. Effective measures can be taken by governments, financial institutions and individuals to help contain environmental destruction in the region and to direct development in sustainable, non-destructive directions.

Note

1. Earlier versions of sections of this paper have been presented at the 'International Conference on the Environmental Future: Maintenance of the Biosphere', 24–27 September 1987, Edinburgh, Scotland (Fearnside, n.d. b); the 'Symposium on the Amazonia: Deforestation and Possible Effects', 46th International Congress of Americanists, 4–8 July 1988, Amsterdam, the Netherlands (Fearnside n.d. c); and the 40ª Reunião Anual da Sociedade Brasileira para o Progresso da Ciência (SBPC), 10–16 July 1988, São Paulo (Fearnside, n.d. d). I thank Summer Wilson and the volume editors for helpful comments on the manuscript.

References

Abelson, P. H. (1975), 'Energy alternatives for Brazil', *Science*, 189, p. 417.

Anderson, A. B. and Benson, W. W. (1980), 'On the number of tree species in Amazonian forests', *Biotropica* 12(3), pp. 235–7.

Beisiegel, W. de R. and de Souza, W. O. (1986), 'Reservas de fosfátos – Panorama nacional e mundial', in Instituto Brasileiro de Fósforo (IBRAFOS), *III Encontro Nacional de Rocha Fosfática, Brasília*, 16–18 June 1986, Brasília: IBRAFOS, pp. 55–67.

Bell, T. I. W. (1973), 'Erosion in the Trinidad teak plantations', *Commonwealth Forestry Review*, 52(3), pp. 223–33.

Brazil, ELETROBRÁS (1986), 'Plano diretor para proteção e melhoria do meio ambiente nas obras e serviços do setor elétrico', Centrais Elétricas Brasileiras SA (ELETROBRÁS), Diretoria de Planejamento e Engenharia, Brasília: Departamento de Recursos Energéticos (mimeo).

Brazil, ELETROBRÁS (1987), *Plano Nacional de Energia Elétrica 1987/2010: Plano 2010: Relatório Geral* (December 1987), Rio de Janeiro: Centrais Elétricas Brasileiros SA (ELETROBRÁS).

Brazil, ELETROBRÁS/CEPEL (1983), *Relatório Técnico Final no. 963/83: Estudo Comparativo de Manutenção nas Usinas de Curuá-Una e Moxotó*, Brasília: ELETROBRÁS/Centro de Pesquisas de Energia Elétrica (CEPEL).

Brazil, ELETRONORTE, n.d. (1984), 'Tucuruí Urgente', Brasília: Centrais Elétricas do Norte do Brasil SA (ELETRONORTE) (leaflet).

Brazil, ELETRONORTE (1985), *Polit-kit. Ano II. no. 3. Abril/85. O Novo Perfil da Amazônia*, Brasília: Centrais Elétricas do Norte do Brasil SA (ELETRONORTE).

Brazil, ELETRONORTE (1987), *UHE Balbina*, Brasília: Centrais Elétricas do Norte do Brasil SA (ELETRONORTE).

Brazil, ELETRONORTE/CNEC n.d. (1986), 'The Altamira Hydroelectric Complex', São Paulo: Centrais Elétricas do Norte do Brasil, SA (ELETRONORTE)/ Consórcio Nacional de Engenheiros Consultores SA (CNEC).

Brazil, ELETRONORTE/CNEC (1987), *Estudos Xingú Contrato DT-1HX-001/75. Estudos de Viabilidade UHE Kararaô: Panorama Atual*, São Paulo: Consórcio Nacional de Engenheiros Consultores SA (CNEC).

Brazil, ELETRONORTE/MONASA/ENGE-RIO (1976), *Estudos Amazônia, Relatório Final Volume IV: Aproveitamento Hidrelétrico do Rio Uatumã em Cachoeira Balbina, Estudos de Viabilidade*, Brasília: Centrais Elétricas do Norte do Brasil (ELETRONORTE)/MONASA Consultoria e Projetos Ltda./ENGE-RIO Engenharia e Consultoria SA.

Brazil, EMBRAPA/IPEAN (1974), *Solos da Rodovia Transamazônica: Trecho Itaituba-Rio Branco. Relatório Preliminar*, Belém: Empresa Brasileira de Pesquisa

Agropecuária (EMBRAPA)/Instituto de Pesquisas Agropecuárias do Norte (IPEAN).

Brazil, IBDF (1983), *Desenvolvimento Florestal no Brasil*, Brasília: Folha Informativa No. 5. Instituto Brasileiro de Desenvolvimento Florestal (IBDF).

Brazil, IBDF (1985), *Monitoramento da Alteração da Cobertura Vegetal da Área do Programa POLONOROESTE nos Estados de Rondônia e Mato Grosso: Relatório Técnico*, Brasília: Instituto Brasileiro de Desenvolvimento Florestal (IBDF).

Brazil, INPA/ELETRONORTE (1982–1984), *Estudos de Ecologia e Controle Ambiental na Região da UHE Tucuruí: Relatórios Setorias*, Manaus: Instituto Nacional de Pesquisas da Amazônia (INPA).

Brazil, Ministério da Agricultura (1983), *Programa Grande Carajás Agrícola, Versão Preliminar*. Brasília: Ministério da Agricultura, 6 vols.

Brazil, MIRAD (1985), *Proposta para a elaboração do 1º Plano Nacional de Reforma Agrária da Nova República – PNRA*, Brasília: Ministério da Reforma e do Desenvolvimento Agrária (MIRAD) (mimeo).

Brazil, Projeto RADAMBRASIL (1973–1982), *Levantamento de Recursos Naturais, vols 1–23*, Rio de Janeiro: Departamento Nacional de Produção Mineral.

Brazil, Universidade Federal do Pará, Núcleo de Altos Estudos Amazônicos (NAEA), Grupo Interdisciplinar de Política Científica e Tecnológica (GIPCT) (1987), *Projeto Calha Norte: Autoritarismo e Sigilo na Nova República*, Belém: Série Documentos GIPCT no. 2, NAEA.

Brünig, E. F. (1977), 'The tropical rain forest – A wasted asset or an essential biospheric resource?', *Ambio* 6(4), pp. 187–91.

Bunker, S. G. (1980), 'Forces of destruction in Amazonia', *Environment* 22(7), pp. 14–43.

Cardoso, F. H. and Müller, G. (1978), *Amazônia: Expansão do Capitalismo*, São Paulo: Editora Brasiliense, 2nd edn.

Caufield, C. (1983), 'Dam the Amazon, full steam ahead', *Natural History* 1983(7), pp. 60–7.

Caufield, C. (1985), *In the Rainforest*, London: Heinemann; Picador (Pan Books).

CEDI (1986), *Povos Indígenas no Brasil – 85/86*, Aconteceu Especial 17, São Paulo: Centro Ecumênico de Documentação e Informação (CEDI).

CEDI/CONAGE (1988), *Empresa de Mineração e Terras Indígenas na Amazônia*, São Paulo: Centro Ecumênico de Documentação e Informação/Coordenação Nacional de Geólogos (CEDI/CONAGE).

CEDI/Museu Nacional (1987), *Terras Indígenas no Brasil*, São Paulo: Centro Ecumênico de Documentação e Informação (CEDI).

CIMI, CEDI, IBASE and GhK (1986), 'Brasil: Areas Indígenas e Grandes Projetos', (map scale 1:5,000,000). Conselho Indigenista Missionária (CIMI), Brasília: Centro Ecumênico de Documentação e Informação (CEDI), IBASE, GhK.

Collins, N. M. and Wood, T. G. (1984), 'Termites and atmosphere gas production', *Science*, 224, pp. 84–6.

Comité Interdisciplinar de Estudos sobre o Projeto Calha Norte (1987), 'Calha Norte: Documento síntese dos posicionamentos aprovados pelo Comité Interdisciplinar de Estudos sobre o Projeto Calha Norte a partir do seminário "O Projeto Calha Norte: A Política de Ocupação de Espaços no País e seus Impactos Ambientais" realizado pelo Comité Interdisciplinar de Estudos sobre o Projeto Calha Norte, nos dias 26, 27 e 28 de agosto de 1987 na Universidade Federal de Santa Catarina – Florianópolis/SC', Florianópolis: Imprensa Universitária.

A Crítica (Manaus) (4 February 1983), 'Capemi em situação difícil vai deixar área de Tucuruí', Caderno 2, p. 4.

A Crítica (Manaus) (8 August 1987), 'Saída da China afunda mercado', p. 8.
Crutzen, P. J., Heidt, L. E., Krasnec, J. P., Pollock, W. H. and Seiler, W. (1979), 'Biomass burning as a source of the atmospheric gases CO, H_2, N_2O, CH_3Cl, and COS', *Nature*, 282, pp. 253–6.
Cunningham, R. K. (1963), 'The effect of clearing a tropical forest soil', *Journal of Soil Sciences*, 14(2), pp. 334–45.
CVRD – revista (1983), 'ALBRÁS ALUNORTE', *CVRD – Revista* 4(14).
Dantas, M. (1979), 'Pastagens da Amazônia Central: Ecologia e fauna de solo', *Acta Amazônica* 9(2), suplemento, pp. 1–54.
De Freitas, M. de L. D. and Smyrski-Shluger, C. M., n.d. (c.1983), 'Projeto Ferro Carajás – Brasil: Aspectos ambientais', paper presented at the *Interciencia* Association International Symposium on Amazonia, Belém, 7–13 July 1983, manuscript, Rio de Janeiro: Companhia Vale do Rio Doce.
De Lima, J. M. G. (1976), *Perfil Analítico dos Fertilizantes Fosfatados*, boletim no. 39, Brasília: Ministério das Minas e Energia, Departamento Nacional de Produção Mineral.
Denevan, W. M., Treacy, J. M., Alcorn, J. B., Padoch, C., Denslow, J. and Paitan, S. F. (1984), 'Indigenous agroforestry in the Peruvian Amazon: Bora Indian management of swidden fallows', *Interciencia* 9(6), pp. 346–50.
Dicks, S. E. (1982), *The Use of LANDSAT Imagery for Monitoring Forest Cover Alteration in Xinguara, Brazil*, master's Thesis in Geography, Gainseville, Florida: University of Florida.
Dickinson, R. E. and R. J. Cicerone (1986), 'Future global warming from atmospheric trace gases', *Nature*, 319, pp. 109–15.
Eagleson, P. S. (1986), 'The emergence of global-scale hydrology', *Water Resources Research*, 22(9), pp. 6s–14s.
Eckholm, E. (1978), 'Disappearing species: the social challenge', *Worldwatch Paper*, 22, Washington, D.C.: Worldwatch Institute.
Ehhalt, D. H. (1985), 'On the rise: Methane in the global atmosphere', *Environment*, 27(10), pp. 6–12, 30–33.
Ehrlich, P. R. and Mooney, H. A. (1983), 'Extinction, substitution, and ecosystem services', *BioScience*, 33(4), pp. 248–54.
Elisabetsky, E. (1987), 'Pesquisas em plantas medicinais', *Ciência e Cultura*, 39(8), pp. 697–702.
Environmental Policy Institute (1987), 'Potential environmental disasters in Latin America: A set of projects the Inter-American Development Bank and the World Bank should not fund', Washington D.C.: Environmental Policy Institute.
Falesi, I. C. (1974), 'O solo na Amazônia e sua relação com a definição de sistemas de produção agrícola', in Empresa Brasileira de Pesquisas Agropecuárias (EMBRAPA), *Reunião do Grupo Interdisciplinar de Trabalho sobre Diretrizes de Pesquisa Agrícola para a Amazônia (Trópico Úmido), Brasília, Maio 6–10, 1974*, vol. 1, Brasília: EMBRAPA. pp. 2.1–2.11.
Fearnside, P. M. (1979a), 'Cattle yield prediction for the Transamazon Highway of Brazil', *Interciencia*, 4(4), pp. 220–25.
Fearnside, P. M. (1979b), 'The development of the Amazon rain forest: Priority problems for the formulation of guidelines', *Interciencia*, 4(6), pp. 338–43.
Fearnside, P. M. (1980a), 'The prediction of soil erosion losses under various land uses in the Transamazon Highway Colonization Area of Brazil', pp. 1287–95, in J. I. Furtado (ed.), *Tropical Ecology and Development: Proceedings of the 5th International Symposium of Tropical Ecology, 16–21 April 1979, Kuala Lumpur, Malaysia*, Kuala Lumpur: International Society for Tropical Ecology–ISTE.
Fearnside, P. M. (1980b), 'The effects of cattle pastures on soil fertility in the

Brazilian Amazon: consequences for beef production sustainability', *Tropical Ecology*, 21(1), pp. 125–37.

Fearnside, P. M. (1980c), 'Black pepper yield prediction for the Transamazon Highway of Brazil', *Turrialba*, 30(1), pp. 35–42.

Fearnside, P. M. (1982), 'Deforestation in the Brazilian Amazon: How fast is it occurring?', *Interciencia*, 7(2), pp. 82–8.

Fearnside, P. M. (1983a), 'Land use trends in the Brazilian Amazon Region as factors in accelerating deforestation', *Environmental Conservation*, 10(2), pp. 141–8.

Fearnside, P. M. (1983b), 'Development Alternatives in the Brazilian Amazon: An Ecological Evaluation', *Interciencia*, 8(2), pp. 65–78.

Fearnside, P. M. (1984a), 'A floresta vai acabar?', *Ciência Hoje*, 2(10), pp. 42–52.

Fearnside, P. M. (1984b), 'Simulation of meteorological parameters for estimating human carrying capacity in Brazil's Transamazon Highway colonization area', *Tropical Ecology*, 25(1), pp. 134–42.

Fearnside, P. M. (1984c), 'Brazil's Amazon settlement schemes: conflicting objectives and human carrying capacity', *Habitat International*, 8(1), pp. 45–61.

Fearnside, P. M. (1985a), 'Deforestation and decision-making in the development of Brazilian Amazonia', *Interciencia*, 10(5), pp. 243–7.

Fearnside, P. M. (1985b), 'Environmental Change and Deforestation in the Brazilian Amazon', in J. Hemming (ed.), *Change in the Amazon Basin: Man's Impact on Forests and Rivers*. Manchester: Manchester University Press,, pp. 70–89.

Fearnside, P. M. (1985c), 'Brazil's Amazon forest and the global carbon problem', *Interciencia*, 10(4), pp. 179–86.

Fearnside, P. M. (1985d), 'Agriculture in Amazonia', in G. T. Prance and T. E. Lovejoy (eds), *Key Environments: Amazonia*, Oxford: Pergamon Press, pp. 393–418.

Fearnside, P. M. (1986a), 'Spatial concentration of deforestation in the Brazilian Amazon', *Ambio*, 15(2), pp. 72–9.

Fearnside, P. M. (1986b), *Human Carrying Capacity of the Brazilian Rainforest*, New York: Columbia University Press.

Fearnside, P. M. (1986c), 'Brazil's Amazon forest and the global carbon problem: Reply to Lugo and Brown', *Interciencia*, 11(2), pp. 58–64.

Fearnside, P. M. (1986d), 'Os planos agrícolas: Desenvolvimento para quem e por quanto tempo?, in J. M. G. de Almeida (ed.), *Carajás: Desafio Político, Ecologia e Desenvolvimento*, São Paulo: Editora Brasiliense, pp. 362–418.

Fearnside, P. M. (1986e), 'Agricultural plans for Brazil's Grande Carajás program: Lost opportunity for sustainable development?', *World Development*, 14(3), pp. 385–409.

Fearnside, P. M. (1986f), 'Settlement in Rondônia and the token role of science and technology in Brazil's Amazonian development planning', *Interciencia*, 11(5), pp. 229–236.

Fearnside, P. M. (1987a), 'Causes of deforestation in the Brazilian Amazon', in R. F. Dickinson (ed.), *The Geophysiology of Amazonia: Vegetation and Climate Interactions*, New York: John Wiley, pp. 37–53.

Fearnside, P. M. (1987b), 'Summary of progress in quantifying the potential contribution of Amazonian deforestation to the global carbon problem', in D. Athié, T. E. Lovejoy and P. de M. Oyens (eds), *Proceedings of the Workshop on Biogeochemistry of Tropical Rain Forests: Problems for Research*, São Paulo: Universidade de São Paulo, Centro de Energia Nuclear na Agricultura (CENA), Piracicaba, pp. 75–82.

Fearnside, P. M. (1987c), 'Rethinking continuous cultivation in Amazonia', *BioScience*, 37(3), pp. 209–14.

Fearnside, P. M. (1987d), 'Deforestation and international economic development projects in Brazilian Amazonia', *Conservation Biology*, 1(3) (in press).
Fearnside, P. M. (1988), 'Jarí at age 19: Lessons for Brazil's silvicultural plans at Carajás', *Interciencia*, 13(1), pp. 12–24.
Fearnside, P. M. (n.d. a), 'Forest management in Amazonia: The need for new criteria in evaluating development options', *Forest Ecology and Management*, (in press).
Fearnside, P. M. (n.d. b), 'Practical targets for sustainable development in Amazonia', in J. Burnett and N. Polunin (eds), *Proceedings of the International Conference on the Environmental Future: Maintenance of the Biosphere*, Edinburgh: Edinburgh University Press (in press).
Fearnside, P. M. (n.d. c), 'Deforestation and agricultural development in Brazilian Amazonia', in P. R. Leopoldo (ed.), *Amazonia: Deforestation and Possible Effects*, The Hague: Elsevier (forthcoming).
Fearnside, P. M. (n.d. d), 'Brazil's Balbina Dam: Environment versus the legacy of the pharaohs in Amazonia', paper presented at the 40[a] Reunião Anual da Sociedade Brasileira para o Progresso da Ciência (SBPC), 10–16 July 1988, São Paulo-SP (forthcoming).
Fearnside, P. M. and Ferreira, G. de L. (1984), 'Roads in Rondônia: Highway construction and the farce of unprotected reserves in Brazil's Amazonian forest', *Environmental Conservation*, 11(4), pp. 358–60.
Fearnside, P. M. and Rankin, J. M. (1980), 'Jarí and development in the Brazilian Amazon', *Interciencia*, 5(3), pp. 146–56.
Fearnside, P. M. and Rankin, J. M. (1982a), 'The New Jarí: Risks and Prospects of a Major Amazonian Development', *Interciencia*, 7(6), pp. 329–39.
Fearnside, P. M. and Rankin, J. M. (1982b), 'Jarí and Carajás: The uncertain future of large silvicultural plantations in the Amazon', *Interciencia*, 7(6), pp. 326–8.
Fearnside, P. M. and Rankin, J. M. (1985), 'Jarí revisited: Changes and the outlook for sustainability in Amazonia's largest silvicultural estate', *Interciencia*, 10(3), pp. 121–9.
Fearnside, P. M. and Salati, E. (1985), 'Explosive deforestation in Rondônia, Brazil', *Environmental Conservation*, 12(4), pp. 355–6.
Gall, N. (1980), 'Why is inflation so virulent?', *Forbes*, 13 (October 1980), pp. 67–71.
Garzon, C. E. (1984), *Water Quality in Hydroelectric Projects: Considerations for Planning in Tropical Forest Regions*, World Bank technical paper no. 20, Washington, D.C.: World Bank.
Goldemberg, J. (1984), 'O gás de Juruá, uma solução para a região de Manaus', *São Paulo Energia*, 2(17), p. 2.
Gómez-Pompa, A., Vásquez-Yanes, C. and Gueriara, S. (1972), 'The tropical rain forest: a non-renewable resource', *Science*, 177, pp. 762–5.
Goodland, R. J. A. (1978), *Environmental Assessment of the Tucuruí Hydroproject, Rio Tocantins, Amazonia, Brazil*, Brasília, Centrais Elétricas do Norte do Brasil SA (ELETRONORTE).
Goreau, T. J. and Mello, W. Z. (1987), 'Effects of deforestation on sources and sinks of atmospheric carbon dioxide, nitrous oxide, and methane from central Amazonian soils and biota during the dry season: A preliminary study', in D. Athie, T. E. Lovejoy and P. de M. Oyens (eds), *Proceedings of the Workshop on Biogeochemistry of Tropical Rainforests: Problems for Research*, Universidade de São Paulo, Centro de Energia Nuclear na Agricultura (CENA), São Paulo: Piracicaba, pp. 51–66.
Greenland, D. J. and Nye, P. H. (1959), 'Increases in the carbon and nitrogen contents of tropical soil under natural fallows', *Journal of Soil Science*, 10(2), pp. 285–99.

Hall, A. (1987), 'Agrarian crisis in Brazilian Amazonia: The Grande Carajás Programme', *The Journal of Development Studies*, 23(4), pp. 522–52.
Hecht, S. B. (1981), 'Deforestation in the Amazon Basin: Practice, theory and soil resource effects', *Studies in Third World Societies*, 13, pp. 61–108.
Hecht, S. B. (1983), 'Cattle ranching in the eastern Amazon: environmental and social implications', in E. F. Moran (ed.), *The Dilemma of Amazonian Development*, Boulder, Col.: Westview Press, pp. 155–88.
Hecht, S. B. (1985), 'Environment, development and politics: Capital accumulation and the livestock sector in eastern Amazonia', *World Development*, 13(6), pp. 663–84.
Iltis, H. H. (1983), 'Tropical forests: What will be their fate?', *Environment*, 25(10), pp. 55–60.
Iltis, H. H., Doebley, J. F., Guzmán, R. and Pazy, B. (1979), '*Zea diploperennis* (Graminae): a new teosinte from Mexico', *Science*, 203, pp. 186–7.
International Bank for Reconstruction and Development (1981), *Brazil: Integrated Development of the Northwest Frontier*, Washington, D.C.: The World Bank, Latin American and Caribbean Regional Office.
Isto É (15 July 1987), 'Fraude Fiscal: Orgia Amazônica. Incentivos desperdiçam bilhões de cruzados', pp. 62–5.
Janzen, D. H. (1973), 'Tropical agroecosystems: Habitats misunderstood by the temperate zones, mismanaged by the tropics', *Science*, 182, pp. 1212–19.
Kleinpenning, J. M. G. (1979), *An Evaluation of the Brazilian Policy for the Integration of the Amazon Basin (1964–1975)*, Publikatie 9, Vakroep Sociale Geografie van de Ontwikkelinsgslanden, Nijmegen: Geografisch en Planologisch Instituut.
Knowles, O. H. (1988), 'Aceleração da regeneração florestal em locais degradados: A experiência da Companhia Mineração Rio do Norte (Porto Trombetas, Estado do Pará, Brasil)', paper presented at the Simpósio Internacional sobre Alternativas para o Desmatamento, 27–30 January 1988, Belém (manuscript).
Kohlhepp, G. (1980), 'Analysis of state and private regional development projects in the Brazilian Amazon Basin', *Applied Geography and Development*, 16, pp. 53–79.
Koster, H. W., Khan, E. J. A. and Bosshart, R. P. (1977), *Programa e Resultados Preliminares dos Estudos de Pastagens na Região de Paragominas, Pará, e nordeste do Mato Grosso junho 1975 – dezembro 1976*, Superintendência do Desenvolvimento da Amazônia (SUDAM), Belém: Convênio SUDAM/Instituto de Pesquisas IRI.
Leopoldo, P. R. and Salati, E. (1987), 'Rondônia: Quando a floresta vai acabar?', *Ciência Hoje*, 6(36), p. 14.
Lewin, R. (1983), 'No dinosaurs this time: calculations from ecological theory indicate that the loss of species through felling of tropical forests will reach mass extinction proportions by next century', *Science*, 221, pp. 1168–9.
O Liberal (Belém) (1 February 1988), 'Desastre ecológico com primeiros dados quase prontos', p. 17.
Lovejoy, T. E., Rankin, J. M., Bierregaard, Jr, R. O., Brown, Jr, K. S., Emmons, L. H. and Van der Voort, M. E. (1984), 'Ecosystem decay of Amazon forest remnants', in M. H. Nitecki (ed.), *Extinctions*, Chicago: University of Chicago Press, pp. 295–325.
Mahar, D. J. (1976), 'Fiscal incentives for regional development: A case study of the western Amazon Basin', *Journal of Interamerican Studies and World Affairs*, 18(3), pp. 357–78.
Mahar, D. J. (1979), *Frontier Development Policy in Brazil: A Study of Amazonia*, New York: Praeger.

Malingreau, J.-P. and Tucker, C. J. (1988), Large-scale deforestation in the southeastern Amazon Basin of Brazil, *Ambio*, 17(1), pp. 49–55.

Malingreau, J.-P. and Tucker, C. J. (1987), 'The contribution of AVHRR data for measuring and understanding global processes: Large-scale deforestation in the Amazon Basin', in *IGRASS' 87*, Ann Arbor, Michigan: IGRASS, pp. 443–8.

Martinelli, L. A., Ferreira, J. R., Forsberg, B. R. and Victoria, R. L. (n.d.), 'Mercury contamination in the Amazon: A gold rush consequence', *Ambio* (in press).

Matias, F. (1987), 'A quem interesse o Projete Calha Norte?', *Enfoque Amazônico*, 2(5), pp. 18–24.

Melchiades Filho (1987), 'Balbina: um escándalo ecológico', *Universidade de São Paulo Jornal do Campus*, 25 November 1987, 59, pp. 4–5.

Monosowski, E. (1986), 'Brazil's Tucuruí Dam: Development at environmental cost', in E. Goldsmith and N. Hildyard (eds), *The Social and Environmental Effects of Large Dams, Vol. 2: Case Studies*, Camelford: Wadebridge Ecological Centre, pp. 191–8.

Mooney, H. A., P. M. Vitousek and P. A. Matson (1987), 'Exchange of materials between terrestrial ecosystems and the atmosphere', *BioScience*, 238, pp. 926–32.

Moraes, V. H. F. and Bastos, T. X. (1972), 'Viabilidade e limitações climáticas para as culturas permanentes, semi permanentes e anuais, com possibilidades de expansão na Amazônia', in *Zoneamento Agrícola da Amazônia (1ª Aproximação)*, Bólétim Técnico do Instituto de Pesquisa Agropecuária do Norte (IPEAN) no. 54, Belém: IPEAN, pp. 123–53.

Mougeot, L. J. A. (1987), 'O reservatório da Usina Hidrelétrica de Tucuruí, Pará, Brasil: Uma avaliação do programa de reassentamento populacional (1976–85)', in G. Kohlhepp and A. Schrader (eds), *Homem e Natureza na Amazônia*, Tübinger Geographische Studien Heft 95/Tübinger Beiträge zur Geographischen Lateinamerika – Forschung 3, Geographisches Institut, Tübingen: Universität Tübingen, pp. 387–404.

Mueller, C. (1980), 'Frontier based agricultural expansion: The case of Rondônia', in F. Barbira-Scazzocchio (ed.), *Land, People and Planning in Contemporary Amazonia*, University of Cambridge: Centre of Latin American Studies, Occasional Publications, 3, pp. 141–53.

Myers, N. (1976), 'An expanded approach to the problem of disappearing species', *Science*, 193, pp. 198–202.

Myers, N. (1979), *The Sinking Ark: a New Look at the Problem of Disappearing Species*, New York: Pergamon.

Myers, N. (1981), 'The hamburger connection: How Central America's forests become North America's hamburgers', *Ambio*, 10(1), pp. 3–8.

Myers, N. (1984), *The Primary Source: Tropical Forests and our Future*, New York: Norton.

Nations, J. D. and Komer, D. I. (1983), 'Rainforests and the hamburger society', *Environment*, 25(3), pp. 12–20.

A Notícia (Manaus) (25 June 1983), 'Governo justificou exportação de peles', Caderno 1, p. 5.

Oldfield, M. L. (1981), 'Tropical deforestation and genetic resources conservation', *Studies in Third World Societies*, 14, pp. 277–345.

Peters, C. (n.d.), 'Population ecology and management of forest fruits in the Peruvian Amazon', in A. Anderson (ed.), *Alternatives to Deforestation: Steps Toward Sustainable Uses of Amazonian Forests*, New York: Columbia University Press (in press).

Pires, F. B. and Vaccari, F. (1986), 'Alta-tensão por um fio', *Ciência Hoje*, 4(23), pp. 49–53.

Posey, D. A. (1983), 'Indigenous ecological knowledge and development of the Amazon', in E. F. Moran (ed.), *The Dilemma of Amazonian Development*, Boulder, Col.: Westview Press, pp. 225–57.

Prance, G. T. (1978), 'The origin and evolution of the Amazon flora', *Interciencia*, 3(4), pp. 207–22.

Prance, G. T., Rodrigues, W. A. and da Silva, M. F. (1976), 'Inventário florestal de um hectare de mata de terra firme km 30 da Estrada Manaus–Itacoatiara', *Acta Amazônica*, 6(1) pp. 9–35.

A Província do Pará (Belém) (15 June 1982), 'Eletronorte não fará desmatamento: Tucuruí', Caderno, 1, p. 9.

Ramanathan, V., Cicerone, R. J., Singh, H. B. and Kiehl, J. T. (1985), 'Trace gas trends and their potential role in climate change', *Journal of Geophysical Research*, 90(D3), pp. 5547–66.

Rasmussen, R. A. and Khalil, M. A. K. (1983), 'Global production of methane by termites', *Nature*, 301, pp. 700–2.

Routley, R. and Routley, V. (1977), 'Destructive forestry in Australia and Melanesia', in J. H. Winslow (ed.), *The Melanesian Environment*, Canberra: Australian National University, pp. 374–97.

SBPC (Sociedade Brasileira para o Progresso da Ciência) (1986), 'A mineração ameaça terras indígenas', *Ciência Hoje*, 4(24); p. 86.

Salati, E., Dall'Olio, A., Matusi, E. and Gat, J. R. (1979), 'Recycling of water in the Brazilian Amazon Basin: An isotopic study', *Water Resources Research*, 15, pp. 1250–8.

Salati, E., Marques, J. and Molion, L. C. B. (1978), 'Origem e distribuição das chuvas na Amazônia', *Interciencia*, 3(4), pp. 200–6.

Salati, E. and Vose, P. B. (1984), 'Amazon Basin: A system in equilibrium', *Science*, 225, pp. 129–38.

Sanders, T. G. (1973), 'Development and environment: Brazil and the Stockholm Conference', *East Coast South America Series (American Universities Field Staff)*, 17(7) pp. 1–9.

SBPC, (1986), Ao leitor, *Bóletim Informativo da Sociedade Brasileira para o Progresso da Ciência (SBPC)*, 60 p. 1.

Schubart, H. O. R., Junk, W. J. and Petrere, Jr, M. (1976), 'Sumário de ecologia Amazônica', *Ciência e Cultura*, 28(5), pp. 507–9.

Scott, G. A. J. (1978), *Grassland Development in the Gran Pajonal of Eastern Peru: a Study of Soil–Vegetation Nutrient Systems*, Hawaii Monographs in Geography, no. 1, Honolulu: University of Hawaii at Manoa, Department of Geography.

Serrão, E. A. S. and Falesi, I. C. (1977), *Pastagens do Trópico Úmido Brasileiro*, Belém: Empresa Brasileira de Pesquisa Agropecuária – Centro de Pesquisa Agropecuária do Trópico Úmido (EMBRAPA-CPATU).

Serrão, E. A. S., Falesi, I. C., Viega, J. B. and Teixeira Neto, J. F. (1979), 'Productivity of cultivated pastures on low fertility soils in the Amazon of Brazil', in P. A. Sánchez and L. E. Tergas (eds), *Pasture Production in Acid Soils of the Tropics: Proceedings of a Seminar held at CIAT, Cali, Colombia 17–21 April, 1978*, Cali. Colombia: CIAT series 03 EG-05 Centro Internacional de Agricultura Tropical (CIAT), pp. 195–225.

Setubal, O. E., Couto, R. C., Sayad, J. and Denys, R. B. (1986), Desenvolvimento e segurança na região ao norte das calhas dos rios Solimões e Amazonas – Projeto Calha Norte, Exposition of Motives (no. 770) to President José Sarney.

Survival International (1987a), 'Brazil today: Dams threaten Xingú', *Survival International News*, 17, pp. 1–2.

Survival International (1987b), 'Military take over Indian lands?, *Survival International News*, 17, p. 2.

Sylvester-Bradley, R., de Oliveira, L. A., de Podestá Filho, J. A. and St John, T. V. (1980), 'Nodulation of legumes, nitrogenase activity of roots and occurrence of nitrogen-fixing *Azospirillum* spp. in representative soils of central Amazonia', *Agro-Ecosystems* 6, pp. 249–66.

Talbot, R. W., Harriss, R. C., Browell, E. V., Gregory, G. L., Sebacher, D. I. and S. M. Beck (1986), 'Distribution and geochemistry of aerosols in the tropical North Atlantic troposphere: Relationship to Saharan dust', *Journal of Geophysical Research*, 86, pp. 5163–71.

Tardin, A. T., dos Santos, A. P., Lee, D. C. L., de Moraes Novo, E. M. L. and Toledo, F. L. (1978), 'Projetos agropecuários da Amazônia: Desmatamento e fiscalização – relatório', *Amazônia Brasileira em Foco*, 12, pp. 7–45.

Tardin, A. T., Lee, D. C. L., Santos, R. J. R., de Assis, O. R., dos Santos Barbosa, M. P., de Lourdes Moreira, M., Pereira, M. T., Silva, D. and dos Santos Filho C. P. (1980), *Subprojeto Desmatamento, Convênio IBDF/CNPq-INPE 1979*, Instituto Nacional de Pesquisas Espaciais (INPE) Relatório no. INPE-1649-RPE/103, São José dos Campos, São Paulo: INPE.

Uhl, C. (1988), 'Barreiras ecológicas à regeneração florestal em pastagens altamente degradadas (Município de Paragominas, Estado do Pará, Brasil)', paper presented at the 'Simpósio Internacional sobre Alternativas para o Desmatamento', 27–30 January 1988, Belém, Pará (manuscript).

Valverde, O., Japiassu, A. M. S., Lopes, A. M. T., Neves, A. M., Egler, F. G., Mesquita, H. M., da Costa, I. B., Garrido Filha, I., de Bulhões, M. G., Mesquita, M. G. G. C. and Ferreira, N. A. (1979), *A Organização do Espaço na Faixa da Transamazônica, Vol. I: Introdução, Sudoeste amazônico, Rondônia Regiões Vizinhas*, Rio de Janeiro: Instituto Brasileiro de Geografia e Estatística (IBGE).

Veja (28 November 1984), 'Os Igarapés de ouro: Uma estrada rasga, no sul do Pará, a reserva de ouro que vai desbancar Serra Pelada', pp. 28–9.

Veja (20 May 1987), 'Os canteiros de obras mais caros do país', p. 30.

Villa Nova, N. A., Salati, E. and Matusi, E. (1976), 'Estimativa da evapotranspiração na Bacia Amazônica', *Acta Amazônica*, 6(2), pp. 215–28.

Webster, C. C. and Wilson, P. N. (1980), *Agriculture in the Tropics*, London: Longman, 2nd edn.

Weiss, R. F. (1981), 'The temporal and spatial distribution of tropospheric nitrous oxide', *Journal of Geophysical Research*, 86, pp. 7185–95.

Wolf, E. C. (1987), 'On the brink of extinction: conserving the diversity of life', *Worldwatch Paper*, 78, Washington, D.C.: Worldwatch Institute.

Zimmerman, P. R., Greenberg, J. P. and Darlington, J. P. E. C. (1984), 'Termites and atmospheric gas production', *Science*, 224, p. 86.

Zimmerman, P. R., Greenberg, J. P., Wandiga, S. O. and Crutzen, P. J. (1982), 'Termites as a source of atmospheric methane, carbon dioxide and molecular hydrogen', *Science*, 218, pp. 563–5.

9 The State and Land Conflicts in Amazonia, 1964–88

Alfredo Wagner Berno de Almeida

I INTRODUCTION

The lag between the intensification of land conflicts and the irregular and uneven character of State intervention has been a notable feature of agrarian structures in Amazonia during the past twenty years. The official representation of these struggles has been decidedly technocratic, as if land conflicts and violence were somehow intrinsic to agricultural modernisation and the development of the productive forces in an agricultural frontier region. Rising social tensions are thus interpreted in a natural, matter of fact way, which implicitly endorses the increasing concentration of land ownership under the dictates of brute force and coercion. Despite public statements of moral outrage, the use of violence to subjugate different segments of the peasantry, known regionally as *posseiros* and *peões*, as well as diverse indigenous groups is presented within this convoluted logic as a 'necessary fact', specific to the economic processes and political structures of frontier regions. Violence thus has been a constant element both in periods of outright dictatorship (1964–85) and those of 'democratic transition' (1985–8). In some respects, this concentrationist tendency reproduces on the frontier the cultural patterns intrinsic to the formation of *latifúndia*, as found in areas of early settlement. The subordination of peasants by coercion and various forms of banditry and *pistolagem* has historical parallel with the consolidation of this form of large landed property. That is, properties whose access to the means of production is based on the destruction of pre-existing tenure systems and mechanisms to immobilise labour, such as debt peonage, which represent extreme forms of repression of the labour force (Esterci, 1987).

Such repressive forms may constitute decisive factors in the development of capitalism on the frontier, as Barrington Moore Jr (1966) suggests. In the case of Amazonia, the presence of these immobilising mechanisms is articulated with the general activities of the State which, besides imposing rigid measures to control formal access to land by Indians and settlers (*posseiros*), provides credit subsidies, fiscal incentives and extensive land grants to corporate groups on the pretext of economic rationality and more advanced technology. However, these technologies clearly have a conservative role since they cannot be disassociated from the monopoly of land, the repressive employment practices and the use of coercion to resolve land conflicts. The generalised violence on the frontier is not, therefore, contingent but rather constitutes a structural 'given' essential to this type of capitalist development (Velho, 1976). State action in Amazonia since 1964 thus can be analysed in terms of 'authoritarian modernisation' and a conservative process of innovation.

The administrative norms and operational practices of State intervention, although informed by the official representation of the frontier settlement process, have not followed a consistent and regular pattern. Between 1964 when the Land Statute (Law 4,504 of 30 November 1964) was formulated and 1988, these variations have been determined by the level of peasant mobilisation and the widely recognised difficulty experienced by public land tenure angencies and landowner interests in assimilating the pressures and demands generated by this mobilisation. Above all, however, shifts in the norms and practices of State intervention have been caused by the oscillations of the prolonged 'democratic transition', and the introduction, following the formal end of the military regime on 15 March 1985, of the National Land Reform Programme (Decree 91,766 of 10 October 1985) which, once the initial reformist impetus had passed, soon became innocuous.

The emergence of rival social movements outside the traditional framework of clientelist control, which since 1973 have demanded the execution of 'an ample and massive land reform', has to some extent broken the rules of the game under which such domination was considered to be 'natural'. Recognising the increasing intensity of peasant mobilisation and land conflicts in Amazonia, the National Confederation of Agricultural Workers (CONTAG) began, even during the Medici regime, to distinguish between 'land reform' and 'colonisation' or settlement. CONTAG criticised the transfer and compulsory removal of peasants to 'distant areas' and instead demanded a land reform, which would allow them to stay in the areas where they normally live and work[1] (CONTAG, 1973, p. 132). In May 1974, CONTAG presented a document to President Geisel, which demanded 'an immediate and ample land reform' with the direct participation of those involved (CONTAG, 1974). Concomitantly, CONTAG sought to increase rural trade union organisation in Amazonia.

In similar fashion, the Churches of *Amazonia Legal* meeting in Goiania in June 1975 and worried by the spread of land conflicts decided, with the support of the National Conference of Brazilian Bishops (CNBB), to create a 'Land Commission', which was intended to 'link, advise and stimulate' support activities for rural social movements. The Pastoral Land Commission (CPT) was thus established to operate in this vast region, where union structures were quite weak and poorly equipped, to work with the most important segment of the frontier peasantry, the *posseiros* (CPT 1983). Transcending the usual measures of social control, peasant mobilisations demanded changes in the practices followed by landowners and 'modern ranchers', as well as the revision of bureaucratic, administrative procedures in official land tenure institutions.

Irrespective of the political conjuncture, the traditional response of the large landowner (*latifundiário*) to the emergence of rural problems invariably has been to use violence. This is the case whether we are speaking of rubber plantation owners in Acre and Amazonas, Brazil-nut growers in southern Pará, cattle ranchers of Lower Amazonas, the island of Marajó and Lower Maranhão (*Baixada Maranhense*), lumber firms, mining interests or business groups from the Centre–South, responsible for hundreds of investment projects benefitting from SUDAM fiscal incentives. These criticise the slowness of judicial litigation in land disputes and, seeking what they euphemistically call an 'immediate solution', choose to use force. The speedy resolution they expect to achieve assumes that force will be 'effective', even though it breaks the law. However, over the past five years, this response has gone beyond limits considered 'tolerable', victimising mainly rural workers, indians, union officials and priests. Faced with the impossibility of increasing the level of violence further and having permanently to apply extreme measures to maintain their domination, the *latifúndio* interests have been forced partly to revise their forms of organisation and strategies. In attempting to impose their prerogatives, landowners have mounted ambushes against union leaders, burnt down houses, destroyed crops, massacred indians and peasants, and turned to guns and violence to resolve any rural problem whatsoever. However, by going to these extremes, *latifundiários* have lost control of the situation, weakening the structures of power and domination.

This extreme situation is further complicated by tensions within the dominant groups created by the emergence of industrial and financial interests, above all in the region of the Grande Carajás Programme. Influenced by the so-called 'modern political ideologies', timid and cautious, these interests reject the use of force to achieve economic development and refuse to regard explicit violence as one of its essential elements. They are opposed to fraudulent practices to obtain land, support the demarcation of indian reserves, reject illegal employment practices, such as debt peonage, and recognise the ecological value of the forest. They are also in favour of a more orderly land market (Almeida, 1985a). Although these tensions should not be exaggerated, they undermine the principle that oppression is 'natural and necessary'.

The broader purpose, however, is to re-cast 'traditional' forms of domination in modern guise in order to neutralise land reform legislation that guarantees usufruct rights of occupancy (*posse*) and permits expropriation in the social interest. In other words, the tensions noted above imply that the antagonisms generated by land disputes are transferred to the courts, and especially legal actions for the restoration of occupancy. These increased rapidly in the late 1970s and even today cause the expulsion of peasants in significant numbers. Landowner interests also have been active in lobbying Congress, notably after 1986 with the election of the Constituent Assembly. The aim is to achieve a degree of legalisation of what earlier had been considered 'natural' and which does not depend on legitimacy for its enforcement. The tension between these efforts to express their interests in a legal form and recourse to violence to resolve land disputes reveals the intrinsic difficulties of the differentiated strategies pursued by the dominant groups.

In the same way tensions permeate the official land tenure agencies, leading to successive changes in their policies. The common-sense explanation of these changes is 'administrative discontinuity', and certainly the existence of many agencies has been as ephemeral as the policies they advocated. This applies not only to national agencies, such as the Brazilian Land Reform Institute (IBRA, 1966–70) the National Institute of Colonisation and Land Reform (INCRA, 1970–87) and the Extraordinary Ministry for Land Affairs (MEAF, 1983–85), but also to those whose activities were confined to Amazonia, such as the Special Coordination for the Araguaia–Tocantins (1976–79), the Executive Group for Araguaia–Tocantins, (GETAT, 1980–87), the Executive Group for Lower Amazonia (GEBAM, 1980–86), and the Special Coordination for Acre (1980–85). Apart from operational ineptitude, the relations between these agencies and different social groups on the frontier have been re-defined over the past two decades. One example involves local authorities and their operational capacity. With the exception of IDAGO, which dates from the mid-1960s, state land tenure agencies in Amazonia were revived after 1978 as part of a policy to strengthen local and regional power structures in order to confront the peasant mobilisations and the activities of the Church. The redefinition of these relations, although directed by central government, coincides with changes in the priority of the different instruments of land tenure policy. Thus priority has shifted between official settlement programmes, recognition of 'spontaneous settlement', private settlement projects in selected areas, as in the north of Mato Grosso, redistribution of land without observing the minimum size requirements (*módulos*) established by law, as in the case of GETAT between 1980 and March 1985, and the demarcation of holdings without reference to settlement policy. These shifts have common origins, including an effective veto on the use of expropriation as a policy instrument, difficulties in diagnosing the conflicts, resulting in attempts at 'crisis management', and the pre-eminence of military and security personnel in the operational direction of land tenure agencies. These constant elements emphasise the authoritarian, not to say colonialist, character of the government projects underway.

II LAND CONFLICT AS A POLICY ISSUE

However, it is important to stress that land conflicts in the Amazon region began to be recognised as a policy issue from the mid-1970s. At that time, despite vigorous repressive action, the growing number and spread of land conflicts impeded the implementation of agricultural, timber and mining projects, which threatened the existing system of settlement. *Posseiros* struggled to hold onto their usufruct rights (*posses*), notably in Maranhão, Pará and Goiás. According to IBGE Agricultural Census data, these states accounted for one-third of the national total of 898 thousand *posseiros* in 1980 (Almeida, 1986). Before this period, the land tenure agencies conceived these disputes merely as typical expressions of what the military called 'rural unrest', which should be stamped out 'energetically', as it was among the peasants located in the regions affected by the *Guerrilha do Araguaia* between 1971 and 1974. These agencies believed that land conflicts originated in the North-East and other areas of early settlement and that the solution was to transfer 'surplus population' to official settlement projects in Amazonia. Conflicts were explained essentially by demographic pressure and climatic factors ('North-Eastern drought') whose effects, according to this technocratic thinking could be reduced, if not completely eliminated, by the existence of unused resources on the agricultural frontier. In this view, rural conflicts necessarily were problems confined to the region of out-migration. 'Internal migration' was the order of the day for State action in this period and many studies provided guidelines on how to direct and manage migratory flows to Amazonia.

Within this framework, INCRA was established in 1970 and its activities concentrated primarily on 'directed colonisation' (*colonização dirigida*). However, the Integrated Colonisation Projects (PICs), with their *agrovilas*, *agrópolis* and *rurópolis*, proved to be inadequate and unsuccessful. The phenomenon of 'spontaneous settlement' far exceeded official planning estimates and could not be ignored. INCRA therefore began to implement Directed Settlement Projects (PADs), consisting of officially demarcated holdings but where settlers received assistance from other public agencies via service contracts with INCRA. In this way, INCRA began to reduce its field of activities and operating costs. It no longer provided transportation for peasants from the South to Amazonia and it slowed down its 'directed colonisation' programme.

Various colonisation projects were located on indian lands on the assumption that these were 'empty spaces'. Such lands were considered to be in the public domain, which generated a huge free-for-all. Settlers and *posseiros* were thrown into conflict with tribal peoples in Acre (Yaminawa, Machineri), Pará (Tembé, Parakanã), Maranhão (Guajá, Urubú, Guajajara), Goiás (Xerente), Rondônia and Roraima. An unsuccessful attempt to resolve this worsening situation was made in June 1976, when the President of INCRA established a Working Group to collaborate with FUNAI in studying ways of resettling those who were illegally occupying indian lands (INCRA, 1978).

The land question continued to be formulated in demographic terms as before but a change of emphasis occurred in the policy instruments used. In an address to the Superior War College in August 1977, the director of INCRA's Department of Projects and Operations, Hélio Palma de Arruda, observed that 'the land problem in Amazonia is characterised generally by the need to demarcate public lands [*terras devolutas*] so that the large human contingents in Brazil who need good, cheap land can be taken there' (Arruda, 1977, p. 15). At that time, land demarcation was still presented in conjunction with colonisation. However, the land tenure agencies could no longer ignore the fact that the number of zones of tension and conflict was multiplying within Amazonia itself, and they were forced once again to change their policy. In contrast to earlier formulations, demographic pressure within Amazonia now was identified as the cause of conflicts, and increasingly it was recognised that these could occur in 'areas of arrival'. This diagnosis led land tenure agencies once more to scale down colonisation activities and give priority to demarcation in areas unrelated to official settlement projects.[2] This change was explained in June 1980 by the director of INCRA's Department of Land Resources, Adair Zanatta, as follows: 'In view of the juridical and land tenure situation, population density, and conflicts over ownership and occupancy, priority was given to the demarcation of land in the most critical areas, principally those in Mato Grosso, southern Amazonas, Rondônia, and the south of Pará (Zanatta, 1980, p. 13).

As a result of this 'priority', the activities of INCRA, GETAT, state tenure agencies and *Projeto Nordeste* led to the demarcation of 139.9 million hectares by the end of 1984. Some 83 per cent of this area was in Acre, Amazonas, Pará, Rondônia, and the Territories of Roraima and Amapá, and this proportion rises to 98 per cent of the total if the states of Mato Grosso and Goiás are included (Ribeiro, 1985). This demarcation policy, although maintained, soon proved inadequate to remove the so-called 'distortions' in the land tenure structure and facilitate free access to land. Although it settled some ownership questions, it failed to offer a solution to conflicts involving the maintenance of occupancy rights (*posse*) or the recovery of these rights by rural workers who had been thrown off their holdings. Furthermore the slowness of demarcation work, which still might be inconclusive after a decade, served only to aggravate conflict and tensions. Yet official policy towards the conflicts remained tied to immediate circumstances and episodic events. It was unrelated to any proposal for basic changes in agrarian structures, such as those demanded by CONTAG at its Third National Congress held in May 1979 (CONTAG, 1981).

III THE 'MANAGEMENT' OF AMAZONIAN LAND CONFLICTS

In the period of authoritarian rule immediately before the advent of the so-called New Republic, the scope of land tenure policy was significantly enlarged by the introduction of mechanisms designed to 'manage' the land conflict in Amazonia. In February 1980, the Figueiredo government launched a series of measures in order to 'expedite pending land tenure questions'.

These included the creation of GETAT on 1 February 1980, GEBAM on 28 February 1980, and the Special Coordination for Acre. Activity to 'regularise' land tenure thus was concentrated in those regions of Amazonia where the expansion of spontaneous peasant settlement had gone beyond official expectations. That is, in regions where the existing systems of occupancy, as in the case of the rubber-tappers, Brazil-nut gatherers and *posseiros* in the *babaçu* groves, was in direct opposition to the pattern of land settlement preferred by the government. These spontaneous flows and the constant conflicts in areas where occupancy was well established clashed with the ideal of 'rational settlement' upheld by the State. Whereas land according to the logic of peasant agriculture is incorporated into the productive process via family labour, the relationship of business groups to the land is juridical and commercial in conception. In defending this view, official agencies began to speak of the 'invasion' of public and private lands' in Amazonia. (CSN–GETAT, 1981, p. 5) *Posseiros* and squatters thus were cloaked in the illegitimacy attributed to 'trespassers'. As the document entitled 'Studies of Projeto Carajás' prepared by the General Secretariat of the National Security Council and GETAT warned, 'the disorderly invasion of lands along the access roads to the Serra dos Carajás and its zone of influence has already started and is intensifying uncontrollably. If this situation is not quickly checked, development projects in the region may be irredeemably compromised' (CSN–GETAT, 1981, p. 5).

As this statement reveals, the national security criteria invoked to deal with agrarian problems and support the creation of GETAT and GEBAM clearly had an economic dimension. After all, these measures were used to extend a protective cordon around private settlement companies: between 1968 and 1984 INCRA approved 71 private colonisation projects, with 66 in Mato Grosso, 3 in Maranhão and 2 in Pará. The hundreds of agricultural projects benefitting from FINAM investment incentives, as well as lumber and mineral projects, were similarly protected.[3] It is worth emphasising that the volume of fiscal incentives granted since 1966 has contributed decisively to the high concentration of land ownership found in the North. According to 1985 INCRA cadastral survey data, this region had 69,987 *latifúndio* properties with 98.9 million hectares, which represented 79.7 per cent and 16.6 per cent of the total area in the Northern region and Brazil, respectively.

The summary demarcation and occupancy decisions taken by GETAT and INCRA constituted, in fact, instruments of land tenure policy whose aim was to guarantee a certain type of capitalist development. The reproduction of this model could be maintained only by counteracting the settlement of *posseiros*, the demarcation of indian lands, and the movements of rubber-tappers, Brazil nut-gatherers and gold prospectors, which have struggled against the implementation of large-scale agricultural projects and undertakings of timber and mining companies (Almeida, 1980, 1985a).

In this sense, by attempting to establish a legal basis for landownership by social groups whose relation to the land is commercial, 'tenure regularisation' measures increasingly clashed with the real interests of the peasant movements and indians. The conflicts which ensued began to take on a new dimension, representing forms of political participation devised by peasants and indians to secure recognition of their civil rights of citizenship. These movements thus established themselves as legitimate representatives and mediators (*interlocutores*) with official agencies. Through this intermediation, the conflicts gradually came to represent, though in a paradoxical way, both a form of organisation as well as a means of securing access to available land and obtaining ownership in areas where *posse* rights were already well-established. Zones where social tensions were critical but hitherto neglected by the bureaucracy of the military regime thus began to gain recognition. Peasant mobilisations also forced land tenure organs to maintain an official register of conflicts, and in September 1981, INCRA formally instituted its survey of Areas of Social Tension. At the same time, a Survey of Problem Areas was introduced within the area covered by GETAT.

By now, the question of land conflicts had already become an important matter of concern in military and security circles. In March 1980, for example, the National Security Council (CSN) reviewed two historic documents dealing with *posse* and property titles (Castelo Branco, 1980). In July, 1980, the President of INCRA, Paulo Yokota, addressed the Chiefs of Staff on the land problem in Amazonia and the North-East, and in the following month the Governor of Bahia, Antonio Carlos Magalhães, called for a 'social pact in the countryside' (*Journal do Brasil*, 19 August 1980). Gradually, land conflicts came to be regarded as a question of national security. This new perception informed the sharp change of direction in the legal and administrative treatment of this issue which occurred under the Figueiredo regime. In the name of 'expediency', the land and indian questions were 'militarised' and policy actions subordinated to the CSN. Decree 87,457 of 16 August 1982 established the National Land Tenure Programme and Extraordinary Ministry of Land Affairs, with powers vested in the General-Secretary of the CSN, General Danilo Venturini. It can be seen that the closer the military rulers came to the date set for the 'change of regime' and 'the beginning of the democratic transition under civilian government', the more they applied authoritarian mechanisms in the countryside, and particularly in Amazonia. These new measures, together with those related to the earlier creation of GETAT and GEBAM, made it clear that democracy would not be allowed to prevail in the countryside, and especially not on the frontier.

Before the end of 1982, GETAT had undertaken no expropriations in the social interest, and this policy instrument was formally defined as merely an auxiliary mechanism of 'tenure regularisation'. Instead, so-called 'deals' and 'swaps' predominated in this period, with GETAT arranging the exchange of areas effectively occupied by *posseiros* but coveted by *latifundiários* or large projects for larger areas of public lands, generally in the river Xingú valley. *Posseiro* families were settled in this region, which was meant to function as a 'safety-valve' to relieve tensions in the areas bordering on the large mining projects of the Grande Carajás Programme.

Following peasant mobilisations and denunciations of violence, the Extra-ordinary Ministry for Land Affairs (MEAF) began to investigate the 'origins' of these conflicts and raised the possibility of expropriation in critical situations. That is, in areas with heavy concentrations of *posseiros*, where the level of violence was high and resistance to expulsion had been prolonged. The criteria adopted by the MEAF, which provided an initial characterisation of the land conflicts, were the following:

(a) doubts over the legality of titles
(b) litigation among *posseiros* on public lands
(c) rural properties that do not fulfil their social function
(d) lack of demarcation of tribal areas and of large-scale extractive properties (rubber and Brazil-nut estates)
(e) disrespect for *posse* rights established before privatisation
(f) resettlement to permit public infrastructure projects, such as dams, to proceed, and
(e) demands by *posseiros* for land in order to attain the minimum size of the rural *módulo* (Venturini, 1985, p. 31).

In its characterisation of the conflicts, MEAF made no explicit reference to the problem of violence, regarding it as the concern of other agencies. Faced with demands for justice made by different entities, such as CONTAG, CPT, CIMI and the Brazilian Law Association (OAB), repeated denunciations of the massacre of the indian population, the murder of over 800 rural workers and indians in rural conflicts in the period 1970–84 (Leme and Pietrafesa, 1985), and the general impunity of those responsible, General Venturini of MEAF alleged that, 'Some institutions, intent on showing the increase in the number of conflicts, confuse problems which are exclusively police matters (including crime) with conflicts over land tenure' (Venturini, 1985, p. 31). However, the separation of 'police' and 'tenure' matters, intended to show that MEAF confined its attention to technical issues, runs counter to the whole idea of conflicts as a national security question and hence the responsibility of the military, and on which the Minister based his own authority.

Expropriation in this period was unconnected with settlement policy and was reserved for situations where conflicts were considered to be insoluble and the source of 'grave social tensions'.[4] Yet such action was the exception. Essentially, it was restricted to those disputes where the would-be landowners had failed to 'clean up the area' and where the number of peasants ruled out normal methods of adjusting competing tenure claims. The first expropriations made by GETAT occurred in November 1982, almost three months after the creation of MEAF, and only five more followed before its abolition in February 1985, amounting to a total area of 396,694 hectares. As can be seen from Table 9.1, none of these expropriations occurred in the area of Maranhão under GETAT's jurisdiction, which was one of the most critical zones of conflict.

IV THE STRUGGLE FOR LAND: SOME EMPIRICAL EVIDENCE

At the Fourth National Congress of Rural Workers held in Brasília on 25 May 1985, the Sarney government announced the Proposal for the National Land Reform Plan (PPNRA), and this was followed by the creation of the Ministry of Land Reform and Agrarian Development (MIRAD) by Decree 91,214 of 30 April 1985. (CONTAG, 1985). For the peasant movements and

Table 9.1 GETAT: areas expropriated in the social interest, November 1982–February 1985

Area/Property	*Municipality*	*State*		*Total (ha)*
		Pará (ha)	*Goiás* (ha)	
Fundação Brasil Central	Xinguara	141,326	—	141,326
Fazenda Tupá Ciretá	Rio Maria/ Xinguara	34,848	—	34,848
Colonia Verde Brasileira	Santana do Araguaia	52,316	—	52,316
Fazenda Extrema	Itacajá	—	159,400	159,400
Fazenda Extrema Norte	Nazaré	—	7,101	7,101
Fazenda Serra (GL.J-L22)	Sitio Novo	—	1,703	1,703
Total		228,490	168,204	396,694

indian groups, these initiatives demonstrated that the State in the so-called New Republic had taken on responsibility to resolve rural conflicts, principally through changes in the structure of land tenure. Official statements began to refer to land reform as a 'social debt' and to emphasise the innocuous changes in tenure made during the authoritarian period. A new hierarchy of policy measures was presented, with priority for expropriation in the social interest over colonisation policy, land taxation or land purchases with PROTERRA funds.

When the files were opened, the Survey of Areas of Social Tension kept by INCRA registered conflicts in Amazonia on only 154 properties covering an area of 3,043,063 hectares and affecting 38,655 families. It was impossible to obtain an immediate picture of the situation from GETAT's Survey of Problem Areas. However, a commission set up by MIRAD in June 1985 estimated that there were 125 conflicts in the GETAT area involving 13,133 families and 1,598,227 hectares (MIRAD, 1985). INCRA documents on rural conflicts, which always were stamped 'reserved' or 'confidential', referred to areas of social tension principally as those where 'ownership litigation' was under way in the courts. For its part, GETAT considered that disputes in areas already titled should be resolved by the courts rather than land tenure agencies. The notion of 'problem area' adopted by GETAT comprised cases of doubtful land title, cases where *posseiros* had already made land improvements (*benfeitorias*) in areas wanted for large-scale projects, and areas designated as '*posseiro* invasions'.

The other cases included in these surveys refer to emergency situations, not necessarily arising from legal problems, which had created such an impact or generated such antagonism that they could not possibly be ignored. The latter cases usually arise from violence, coercion and the maltreatment of rural workers. The data presented below in Tables 9.2 and 9.3 are taken from the file cards used in the surveys mentioned earlier. The INCRA cards were completed regularly in the Regional Coordinating Offices and forwarded to the Directorate of Land Resources in Brasília. These data refer to 'areas (*focos*) of social tension', which are divided into those which have been resolved (*focos solucionados*) by demarcation, purchase or expropriation and

Table 9.2 Conflicts in the area administered by GETAT

State	*No. of Properties*	*No. of Families*	*Total area* (ha)
Pará	66	8,084	98,075
Goiás	46	2,092	412,967
Maranhão	12	2,957	268,185

Source: MIRAD, Estudo da situação das areas do conflito da Regiao Araguaia–Tocantins, Brasilia (August 1985) p.26.

Table 9.3 Land conflicts according to the INCRA survey of Areas of Social Tension, May 1985

State	*'focos pendentes'*			*'focos solucionados'*		
	No. of properties	*Area (ha)*	*No. of families*	*No. of properties*	*Area (ha)*	*No. of families*
Maranhão	54	533, 250	17, 491	13	234,812	5,989
Pará	10	312,844	1,865	1	12,539	243
Amapá	—	—	—	2	21,051	417
Acre	2	155,000	327	9	304,554	1,609
Roraima	3	288,500	496	4	73,741	1,527
Mato Grosso	22	616,226	2,855	16	366,843	5,142
Goiás	4	90,242	210	3	34,460	404

Source: INCRA.

those still to be cleared up ('*focos pendentes*'). It appears that the course of action followed in these emergency situations was based on the experience of the military bureaucrats and the rules set informally by the so-called 'fire-fighters', who constantly moved about 'putting out fires'. The conflicts were managed crisis by crisis, ruling out any possibility of making interventions by land tenure agencies more systematic (MIRAD/CCA, 1986). This spasmodic action explains the under-estimated number of conflicts, the disregard for any continued reporting of the disputes, and the lack of detailed documentation on the position of the different sides involved.

Some two months later, when the initial draft of the Regional Land Reform Plans began to be elaborated, it became obvious that data drawn from the INCRA and GETAT Surveys seriously under-estimated the number of conflicts in Amazonia. The evidence collected from rural trade unions, Church groups, landowners' organisations and voluntary institutions provided quantitatively and qualitatively more accurate information on each conflict. These sources reveal that in only 4 states – Maranhão, Mato Grosso, Pará and Rondônia – there were 371 conflicts, embracing 82,447 families and 5,666,430 hectares (Table 9.4). The difference between these data and the Surveys would have been even greater had it been possible to consult the original version of the Regional Land Reform Plans. However, these documents enjoyed only a brief ephemeral existence and were not formally released. By order of the National Security Council, which advised the President and directed the elaboration of the final version of the National Land Reform Plan (Decree 91,7666 of 19 October 1985), these regional versions were changed and reformulated, with substantial alterations to the data on land conflicts (Silva, 1987). For example, the Regional Land Reform

Table 9.4 Data on land conflicts from the regional land reform plans

	Land conflicts				
State	*No. of municipalities*	*No. of properties*	*No. of conflicts*	*Area (ha)*	*No. of families*
Maranhão	45	—	—	1,260,756	28,487
Mato Grosso	30	144	—	—	24,297
Pará	29	—	122	1,668,610	21,727
Ronônia	—	60	—	2,737,064	7,926

Note:
1. Data from the original drafts of these plans before being modified for publication in the *Diaria Official*.

Source: Planos Regionais de Reforma Agrária (December 1985).

Plans for Acre and Goiás, published in the *Diário Oficial* in May 1986, make no reference to land conflicts, while the plan for Mato Grosso presents a higher number of conflicts but reduces the estimated number of families involved (MIRAD/INCRA, 1986a). Almost all the rural properties in these critical zones of conflict are formally classified by INCRA criteria as *latifúndia*. In many cases, these large estates were established by transferring public lands to business groups and by public auctions of units ranging from 500 to 3,00 hectares. The latter were particularly important in southern Pará and western Maranhão (Ribeiro, 1987).

V THE FAILURE OF LAND REFORM

Initially, recourse to expropriation in the social interest was not inhibited by the prolonged debate on the final provisions of the National Agrarian Reform Plan (PNRA), the delays in reformulating regional plans, nor the counter-offensive mounted by conservative landed interests. This was strengthened by the creation in May, 1985 of the Rural Democratic Union (UDR), which imparted a fierce, warlike militancy to landowners' organisations. However, until the Rural Commissions (*Comissões Agrárias*) were established in August 1986, emergency procedures prevailed and expropriations were made in Amazonia. Between 25 October 1985 and 1 July 1986, expropriations occurred on 67,694 hectares in Maranhão, 67,246 in Pará, 33,028 in Rondônia, 65,940 in Mato Grosso, 56,084 in Goiás, and 2,984 in Acre. Land conflicts intensified in this period. Landowners increased their private militias and, without even the cover of a legal order, proceeded to expel *posseiros* and apply a 'scorched earth' strategy. Entire peasant villages were destroyed, notably in the valleys of the River Mearim (São Manuel, Serraria, Palmeira Torta) and River Itapecurú in Maranhão and in the northern region of Mato Grosso. In July 1986, the Office of Land Conflicts in MIRAD estimated that there were 892 conflicts in Amazonia, with 778 in Mato Grosso, Pará, Maranhão and Goiás.

This period of emergency administrative procedures ended in July 1986 as the Rural Commissions came into operation. These, paradoxically, had been set up to create more democratic mechanisms of land reform but henceforth MIRAD and INCRA were unable to surmount the obstacles and delays caused by counter-reform forces. In the absence of clear official priorities, expropriation occurred only on properties where this was in the landowner's interest. The *latifundiários* took advantage of the situation, getting rid of poor land which was then transferred to the peasants via the PNRA. So-called 'friendly expropriations' became common, fostering an illusion that divergent interests were being reconciled. In Amazonia, the Rural Commissions were an important factor in slowing down the rhythm achieved by the process of expropriation. For example, the Commission for the State of Pará met only once in 1987. In various states, legal expropriations considered a priority by social movements were suspended and filed away. The working of these commissions was marked by every conceivable kind of bureaucratic obstacle, undermining the support initially mobilised by creating the illusion of participation and the equal representation of peasants' and landowners' interests. In turn, MIRAD failed to take emergency action which would have 'corrected' in part the problem of implementing the land reform in accordance with the demands of the social movements. Such action occurred only twice between August 1986 and May 1987, in the cases of Gleba Aymorés (Mato Grosso) and Castanhal Araras (Pará).

Furthermore, demands by the rural trade union movement for the immediate abolition of GETAT and GEBAM also were not met. These agencies were maintained under their conservative leadership and without major changes in orientation.[5] Yet the pressure of peasant mobilisations was very intense in this region and some expropriations were achieved. Between 29 November 1985, when the first one occurred in the GETAT area, and 15 January 1987, 15 *latifúndios* with a total area of 77,674 hectares were expropriated for 1,208 peasant families. The rest of the area expropriated by MIRAD in Amazonia following the end of the period when emergency procedures were applied is explained by the so-called 'friendly expropriations' and the run-up to the November 1986 elections. Political compromises agreed at the regional level and effected through the Rural Commissions led to the 'de-politicisation' of land reform. In short, the Commissions gave preference to the so-called 'agreements' over the demands of the rural unions. This qualification must be borne in mind when considering the fact that, by 15 December 1987, some 1.5 million hectares had been expropriated in Amazonia.

The gradual loss of momentum behind expropriation and the dismissal of the most active group in the land reform agencies, together with successive changes in the leadership of MIRAD, created a general situation of stagnation and immobility. Without their principal policy instrument, these agencies assumed a passivity much to the taste of the counter-reform interests. Analyses and statistics of the conflicts and number of murders committed in this period, which had been compiled for 1985 and 1986, were no longer prepared for wide distribution. This return to confidentiality and the restricted circulation of these data also reflected a loss of technical competence. Investigations of conflicts *in loco* were equally limited.[6] In the case of Amazonia, this generalised inertia implicitly meant the tacit endorsement of forms of domination based on force, which continued to be used by landowners and *grileiros*, safe in the knowledge that now not even the sanctions of the Land Statute would be applied.

The abolition of GEBAM in May 1986 and GETAT in May 1987 brought no significant changes in this general picture. However, a new blow against the use of expropriation in the social interest came with the closure of

INCRA (Decree-Law 2,363 of 21 October 1987), and the provision that rural properties in the area administered by SUDAM with 1,500 hectares or less would not be expropriated. The repeal of Decree 1.164 of 1 April 1971 on 24 October 1987 also left vast areas, such as the municipalities of Itaituba Altamira and Marabá in Pará, provisionally under federal jurisdiction, awaiting a policy decision from the Ministry of the Army. These various legal changes relating to Amazonia continue to ignore the demands made by social movements and fail to counteract anti-reformist tendencies. On the contrary, policy inertia has strengthened these by allowing the *latifúndio* faction to make considerable gains both in the courts and in the national Constituent Assembly. Thus MIRAD has succeeded in obtaining titles only for usufruct rights (*posse*) on little more than one-third of the properties for which decrees of expropriation have been approved. Secondly, landowning interests have managed to win acceptance of the notion of 'productive land' in the new Constitution.

However, the *coup de grâce* for the already discredited land reform did not really come with the decisions made in the National Constituent Assembly to exclude 'productive' land from expropriation but, rather, from the re-introduction of land purchase. On 24 May 1988 in Pará, the Minister of MIRAD signed deeds of purchase for public lands (*terras públicas aforadas*), marking the return to land purchases as the solution to land conflicts, benefitting the *latifundiários* who had rented the land and ruling out expropriation. MIRAD purchased 56 rural properties, of which 53 were rented (*aforados*) and 3 were titled, in the so-called *Polígono dos Castanhais* in southern Pará, with a combined area of 205,303 hectares and 2,670 families of *posseiros*. MIRAD issued 404,613 Agrarian Debt bonds (*Títulos de Dívida Agrária*), redeemable in five years, with a period of two years' grace representing the equivalent of roughly CZ$2.2 billion (Silva, 1988). The areas acquired by MIRAD were then transferred to the state government and its land tenure agency, ITERPA, which assumed responsibility for the settlement of the *posseiros*. The transaction clearly reflected the pressure exerted by the landowners and large tenants who control Brazil-nut production in this area. In effect, they took advantage of the opportunity to sell off certain areas which were already in the hands of the *posseiros*. Moreover, other areas in this polygon where conflicts were more critical remained free of official intervention. Bluntly speaking, the land tenure agencies are in danger of becoming little more than real estate offices, offering no alternative measures and casting doubt on the outcome of the land conflicts, which remain intense and without any prospect of a solution.

VI CONCLUSION

A more careful analysis of the concrete results of this unsuccessful experience of land reform would suggest perhaps that the so-called 'democratic transition' and, by extension, the practice of democracy, which ensures the basic rights of citizenship, has had only a tangential impact on rural areas. There is absolutely no doubt that these rights are not enjoyed by the peasants and indian groups on the frontier. Popular mobilisation continues to be intense in Amazonia. The emasculation of the basic instruments of land reform and the running down of MIRAD have left an open space, if not a vacuum, in terms of institutional intermediation. Since the end of MIRAD's transitional phase when emergency procedures were used, the official intermediaries were again the agencies subordinated, directly or indirectly, to the National Security Council (CSN), which is now designated as the President's Advisory Secretariat on National Defence. Measures along the lines used by GETAT have been re-introduced by strengthening the Special Projects of Calha Norte, a process which has been going on since mid-1986. These same policy interests have taken over the coordination of the new environmental initiative, 'Our Nature' Programme (*Programa Nossa Natureza*), launched on 12 October 1988, orientating its executive commission and the interministerial working groups. Similarly, these interests now wield direct control over the Brazilian Institute of Forestry Development (IBDF) and its policies. They also influence decisions on mining policy. In September 1988, a former director of GETAT, Iris Pedro, who is sympathetic to the aims of these ex-CSN policy groups, was designated President of the National Indian Foundation (FUNAI), with formal and explicit control over indian policy.

Once again, albeit within a different framework, the crucial problems in Amazonia have been elevated to the status of 'security matters'. Questions concerning the environment, mining, indians, peasants and even industrialisation have clearly become problems under the control, directly or indirectly, of the military bureaucracy and its specialised cadres. (Indeed, this control has always been maintained, despite the formal end of the dictatorial military regime). This defines the context within which the conflicts and struggles in Amazonia exist, and which, tragically, will continue to break out in the anguished search for democratic and non-coercive solutions – an ever more improbable prospect given the levels of violence in rural conflicts recently recorded. These continuities are accompanied by measures to reduce the intensity of social conflict by pursuing economic solutions which are thought to prevent change. This logic perhaps can be understood by attempting to answer the following question: can the centralising sources of power within the State relinquish their frontier strategies without erecting insurmountable obstacles to the reproduction of this authoritarian mode of capitalism?

Notes

1. The Second National Congress of Rural Workers was held between 21 and 25 May 1973. The Congress Proceedings (*Anais*) published in August 1973 by CONTAG

reveal the weakness of union structures in Amazonia. Whereas the North-East, excluding Maranhão, had 678 rural unions (STRs) and over 1,255 *delegacias*, the North and Central–West had only 169 STRs, distributed as follows: Maranhão (87), Pará (34), Goiás (26) and Mato Grosso (22). Pará and Mato Grosso had 71 and 93 *delegacias sindicais*, respectively, and CONTAG had only one *delegacia* in Amazonia.

2. According to INCRA data collected in 1985, 64 public settlement projects were established in the period 1970–84, covering an area of over 12 million hectares and providing land for 85,503 families of which 65,435 were in Amazonia, defined as Acre, Amazonas, Pará, Rondônia and Roraima.
3. According to the report of the Commission of Evaluation of Tax Incentives (COMIF), established in 1985, 621 agricultural and industrial projects had been approved by FINAM during the previous 20 years. Despite the laxity of SUDAM project review procedures, 90 projects were cancelled with estimated losses to the federal government of 4.5 million ORTN which were not recovered. For further details, see Gasques and Yokomizo (1985).
4. Between 1964 and February 1985, 13.6 million hectares were expropriated in Brazil, with 10.5 million in the North region (Acre, Amazonas, Pará, and Rondônia). It is worth noting that 6.4 million hectares of this sub-total were expropriated in 1971 in the *Polígono de Altamira* as a result of the construction of the Transamazonian Highway and *agrovilas* for settlers. Overall, only 5 per cent of the area was expropriated in order to resolve land conflicts. See Yokota (1981). A detailed survey of the expropriations undertaken in this period can be found in MEAF (1984), especially in the Appendices, pp. 702–3 and 714–15.
5. To understand this continuity, see *Pronunciamento do President do GETAT, Sr Asdrubal Mendes Bentes, perante a Comissao do Interior da Camara Federal*, Brasília: MIRAD/GETAT (September 1985).
6. Between 1985 and 1987, 479 people were murdered in land conflicts in Amazonia. However, this figure is probably an under-estimate as the data for 1987 were obtained principally from rural unions (STRs) and other entities, whereas those for 1985 and 1986 were collected by the Office of Land Conflicts (*Coordenadoria de Conflitos Agrários*) and include the number of gunmen and fraudulent operators (*grileiros*) who were killed. This Office was abolished in December 1986.

References

Almeida, A. W. B. (1980), 'GETAT – A segurança nacional e o revigoramento do poder regional', *Revista FIPES*, São Luís, (July–December).

Almeida, A. W. B. (1985a), 'As áreas indígenas e o mercado de terras' in CEDI, *Povos Indígenas do Brasil*, São Paulo: CEDI.

Almeida, A. W. B. (1985b), *O GETAT e a arrecadação de áreas rurais como terra devoluta*, Belém: IDESP.

Almeida, A. W. B. (1986), 'Estrutura Fundiária e Expansão Camponesa', in J. M. Gonçalves de Almeida Jr (ed.), *Carajás: desafio político, ecologia e desenvolvimento*, São Paulo: Brasilense.

Arruda, H. P. de (1977), *Os problemas fundiários na estrategia do desenvolvimento e da segurança*, Rio de Janeiro: ADESG (mimeo).

Barrington Moore, Jr (1975), *As origens sociais da ditadura e da democracia: senhores e camponeses na construção do mundo moderno*, Lisbon: Cosmos/M. Fontes.

Campos, R. (1988), 'Buraco Branco', São Paulo: *Folha de São Paulo* (10 May).

Castelo Branco, C. (1980), 'INCRA legisla com Dom João VI', Rio de Janeiro: *Jornal do Brasil* (19 March).

CONTAG (1974), *Memorial*, Brasília: CONTAG (18 April).
CONTAG (1981), *As lutas camponesas no Brasil – 1980*, Rio de Janeiro: Marco Zero.
CONTAG (1985), *Anais do IV Congresso Nacional dos Trabalhadores Rurais*, Brasília: CONTAG.
CPT (Comissão Pastoral da Terra) – (1973), *CPT – Pastoral e Compromisso* Petrópolis: Vozes.
CSN–GETAT (1981), 'Estudos sobre o Projeto Carajás', Marabá: SG/CSN–GETAT (September).
Esterci, N. (1987), *Conflito no Araguaia – Peões e Posseiros contra a Grande Empresa*, Petrópolis: Vozes.
Gasques, J. G. and Yokomizo, C. (1985), *Avaliação dos incentivos fiscais na Amazônia*, Brasilia: IPEA (December).
INCRA (1978), *Desenvolvimento e Integraçâo – GT/Port. 724/76* Brasília.
Leme, M. C. V. and Pietrafesa W. M. de A. (1985), *Assassinatos no campo – crime e impunidade*, São Paulo: Movimento dos Trabalhadores Rurais sem Terra.
MEAF (1984), *Anais do Simpósio Internacional de Experiencia Fundiária*, Salvador: MEAF (August).
MIRAD (1985), *Estudo da situação das áreas de conflito da região Araguaia – Tocantins*, Brasília.
MIRAD/CCA (1986), *Conflitos de Terra*, Brasília.
MIRAD/INCRA (1986a), *Plano Regional de Reforma Agrária de Mato Grosso*, Brasília.
MIRAD/INCRA (1986b), *Plano Regional de Reforma Agrária de Maranhão*, Brasília.
Ribeiro, N. de F. (1985), *Pronunciamento do Ministro da Reforma Agrária e do Desenvolvimento Agrário no Plenario da Camera dos Deputados sobre o Plano Nacional de Reforma Agrária*, Brasília: MIRAD.
Ribeiro, N. de F. (1987), 'A questão agrária na Amazônia', in *Caminhada e esperança da reforma agrária*, Rio de Janeiro: Paz e Terra.
Silva, F. C. da (1988), *Desapropriação negociada: a exceção que virou regra*, Bélem (mimeo).
Silva, J. Gomes da (1987), *Caindo por Terra*, São Paulo: Editora Busca Vida.
Velho, O. G. (1976), *Capitalismo Autoritário e Campesinato – um estudo comparativo a partir da fronteira em movimento*, São Paulo: Difel.
Venturini General Danilo (1985), *A questão fundiária no Brasil*, Brasília: MEAF.
Yokota, P. (1981), *Questao Fundária Brasileira*, Brasília: INCRA.
Zanatta, A. (1980), *A legislação e a ocupação do territorio nacional*, Brasília: INCRA.

10 The Political Impasses of Rural Social Movements in Amazonia

José de Souza Martins

It is easy, though misleading, to take the large number of land conflicts in Amazonia as a sign of the vitality of peasant movements. It is just as wrong to consider such conflicts, generically and indiscriminately, as *social movements*. The opposite mistake can also be made. The predominantly local character of these struggles may suggest their fragility and reveal the absence of true social movements or, a more generous hypothesis, indicate their pre-political nature.

Indeed, the characterisation of these struggles by different social groups has been a factor responsible for the 'pre-politicisation' of these conflicts. It is possible to show that different groups, political and 'non-political' (or para-political), such as the political parties and the Church have contributed to the relative political importance of social struggles in the countryside. This creates a very complex situation since, at the same time, the struggles of rural workers are unable to develop and mature, under current conditions, without the mediation of political parties or even para-political groups.

The fragmented local bases of the many conflicts are frequently linked to the various groups which seek to mediate these struggles. Nevertheless, the distance and differences between those caught up in the conflicts, generally as victims, and the mediating groups, which give peasant struggle a wider historical and political dimension, are always striking.

This division has its genesis in the spontaneous and non-political origins of most conflicts. This is explained, at least in part, by the political conjuncture which gave rise to the contemporary and Amazonian phase in the history of peasant struggle in Brazil. This conjuncture began to take shape after the military *coup* of 1964, when the State unleashed widespread and harsh repression against left-wing elements in rural areas, including groups active in Goiás on the edge of what was later defined as *Amazônia Legal*. Here I refer specifically to the peasant 'republic' of Trombas and Formosa in the municipality of Uruaçú.[1] Although this area was invaded by the military only in the early 1970s, the leaders had already dispersed soon after the *coup*. Nevertheless, tangible signs remained of the political work undertaken by the Communist Party of Brazil over a period of 15 years, as well as that of other parties and groups formed by the fragmentation of the Left in the early

1960s. Political militants, although few in number, remained scattered in northern Goiás and neighbouring regions.

In the different forms of organising resistance against the violence of *latifundiários* in the 1970s, it was still possible to find traces in Goiás, Mato Grosso, Pará and Maranhão of the same defensive strategy used by the *posseiros* of Trombas. In addition, the peasantry rapidly assimilated some of the counter-insurgency tactics utilised by the Army during the violent repression of the guerrilla movement in the Araguaia region in the early 1970s. Some groups of workers, directly affected by this repression, incorporated similar tactics in their arsenal of defensive operations against the gunmen and police forces serving landowners and *grileiros*. Despite these residual signs, however, no organised groups were able to operate in this region with any continuity. The peasant struggles in Amazonia thus did not arise out of political organisation nor were they characterised by a historical process capable of binding together the many local and dispersed conflicts. On the contrary, for much of the military dictatorship, these struggles emerged spontaneously and defensively from resistance to violence: eviction from the land, assassination, the burning and destruction of houses, crops and harvests. In the vast majority of cases, rural workers had no real alternative: either they accepted eviction or they reacted to save their own lives.

The separation between the workers directly involved in the conflicts and mediating groups is also seen in the gradual and slow arrival on the scene of remnants of the clandestine parties and party factions, the so-called '*tendencias*', which had been decimated by violent military repression. These elements and parties, which had attempted to hasten the revolutionary process but had found little support in the mass of urban and rural workers, were composed mainly of middle-class militants, above all, students. For this reason, they remained politically isolated, divided by the ideological crisis of the Left, between the '*foquismo*' and political voluntarism characteristic of the period. After their defeat, some of these groups were attracted by the fact that, although the number of conflicts and struggles for land was increasing, particularly in Amazonia, the only channel of mediation and politicisation was the Church. Defeated as active agents in the political process and with a very limited following among the mass of workers, these party factions felt directly challenged by the growing number of rural workers who entered the fray without awaiting the messianic arrival of a vanguard party.

This division is therefore clear. On the one hand, the struggle for land is initially not politicised. On the other, the parties and 'tendencies' usually arrive in rural areas long after the struggles have begun and with different aims, influenced by a defeated revolutionary programme. They arrive quietly, attempting to 'equip' existing organisations, such as the Church and the unions, competing within these entities for political hegemony over the workers. However, this work is undertaken in the name of the same programmes already defeated by repression, and by lack of political support, and reflects the same ideological and political divisions[2] – indeed, even to the point of seeking to direct the struggle in rural areas on the basis of theories produced in other social and historical contexts, such as those which assign the leading role in the revolution to the working class or to the peasantry. As a result, the articulation of local movements and confrontations between peasants and landowners comes to depend on groups which are in fact involved in huge disputes among themselves in their efforts to gain the loyalty of rural workers. These groups are guided by conflicting 'theories' and ideologies in which, frequently the peasant does not appear as a historical subject with a recognised historical mission as an agent of social change.

In the centre is the Church which, although not a political agent, has had to face up to the political task of acting as a channel for mediating peasant struggles and expressing their programme. Consequently this remains an implicit, barely delineated and unfinished formulation of the social and political potential implicit in these conflicts. For this reason, the Church is the favourite target of the '*aparelhismo*' of different political groups, which regard it as a party and, very often, as a competing party and adversary. These groups therefore deliberately seek to weaken this non-party institution, which gives expression and unity to the diverse kinds of popular struggle in the countryside, and particularly in Amazonia. Today in Brazil there are political groups which are actively working to disrupt this non-party mediation and, consequently, to undermine rural social movements. Although this phenomenon is particularly clear in the cities, it is also plain in rural areas. Since 40 per cent of all land conflicts occur in Amazonia, it is precisely here that these political impasses are most intense.

The core of these difficulties, both of action as well as in the understanding of rural struggles in Amazonia, lies in the historical conjuncture of which these conflicts and social movement are part. In general, these struggles are treated as events that coincide chronologically with the period of military dictatorship, which is identified as their main cause and which will disappear when civilian rule is restored. Some observers are even tempted to see the dictatorial and repressive character of the military regime, essentially rightist and anti-popular, as the cause of concentrated landownership, fraudulent land sales (*grilagem*), peasant evictions and rural violence. Along similar lines, some groups believe that the end of the military dictatorship *per se* will lead to land reform. A civilian regime, whichever it may be, would necessarily create conditions in which the land tenure question, and the rural violence it engenders, would be resolved. However, such a chronological analysis fails to grasp the true historical amplitude of the processes under way in Amazonia.

These take place far more slowly than is usually thought. In fact, current analyses, such as this one, are dealing with processes as yet incomplete and situations which are still evolving. In this respect, it is presumptuous to think that the apparently rapid transformations occurring in Amazonia, such as the destruction of the rainforest, the massive eviction of *posserios*, and the fencing in and reduction of indian lands, are happening with such speed that analyses made ten years ago are now outdated. Nothing would be further from the truth. In studying Amazonia, we must be careful to avoid the trap of thinking that the tempo of change is faster than it has been in reality.

The classical thesis that the eviction of peasants is the first step towards proletarianisation and the demise of the peasantry is being disproved daily in Amazonia. At the same time as *posseiros* are being evicted, the new estates (*fazendas*) are demonstrating a remarkable capacity to revive 'archaic' forms of labour exploitation, such as debt service tenure (*peonagem*). Similarly, *posseiros* reveal great tenacity in re-establishing family agriculture and petty commodity production. However, evictions do result in net out-migration; witness the *favela* populations in such regional centres as Goiânia, Cuiabá, Belém, São Luís and Rio Branco. This process thus proceeds at a different pace, with its own complex internal contradictions.

POLITICAL CHANGES IN THE COUNTRYSIDE

In historical terms, it is important to note that migration to Amazonia and conflicts between *grileiros* and *posseiros* (which have so bloodily marked the past 25 years) represent merely an acceleration of processes already under way in Brazil. In the 1950s, the construction of the Belém–Brasilia highway caused a significant movement of North-Eastern migrants from Piauí and Maranhão into Goiás and Mato Grosso. The bloody battles between *grileiros* and *posseiros* left deep marks on the rural scene in Goiás. The peasant 'republic' of Trombas mentioned earlier was formed by *posseiros* from Maranhão. These migrants followed the course of the River Tocantins in the late 1940s, when construction of a Trans-Brazilian road was first proposed, and settled in the Trombas area (Valverde and Dias, 1967). The 1940s also witnessed the first migrations of followers of the 'Green Flag', a millenial movement which persuaded many North-Easterners to cross the Araguaia river to escape the Final Judgement. This movement still survives in Mato Grosso, Goiás and Pará.

In the same way, the famous *grileiros* of Paraná, who had taken part in the violent evictions of *posseiros* and *colonos* in the south–east and northern regions of the state during the black days of *grilagem* which characterised the two periods of government of Moises Lupion, were already shifting their activities to northern Mato Grosso in the early 1960s and even earlier (Rivère d'Arc, n.d.; Asselin, 1982; Branford and Glock, 1985). There they used the same techniques for monopolising and selling land which had made their fortunes in the south of Brazil. Some of these same *grileiros* have been behind major current land settlement projects and the creation of huge estates in the years since 1964.

Even before the military dictatorship emerged as a real possibility, the map of Amazonia already bore the marks of the branding iron of experienced *grileiros*, who had begun to sell off large sections of land. Even today, the leading newspapers in São Paulo frequently carry advertisements announcing the sale of titles to land in Amazonia which were bought in the pre-1964 period. These titles were originally sold either by *grileiros* or directly by state governments, as in the case of Mato Grosso, in an attempt to transform land, which in those days had no market, into a source of funds for public coffers. Titles to lands which the owners had never seen nor even knew where they were. Indeed, in Mato Grosso, the *same* piece of land was sold by different *grileiros*, as well as by the state government, to different buyers.

These are the same processes which occur today: poor North-Eastern migrants practising slash-and-burn agriculture moved westwards searching for so-called 'free land' (*terras livres*) and large and medium-size buyers of land living in the South, acquired property titles which often had no legal value whatsoever. With or without the military regime, a large number of land conflicts occurred in Amazonia in the 1960s. It is clear, however, that the development of these conflicts and their consequences would probably have been very different from those we are now observing. In the 1950s, landgrabbing by *grileiros* led to peasant uprisings in Paraná (Foweraker, 1971, 1974 and 1982; Colnaghi, 1984; Gomes, 1986). The peasant revolt at Trombas in Goiás occurred in the same period. In both cases, these struggles were politicised by the Communist Party of Brazil, resulting in the creation of popular governments and free, self-governing enclaves. As migrants and *grileiros* turned towards Amazonia, they took with them the peasant struggles and the social movement of which they are part: their tactics, instruments and their implicit historical aims.

The military *coup* of 1964 thus did not create a situation of conflict. On the contrary. The *coup* and land tenure policy of the military regime also struck down the peasant movement, the revolt in the countryside. They also tried to de-politicise it. The message sent to Congress by General Castelo Branco proposing constitutional reform to allow approval of the Land Statute clearly established what the aim of the land reform would be under the military; to cut off the head of the peasant movement. That is, to separate rural workers from left-wing political groups, such as the Communist Party (PCB) and the Peasant Leagues, which were deeply and actively involved in the conflicts, providing them with political leadership and a sense of historical perspective. The dictatorship gave a new military and geopolitical direction to the struggle, in an attempt to by-pass the rural workers, while at the same time via fiscal incentives, transforming those who control large-scale capital into landowners.

Military involvement in the agrarian question is directly related to the rupture of dependency links and the crisis of personal domination, which had subjugated rural workers to the landowner (*patrão*). These relations, although fragmented, still persist in many regions. This clientelism (*tutela clientalística*), the basis of the Brazilian political system, existed in the sugarcane plantations of the North–East, the coffee estates of São Paulo, and the rubber plantations in Amazonia. The economic changes which began to affect large estates in the 1950s gradually undermined the bonds of personal domination: *colonos* were evicted from the coffee estates and *moradores* from the sugarcane plantations. Concomitantly, the increasing separation of workplace and residence began to create perceptions among rural workers of other relationships, between strong and weak, rich and poor. These changes were also occurring in other sectors of the economy. Although clientelism was not always destroyed, it was modified. In Amazonia, this change began to take place in the late 1960s with the opening of roads, diverting traffic and the flow of merchandise from river navigation. As a result, Amazonian traders (*feitorias*), whose trading posts (*barracões*) were the basis of the enslavement of the rubber tappers and Brazil-nut gatherers lost control of these flows.

In general, these trends do not reflect a transformation of relations of production arising from changes in the rural labour process, which has remained virtually the same. This is true both of the North–Eastern sugarcane plantations (Sigaud, 1979) and the São Paulo coffee estates. The only difference is that the expulsion of *moradores* and *colonos* from the plantations led to the imposition of wage relations and the emergence of casual workers, the '*clandestino*' and the '*boia-fria*'. In practice, the significant change was that the land previously used by resident workers for subsistence production was now available for landowners to plant commercial crops. This change resulted not from the transformation of the labour process but from an increase in land prices and rent. That is, the change occurred not in the *production* of surplus value but in its *distribution*, in the form of rent.

Similar fundamental changes of this kind began to occur in Amazonia in the 1960s and 1970s. In this region, the key to the agrarian question is not between capital and labour in the production of surplus value but in its distribution as rent. This is the essential point to grasp if we are to understand the recent transformations and conflicts in Amazonia, and the real nature of its social movements and the obstacles which prevent them from becoming a true force for change in social and political relations.

Military intervention in the Brazilian political process and the concrete form it assumed in Amazonia was an explicit attempt to prevent the 'empty space' left by the decline of personal relations of dependence from being filled by a new historical figure: the rural worker. Traditionally excluded from the political arena by personal subjugation, which made the rural worker a client of the landowner, a member of the property system but not a citizen, this citizenship could be won only by groups or political parties which were also 'outside' the political system, as was the case of clandestine parties and left-wing groups. This new historical figure, despite its diversity, undermined the political pact supporting the Republic, which in a precarious way had 'stitched together' the centralising, developmentalist military tradition with the federalist, parochial and commercial–agrarian tradition of the oligarchy. The leaders of the Revolution of 1930, after defeating the rural oligarchies and replacing them in the states with new ruling groups, generally of military origin, revived and consolidated a new political pact between the military and civilians, which remained the basis of power. The peasant revolt of the 1950s threatened this pact. Indeed, when this struggle was transferred to Amazonia in the 1960s, not only was the existing model of capitalist development threatened but also, and more fundamentally, the authoritarian political model based on landed property and rent.

RENTIER CAPITALISM AND AUTHORITARIANISM

It is important to distinguish this conception of the authoritarian political model from others which have different connotations.[3] The basis of *rentier capitalism* is not circumscribed by Amazonia, nor is it explained by the abundance of free land, which supposedly demands recourse to coercive forms of labour exploitation, creating conditions for clientelism and authoritarianism. The lack of viability of the liberal political model in Brazil arises above all from the taxative (*tributário*), anti-liberal character of land rent (*renda territorial*), which generates and requires a clientelistic, cartorial State. This rentier capitalism has its base throughout the whole country and its various regions. In this sense, Amazonia has become a kind of colony for large capital, a source of speculative gains produced by the increase in rent which follows the opening of roads and investment in productive infrastructure by the government. Furthermore these gains do not arise directly from productive activities but from fiscal incentives which are used by the State to transfer income from other sectors to those with a base in agriculture.

It is easy to see that we are not dealing with pre-capitalism or, more simply, with backward capitalism which depends, let us say, on primitive accumulation. On the contrary, since in this case the *expropriation* of land is combined with *taxation* through rent (*renda fundiária*). Large modern capitalists may dispense with landownership in the South but they strive hard to become landowners in Amazonia. This model of capitalism thus differs from the classical European or American model in which the expanded reproduction of capital involves the appropriation and realisation of rent. In this respect, the balance sheets of rural enterprises in *Amazônia Legal* almost invariably show accounting losses or negligible profits (Branford and Glock, 1985).

These losses are offset by fiscal incentives, that is, subsidies, and the speculative rise of land prices. These firms are able to survive because they belong to large conglomerates, whose profitability depends on other activities. In other words, the rationale of agricultural firms in Amazonia is not found in *production*. This lends an apparent ambiguity to the behaviour of rural businessmen and landowners. While strongly opposing *posseiros*, they support colonisation policies and private colonisation schemes, in which they are frequently involved.

For this reason, the customers (*clientela*) for such schemes are not poor North-Eastern migrants who have been moving slowly towards Amazonia over the past 50 years but, rather, small producers from the South who have migrated to Western Amazonia only in the past 20 years. The latter can sell their holdings in Rio Grande do Sul, Santa Caterina and Paraná to the cooperatives and large landowners, as they did, above all, in the 1960s when soybeans provided high and rising gains. Indeed soybeans and especially the price fluctuations caused by the speculative activities of American groups, forced many small Southern producers to take the road to Mato Grosso and Rondônia in the following decade.

Small producers from the South migrated to Amazonia not only because they could increase the size of their holdings and so provide their children with the chance to remain in farming. They also went because technological innovation was increasing the scale of family production. The reproduction of family-based agriculture now depended on a higher level of output due to the deterioration of the terms of trade between agriculture and industry, with the latter swallowing an increasing share of the value created by small producers.

On the other hand, although technical change led to increased scales of production, it also offered opportunities for the survival of those family producers who used machinery, fertilisers, insecticides and improved seeds. In practice, however, small producers faced rising prices of industrial inputs and the declining trend of real product prices. This combination led to the temporary or definite suppression of rent (*renda fundiária*) for the small producer who owned his land because its equivalent was not expressed in the price received for his output. When he bought the *land*, he paid the rent incorporated in the price.[4] When selling his output, rent was not included. It could be recovered only if he sold the land and left agriculture. In extreme cases, such as the Canarana Project in Mato Grosso, the loss of soil fertility led many *colonos* simply to abandon their holdings (*lotes*). In such cases, we have a concrete example of the suppression and loss of rent, while at the same time rent is one of the bases underlying the accumulation of capital by the large capitalist landowners.

Even so, this does not imply that there is an inexorable tendency towards the concentration of landownership and the *latifundização* of the countryside. Although landownership is highly concentrated in Amazonia and elsewhere in Brazil, the place of rent (*renda fundiária*) in the accumulation process depends on the pendular movement between an unconcentrated tenure structure and private settlement. Rent is transformed into capital through the property market. The creation of a rural middle class, which is the guiding principle behind the military regime's Land Statute and World Bank policies supporting the small rural enterprise, is one of the factors cementing the relationship between capital and rent.

The other main ingredient was the fiscal incentives policy, introduced to stimulate large capitalists from the South-East to establish enterprises in Amazonia, particularly agricultural firms. The policy of fiscal incentives established in 1966 gave real *political* expression to the agrarian orientation of the military regime. Moreover, it should be noted that the civilian government of the New Republic which followed the military dictatorship has restricted the implementation of the Land Statute, and hence the possibility of land reform in areas of social conflict. At the same time, it has maintained the policy of fiscal incentives, involving free transfers of public resources to firms established in Amazonia. In fact, the incentives policy was clearly intended to protect landowners, ensuring their place in the power structure. Although this policy is almost always seen simply as a measure of 'implanting' large capital in agriculture, it has been used to *force* these capitals into landownership, maintaining and modernising property without undertaking social reform.

Rather than implementing the classical model, where capital eliminates or reduces rent in order to be able to expand in agriculture, the military regime has subsidised capital to offset the losses arising from its acquisition of land. The economic irrationality represented by rent, instead of being eliminated by the nationalisation of land or land reform, which would have been a mortal blow to the landed oligarchy (as the military did in Peru), was removed by financial subsidy via fiscal incentives, which transferred the burden of maintaining the class of large landowners to society as a whole. As a result, the great traditional landlords (*os grandes senhores territoriais*) of Amazonia and the *grileiros* from the South were not expropriated. However, at the same time, the entry of large capital opened up divisions in the structure of domination, the niches of local power and in the social bases of political clientelism. The violence unleashed against *posseiros* and rural workers broke down traditional loyalties, as well as practices of reciprocity, protection and *tutela*, and the moral obligations of the traditional *latifundista* towards his workers (*agregados*).

In order to cement the partnership between land and capital, and so maintain the conservative foundations of the Brazilian State, it was necessary to institute military protection (*tutela militar*) over those regions where social tensions were most acute, particularly in the Araguaia region but also elsewhere. This was the form used to contain the social forces which were growing in substance in the peasant struggle, and which threatened to create in Amazonia a social reality based on small-scale production supported, above all, by a widespread nationalisation of landed property. This is not fantasy. Many different peasant groups have campaigned vigorously for legal recognition of communal landownership as the basis for small family agriculture, and even family extractive activities. The resistance of the rubber-tappers of Acre in recent years against the felling of rubber trees in areas sold to São Paulo businessmen is an illustration of this demand. Not only have they prevented the clearance of rubber trees by organising new forms of resistance (the '*empates*') against landowners and ranchers from the South, but they have also impeded destruction of the forest. In addition, they have demanded recognition of their collective rights of usufruct (*posse*) of the rubber plantations, and rejected their division into family holdings following expropriation for land reform, as official policy would dictate.

Collective landownership is also part of the logic of family agriculture (*agricultura de roça*) and informs the resistance struggle of different *posseiro* groups in Mato Grosso, Pará, Goiás and Maranhão. By tradition, a collective right to the use of land exists which conflicts with the legal concept of private property, or '*terra de dono*' as the workers say. The shifting agriculture *de roça* occupies the land only for two to three years before moving on to virgin lands, while the old areas recover their fertility. There is a perfect logic and concept of mutual rights in this traditional form of land use. The experience in Canabrava, Mato Grosso, where the workers drew up a 'law of the land', a kind of collective contract between the local peasant population covering the communal use of land, was repeated in the village of Anilzinho in Pará and disseminated through the famous 'land laws' or '*lei Anilzinho*' – a popular 'law' for popular use.

THE CHURCH AND THE CONFLICTS: REFUGE FOR SUFFERERS

As observed earlier, the recent settlement of Amazonia by large capital sealed the political pact introduced by the military *coup* of 1964. This pact overcame the basic antagonisms between the military and the rural oligarchies, which had marked the entire republican history of Brazil, and even earlier. The heavy subsidies conceded by the State to large enterprises opened the way for capitalist investment, protected and endorsed land rent and property speculation, and included landed interests (*a grande propriedade fundiária*) in a process of capitalist development which, in a contradictory fashion tried to organise a modern society on the basis of a rentier and export economy. In rural areas, and in Amazonia in particular, this pact has brought profound changes. The traditional *latifundista* economy was based on the sale of surpluses, which meant that small producers had to be assured some form of subsistence. When this was impossible, as in the rubber economy, it was necessary to enforce the extortionary *aviamento* system and coercive labour relations. Now, however, the incidence of *aviamento* has expanded on a much larger scale, and landed property is not the only factor behind this growth. Now there is large capital: banks, industry, large distributors, the market. Investment now has to produce profit at a certain rate and in a given period, as defined by the expanded reproduction of capital. This is the case even when the new landowners adopt relations of exploitation involving *peonagem* or debt service, as they frequently do. In fact, the level of violence is much greater than under the old relations of *aviamento* in the rubber plantations. There are many denunciations of workers being murdered with impunity, as well as frequent references to the traffic in workers consigned to the slavery of debt service on the new estates of modern enterprises.

This is the context in which the 'new' Church is working in Amazonia – a Church which extends its pastoral services to provide refuge to the afflicted and oppressed. The first signs of this change appeared in 1971 in the pastoral letter of Dom Pedro Casaldáliga, Bishop of Sao Félix in Mato Grosso. (Casaldáliga, 1971, 1978). Peons fleeing from the new estates came to the Church in search of help and asylum. Some died on arrival, bearing the signs of torture by gunmen and weakened by malaria. Many died on the roadside and in the jungle before reaching safety. These escaped peons were joined by *posseiros*, some having lived for several generations on the same land, who began to be violently evicted from their holdings (*roçados*). Fazenda Codeara, owned by the *Banco de Crédito Nacional*, completely surrounded the old village of Santa Terezinha, transforming it into an enclave. It tried to implant a new urban structure on the old one in an attempt to sell the land to those already occupying it, and thereby extract land rent. This provoked an armed uprising in 1973, which led to the imprisonment, prosecution and expulsion from Brazil of Father Francisco Jentel (Branford and Glock, 1985).

The same story was repeated everywhere. As the capitalist front expanded in Amazonia, it ran up against the local Church. When Dom José Patrício Hanrahan arrived in Conceiçao do Araguaia in Pará to take over the Diocese, he was met by a huge crowd of *posseiros* who had just been evicted from their land by a large landowner. Since then, the Bishop has been constantly sought after by evicted, persecuted and tortured workers. It was in this diocese that two French priests, Aristides Camio and Francisco Goriou, were imprisoned and prosecuted by the Military Court, accused of inciting subversion and of being involved in an ambush in which a gunman who had been persecuting rural workers was killed (Figueira, 1986).

The Church became a refuge basically for two reasons. First, because victims and the persecuted have nowhere else to turn. The local police and judicial system are controlled by large landowners, and it is not uncommon for the judge, police chief, militia commander and all other public officials to have been nominated by the state governor on the advice of politicians supported and elected by the landowners. Frequently, runaway peons, threatened with torture and death, have turned to the local police for help, only to be returned to the overseer (*capataz*) and gunmen from the estate from which they had escaped. *Posseiros* also are well aware of the futility of relying on the police for help and protection. The second reason is because the Church in the 1960s opted to give priority to the poor and its social pastoral work. The violence which accompanies capitalist expansion in Amazonia is confronted by a vigilant church, aware of human rights, committed to the poor and oppressed, the voice of those who have no voice. A Church committed to the dignity of man and conscious that in order to free the poor the chains which enslave them must be broken, not only in the work place but also in the mind (CNBB, 1976).

Texts which take a superficial view of the Church's involvement in social conflicts, and particularly in the agrarian question, frequently distinguish between 'progressive' and 'conservative' bishops. This distinction explains nothing. During the 1970 sessions of the Parliamentary Commission of Inquiry into land tenure problems, various bishops, both 'progressive' and 'conservative', were called to appear. All of them denounced the gravity of the situation and mentioned the steps being taken in the different dioceses to help the victims of violence (SEDOC, 1977). In fact, these circumstances, (together with the development of a rural pastoral programme by the Lutherans), persuaded the Church of the urgency of social and political mediation in the countryside. It should be remembered that this period witnessed the violent political repression of left-wing parties, including the Communist Party of Brazil, which had played an important part in organising rural workers in the struggle for land. In addition, the Left was deeply divided on what position to take on this issue.

At a series of regional meetings held in 1973, the bishops declared their commitment to those suffering wrongs in the countryside, denouncing the serious crimes against indians and peasants. Two years later, in 1975, the assistance and pastoral care given by the Church began to be coordinated by the Pastoral Land Commission (CPT), which today includes Catholics and Lutherans. The CPT acquired distinctive characteristics within the context of the Church's activities because the land question was and is clearly political. It arises from a political pact which does not merely exclude rural workers but involves their widespread expulsion from the land. This pact transformed workers into victims, condemning millions to poverty, and even to summary disappearance with the connivance, participation and promotion of the State. This became absolutely clear in the case of tribal peoples which convinced the Church, in 1972, to create the Indian Missionary Council (CIMI) and define a new programme of pastoral care for indians.

Some dramatic cases have occurred, including that of Kreenakarore in Mato Grosso, the so-called 'giant indians'. They had begun to be contacted in 1972 because their territory was being crossed by the Cuiabá-Santarém road, which would open the area to large ranches and private settlement projects. When the first friendly contacts were made in October 1972 and in January 1973, they numbered 350. By January 1975, this number had fallen to 79, all of whom displayed clear signs of tuberculosis. In this period they were initiated into homosexual practices by an official of FUNAI, who also confiscated their bows and arrows, their means of survival. By 1974, they had become beggars, wandering along the new road, prostituting their wives and daughters, drinking alcohol, dirty, fighting over left-overs. In 1975, only three of them were over 39 years old. At this point, they accepted the invitation from the Txukarramei, their traditional enemies, to live with them in the Xingú National Park (*Coojornal*, November 1980, p. 16; *O Estado de São Paulo*, 2 October, 1982, p. 11).

A similar situation has occurred with the Parakanã of Pará, and in 1971 there were only 92 survivors. Their lands were violated and mutilated twice: by the opening of the Transamazon Highway and the construction of the Tucuruí Dam, which led to their displacement, with harmful results (*O Estado de São Paulo*, 12 June 1977, p. 27). The story of the Waimiri–Atroari of Amazonas is the same. Their lands were opened by the Manaus–Caraçaraí road, despite vigorous resistance. This was followed by the beginning of construction of the Balbina Dam, which will flood large areas of their territory. Finally, a large mining enterprise has been given a concession to extract cassiterite on their lands. The Waimiri–Atroari, who numbered 3,000 in 1968, had fallen to 600 by 1982 (*O Estado de São Paulo*, 20 October, 1982, p. 11).

Large enterprises and the so-called *grandes projetos* (roads, hydroelectric projects, settlement schemes and mining) have had a mortal effect on Amazonia. They do more than monopolise the land. They destroy ways of life and corrupt whole peoples, as the case of the indians so cruelly demonstrates.

These reasons explain why peasant resistance in these areas, whether articulated by the Church or not, is more than a struggle for land. For example, in the case of the rubber-tappers in Acre, where the Church has played an active but not a leading role, the struggle has been for the expropriation and *preservation* of the rubber trees – that is, for the preservation of a way of life. This also characterises the work of the Church in various regions. However, this is not a preconceived policy of the Church in the sense that it is not part of Catholic dogma intended, as some suggest, to support family-based agriculture as a bulwark against Communism.

In fact, the organisation of the rural workers' struggle for survival on the basis of collective labour (*mutirão*) and family production is the only real alternative to poverty and degradation. It is certainly true that the Church has taken up this alternative but only in a potential sense has it acquired the dimensions of a project that would give direction to the workers' struggle. This project emphasises the importance of the family and the abundance of independent family production as a defence against the exploitation of unequal exchange, recognising the role of intermediaries in maintaing rural backwardness and poverty. It is not by chance that there is a cooperativist tendency in many rural social movements, as well as awareness of the importance of cooperative forms of production.

One of the basic characteristics of rural social movements is the emphasis on *labour* as the key to relations with society at large. What lies behind the deterioration of the terms of trade is the depreciation of the value of labour and the worker. Workers discover the universality of labour, revealing the real links that unite them with fellow workers and separate them from their exploiters and oppressors. However, labour cannot appear in this context as *abstract labour*, which has merely a monetary equivalent, as in the case of wage labour. For the peasant, labour can be manifest only as *concrete labour*, which takes tangible form as his harvest and the work of his family. A part of this output is essential for his survival. For this reason, his eviction from the land, although often presented as a legal decision, appears unfair because it is always violent and threatens his survival. It robs him of what is his – his labour, the means and instruments of his dignity and self-image. It is on this level where the morality of the rural workers and the Church meet and come together. Both the social doctrine of the Church and peasant ideology conceive of the *person*, rather than the abstract conception of the individual, which appears more clearly in the condition of the worker or employee.

In this context, peasant movements draw support from a political programme which remains only implicit. For the *person*, the Church has a *religious project* and not a *political project*. And left-wing parties have a project for the working class but not the person. Both in theory and practice, the parties still have to rediscover the historic importance of *individuality*, in which internal harmony and overcoming alienation are essential elements. The lack of this understanding weakens the political mediation of the Church, in the same way that it distorts the political parties' view of the relationship between the Church and rural workers in peasant social movements.

THE DISCOURSE OF LAND REFORM AND THE IMPASSE

The end of the military dictatorship did not arrest the decline of peasant struggles in Amazonia nor increase the political possibility that rural workers there would be able to implement an alternative agricultural model. The political weakness of the rural workers and the private and public repression during the military regime have left wounds that will take a long time to heal. This weakness was aggravated by the divorce between the peasant struggle and political groups. Rural workers were further weakened politically by the deliberate policy of consolidating the alliance between landownership and large capital, creating the social and political foundations for a new elite in

Amazonia. The aggressive, right-wing Rural Democratic Union (UDR), which is active throughout Brazil, is a direct result of these changes. Significantly, some of its most important branches are in Legal Amazonia and in states outside the region where businessmen with ranches in Amazonia are well represented (Bruno, 1987); Furthermore, the strong political presence of the regional oligarchies in the new Parliament and the National Constituent Assembly illustrates a recurrent phenomenon in the history of landed oligarchies in Brazil: their great capacity for regeneration, as occured in 1930 (Dantas, 1987).[5]

These conditions were not conducive to real involvement in the peasant struggle of mediating groups, which would give political expression to peasant *praxis*. Rather, these groups are the political expression of middle-class *praxis*. Their ideological position and their programme are guided by political possibilities defined by the State and government which emerged from the pact of 1984. Without realising it, these groups have returned to the old tradition of high-minded liberalism which, in the nineteenth century, believed that the masses could be emancipated without making any commitment to a popular programme of political emancipation. In this sense, contemporary political groups tend to reimpose the ideological and political patronage (*tutela*) of the Left over the peasant organisations involved in land conflicts, a kind of 'progressive *coronelismo*', which is corporatist and which reduces the prospects of political emancipation of the rural poor.[6]

As a result, these political groups have created a discourse between themselves on the theme of *land reform*, which supposedly expresses the aspirations of the popular movements in the countryside. On this basis, these groups have unleashed *a struggle for land reform* which is historically out of context and divorced from peasant *praxis* – that is, from the struggle for land, the bloody experience of eviction, the violence and the violation of human rights, and the lack of citizenship. These political groups found it easy to develop ways of supporting the pact which gave rise to the New Republic. And it was with the same ease that the State weakened and distorted their struggle, coopting them and depriving them of legitimacy because they accentuated the gap separating them from the popular struggles in the countryside.

These groups made the fatal mistake of putting the struggle for land reform in the place of the struggle for land (*luta pela terra*). This forced them to adopt the discourse of the rentier bourgeoisie and landed oligarchies and to defend the idea of a necessary relationship between property and production. By adopting this discourse, they inevitably legitimised the defence of the present system of property, with its concentrationist characteristics, and had to accept the possibility that land reform would be confined to those lands not included in this conceptual trap. In practice, these groups themselves destroyed the legitimacy of the land reform proposal whose discourse was appropriated by the *latifundiários*. Victims of an impoverished 'economicism', so dominant in the 'theories' of different left-wing groups, they failed to comprehend that they were facing not a question of productivity nor of production but rather a political problem represented by the current property system, which continually renewed the economic and class foundations of political conservatism and authoritarianism. Unable to contest the dominant discourse within the government, they were unable to mediate the struggle for land and express its historical and political dimensions.

Even the Church, which through its pastoral social work, particularly significant in Amazonia, had become a channel of expression and political mediation for peasant movements, fell into the trap of believing itself to be the spokesman of the rural workers, thus turning away from the richer, more profound alternative of being the voice of those without voice. Through the activities of some of its members, its voice spoke against that of the workers, substituting *its* struggle for land reform for the workers' struggle for land. It accepted its cooption by the new political regime. Supposedly speaking in the Church's name, some members gave their views on the appointment of successive ministers of land reform and, in some regions, high-level officials of the Ministry of Land Reform and INCRA were informally nominated and supported. In this way, the only organisation with local roots and a deep involvement in the social reality and conflicts of rural workers wavered over its role and its commitment, contributing further to the impasses and limitations facing rural social movements.

Equally, a fatal error of interpretation compromised political action by these different groups: a proposal for land reform made outside the context of peasant revolution and within the ambit of a State which had just sealed a clear, and probably long-lasting, pact between capital and landed property. In these circumstances, no faction within the dominant class was prepared to take up the land reform proposal. Furthermore, at the present time, dependent capitalism is not an ally in the struggle for land reform because it does not need one, at least in the short term.

At the same time, important currents and themes detectable within rural social movements have been excluded from the political agenda of the parties, party factions and para-political groups. These have neglected the fact that capitalist expansion in Amazonia has created a huge political vacuum, replacing the old political leaders with absentee landlords and businessmen, who are ignorant of the loyalties and political commitments of protection (*tutela*) and paternalism incorporated in the traditional conception of landed property. It has been forgotten that the struggle for land is not only about the problem of access to land *per se* but also about the emancipation of rural workers from the old systems of personal domination. The empty space of *local power* has also been neglected, together with the implicit, and often even explicit, political programme of the peasant struggle, involving political participation and direct democracy, which is so much part of the organisation of the centres of peasant resistance. Few regions and groups understood that the struggle for land had aroused and formed civil society in rural areas, and could provide the basis to dominate the local State via the organisation and direction of municipal governments, establishing social management of public resources.

Few have understood that the discourse which would unite the rural struggles was not only and, perhaps, not mainly the disccourse of land reform but rather the discourse of the *relationship between social needs and power*, including local power. Even with all the difficulties and limitations that can be pointed out, it is worth recalling the experience of Araguaia in Mato Grosso, where peasant struggles created the historic prospect of political emancipation at the local level, including the election of mayors and municipal councils. (A similar experience occurred at Trombas, where land freed by the peasants was organised as a municipality, and in south-western Paraná, where the peasants and the political leaders of the movement in 1957 took over the local public administration.)

The struggle for land involves more than property; it concerns the possibility of revolutionising and reorganising local bases of power through the introduction of forms of participatory democracy, which are frequently found, developed and understood in the struggle for land. It is on this level that the peasant struggles in Amazonia have the dimensions of a social movement. They are weakened and drained, however, because their most deeply political themes have been taken off the political agenda by mediating groups. And because, on the other hand, the political treatment of the agrarian question and the peasant question, during the dictatorship, involved the separation of those who *do* and those who *think*. This created the illusion that those who act cannot think and, consequently, that they can act politically only under the tutelage of others. For this reason, all attempts to give political direction to the agrarian question have ended up invoking and privileging the State and have weakened rural social movements.

Notes

1. The peasant revolt in Goiás and the 'liberation' of the territory of Trombas has been examined in a number of texts which, although small, represents a significant contribution to the study of the shift of peasant movements towards Central Brazil and the Amazonia before the *coup* of 1964. These works include Garcia (1966), Carvalho (1978), Amado (1980), Carneiro (1981) and Abreu (1985).
2. A systematic treatment of the fragmentation of the original party 'trees' (Communist, Catholic Left and Trotskyist) can be found in Silva (n.d.). This book also indicates how the different party groups included rural workers in their projects, particularly when dealing with armed struggle; that is, when considering only the overthrow of the political order. A generic study of armed struggle in the countryside and of the *foquista* conception held by the various political groupings on the Left can be found in Gorender (1987). Particular attention is drawn to

references to the following organisations: *ALN – Ação Libertadora Nacional, Partido Comunista do Brasil*, Ala Vermelha, *PCR – Partido Comunista Revolucionário* (the latter two groups were formed by dissidents from the *Partido Comunista do Brasil*); *PRT – Partido Revolucionário dos Trabalhadores* (dissidents from *AP–Ação Popular*, of Catholic origin), *COLINA – Comando de Libertação Nacional, VPR – Vanquarda Popular Revolucionária*, and *VAR–Palmares-Vanguarda Armada Revolucionária*. Although they differed in the form that the armed struggle should take, the different left-wing organisations were unclear about what was happening in rural areas and had no place in their respective political projects for rural workers. The countryside (*o campo*) was only a strategic reference point in the *destruction* of the political system but it had no role in the *construction* of a new political order. Gorender (1987) reveals that these organisations comprised mainly students from the urban middle class. Various studies by Foracchi (1965, 1972, and 1985) which examine student *praxis* emphasise the petty bourgeois motivation of the then young students and the middle class character of their revolutionary impulse.

3. This view differs from the concept of 'authoritarian capitalism' presented by Velho (1976) and Foweraker (1982). These authors both emphasise the abundance of 'free lands' as the factor behind coercive forms of labour *in these regions*. My interpretation does not exclude this. However, I think that political authoritarianism, in Brazil, is supported by a wider mediation which blends together profit and ground rent. In this respect, it is not a regional problem but a national one. For this reason, the expanded reproduction of capital does not occur in a context of economic liberalism (nor, consequently, of political liberalism). The expanded reproduction of capital depends not only on coercive forms of appropriation of *surplus value* but also coercive forms of extraction of *profit*. That is, it involves not only the *production* of value but also its *circulation* and *distribution*. In this sense, it involves both *private violence* and the *public violence* of the State, including police repression, tax subsidies, economic retrenchment policies, etc.

   My conception of authoritarian capitalism also is different in that it is not based on the *structuralist* conception of economic and social formations nor on the related notion of the articulation of modes of production. Rather, it is based on a *historical* conception of economic and social formations.
4. The interesting and well-known study by Vergopoulos (1974) of 'distorted capitalism' (*capitalisme disforme*) fails to consider capitalist expansion in agriculture via peasant production and the systematic use of mechanisms of plunder (*pilhagem*), which suppress rent in product prices but do not suppress it *in practice*. Capital brings rent under its domination, developing means, such as colonisation, which allow it to concentrate and deconcentrate property in a cyclical fashion permitting capitalists to realise rent and reconvert it into capital. In this respect, Brazil is an exemplary case, although not unique.
5. Dantas (1987) analyses the mutations of *coronelismo*, its historical anti-democratic role and its capacity to adapt to different political circumstances.
6. Galjart (1964) had already noted this phenomenon in relation to the Peasant Leagues.

References

Abreu, S. de Barros (1985), *Trombas – A Guerrilha de Zé Porfírio*, Brasilia: Ed. Goethe.

Amado, J, (1980), *Movementos Sociais no Campo: A Revolta de Formoso, Goiás, 1948–64*, Rio de Janeiro (mimeo).
Asselin, V. (1982), *Grilagem–Corrupção e Violência em Terras de Carajás*, Petrópolis: Vozes.
Branford, S. and Glock, O. (1985), *The Last Frontier: Fighting Over Land in the Amazon*, London: Zed Books.
Bruno, R. (1987), *UDR: Crise de Representação e Novas Formas de Poder das Oligarquias Rurais*, Botucatu (mimeo).
Carneiro, M. E. F. (1981), *A Revolta Camponesa de Formoso e Trombas*, Goiânia: Universidade de Goiás.
Carvalho, M. (1978) 'A guerra camponesa de Trombas de Formoso', *Movimento*, 164 (21 August).
Casaldáliga, P. (1971), *Uma Igreja da Amazônia em Conflito com o Latifúndio e a Marginalização Social*, Sao Félix do Araguaia.
Casaldáliga, P. (1978), *Creio na Justiça e na Esperança*, Rio de Janeiro: Civilização Brasileira.
Colnaghi, M. C. (1984), *Colonos e Poder – A Luta pela Terra no Sudoeste do Paraná*, Curitiba: Departamento de Historia da Universidade Federal de Paraná.
CNBB: Conselho Nacional dos Bispos Brasileiros (1976), *Pastoral da Terra*, São Paulo: Edições Paulinas.
Dantas, I. (1987), *Coronelismo e dominação*, Aracajú: Gráfica Diplomata Ltda.
Figueira, R. R. (1986), *A Justiça do Lobo*, Petrópolis: Vozes.
Foracchi, M. M. (1965), *O Estudante e a Transformação da Sociedade Brasilera*, São Paulo: Editora Nacional.
Foracchi, M. M. (1972), *A Juventude na Sociedade Moderna*, São Paulo: Livraria Pioneira Editora.
Foracchi, M. M. (1982), *A Participação Social dos Excluidos*, São Paulo: Editora Hucitec.
Foweraker, J. A. (1971), *The Frontier in South-West Paraná from 1940*, B.Phil, thesis, Oxford.
Foweraker, J. A. (1974), *Political Conflict on the Frontier: a case-study of the land problem in the West of Paraná*, Oxford: University of Oxford (mimeo).
Foweraker, J. A. (1982), *A Luta pela Terra*, Rio de Janeiro: Zahar.
Galjart, B. (1964), 'Class and "Following" in Rural Brazil', *America Latina*, 7 (3) July–September.
Garcia, J. G. (1966), *O Caminho de Trombas*, Rio de Janeiro: Civilização Brasileira.
Gomes, I. Z. (1986), *1957 – Revolta dos Posseiros*, Curitiba: Edições Criar.
Gorender, J. (1987), *Combate das Trevas – A Esquerda Brasileira: das Ilusões Perdidas à Luta Armada*, São Paulo: Editora Ática (3rd edn).
Rivière d'Arc, H. (n.d.), 'Le Nord du Mato Grosso: colonisation et nouveau "Bandeirismo" ', *Annales de Géographie*, LXXXVI, Paris: Librairie Amand Colin.
SEDOC, 1977, vol. 10 (105) (October–November).
Sigaud, L. (1979), *Os Clandestinos e o Direito*, São Paulo: Livraria Duas Cidades.
Silva, A. O. da (n.d.), *História das Tendências no Brasil*, Saõ Paulo: Dag Gráfica e Editorial.
Valverde, O. and Vergolino Dias, C. (1967), *A Rodovia Belém–Brasília*, Rio de Janeiro: IBGE.
Velho, O. G. (1976), *Capitalismo Autoritário e Campesinato*, Saõ Paulo: Difel.
Vergopoulos, K. (1974), 'Capitalisme disforme (le cas de l'agriculture dans le capitalisme)', in S. Amin and K. Vergopoulos, *La Question Paysanne et le Capitalisme*, Paris: Editions Anthropos.

11 Indigenous Peoples in Brazilian Amazonia and the Expansion of the Economic Frontier

David Treece

INTRODUCTION

> The indian economic world is not a closed world. The indian communities are only apparently isolated. On the contrary, they participate in regional systems and in the national economy. Markets and commercial relations are the principal link between the indians and the world of *ladinos*, between the subsistence economy and the national economy. It is true that the largest part of the indians' agricultural production is consumed by themselves. It is also true that the income produced by the indians represents only a very small share of the national product . . . But the importance of these relations does not lie in the amount of the product sold or in the value of the goods bought; it lies rather in the quality of the commercial relations. It is these relations which have transformed the indians into a 'minority' and which have put them in the state of dependence in which they now find themselves (Stavenhagen, 1963, p. 78).

When written, these words were intended to apply, not to the semi-nomadic, forest tribal peoples of the Amazon basin, but to those indians of Central America whose societies and cultures have long since been riven and shaped by the violent forces of market competition and international commodity price fluctuations. Reproducing the experience of the continent as a whole, it is their very integration into the world economy, rather than their marginalisation, which explains the increasingly intolerable conditions of life they are forced to endure. What marginalisation they continue to share with their Amazonian counterparts, it would seem, is of a cultural, social and political nature.

Today, however, with a swiftness and brutality that are breathtaking, the last isolated communities of Amazonian indians are being taught those same laws of world economics. The 10,000 Yanomami of Roraima, barely contacted until recently, are daily suffering shootings, deadly viral infection and the pollution of their rivers, the consequences of a gold-rush that has swept at least 40,000 desperate prospectors into their lands from other exhausted diggings. Whatever is done, and must now be done, to provide the Yanomami with the territorial security which can enable them to withstand and resist this onslaught, the fragile boundaries of a world which seemed closed have been irrevocably opened up. Only the struggle which they and others are forced to wage against the development of capitalism in this and other parts of the globe will determine how far their future repeats the experience of Peru's indian communities; the description below of their disintegration over many decades now threatens to become a familiar pattern in the heart of Amazonia:

> For the indians who continue as farmers, the disappearance of communal organization would bring with it not only their easy exploitation by unscrupulous people – whom experience has shown to be in the majority; but also the indian farmers themselves would enter into competition which would be ruinous for them, given their limited financial capacity, their poor agricultural techniques, and the uneconomic size of their plots. To have the indians, with their limited resources, incorporate themselves to compete actively in an individualistic system would mean sinking them into even greater misery. It is therefore necessary to find new organisational forms to replace the community which will inevitably disappear ... This process, left to its own, can finally achieve the incorporation of the indian into Western ways and the disappearance of the subsistence economy, but at a terrible price in misery, massive tuberculosis, terrible infant mortality, unemployment, crime, etc. The problem is thus changing stage but not actors. The same human masses which cease to be the object of the 'indian problem' become the object of the 'slum problem' which is the problem of the urban subproletariat which lives in extreme misery and which is ever growing (Quintanilla, n.d. pp. 19–20).

Twenty years ago, when the Brazilian government was accused by a group of nations at the UN Conference of Human Rights of permitting the wholesale massacre, enslavement and torture of the country's tribal peoples, anthropologist Darcy Ribeiro predicted that not a single indian would be left alive by 1980. The survival until today of 220,000 indians, the majority of them in the Amazon basin, does not contradict or diminish the gravity of the crisis which Ribeiro was expressing then, nor should it lead us to expect anything short of genocidal catastrophe if the situation of the Yanomami is left unresolved.

What it must force us to do though, is to see the question no longer solely in terms of human annihilation and therefore, by implication, of the search for means of protecting or 'preserving' peoples and ways of life as if they were endangered species or banks of rainforest. To do so would be not only to fail to recognise the profundity and nature of the changes which the region has been experiencing, but also to ignore the demonstrated capacity of the tribal peoples themselves to organise and formulate their own response to the enormous challenges facing them. For the years since 1968 have seen the establishment of major transport infrastructures in the region, often bissecting or passing close to indian lands. Settler immigration along these routes, as well as their use by logging companies and cattle producers, has been intense, bringing not only environmental problems but also demographic and commercial pressures to bear upon the indians' traditional way of life. More recently, however, those same transport infrastructures, aided by the production of energy from ambitious hydroelectricity projects, have made possible the introduction of heavy industries, such as mining and iron and steel smelting. The industrialisation of the Amazon, more so than its agricultural development, will signify profound social, as well as environmental, transformations, many of them irreversible, such as large-scale urbanisation and the proletarianisation of the local rural population.

These changes cannot fail to affect the indian communities in ways other than the simple destruction of the forest and game upon which they depend, or their exposure to disease and violent conflict. Even if such problems can be resolved, the permanence and expansion of urban, industrial complexes in the heart of the region will exert, indeed already has exerted, pressure on those communities to participate in the process. Thus, while the Kayapó of the middle Xingú are actively organising opposition to the installation of a major hydroelectric complex for the river basin, they have for some years been negotiating with gold miners for a percentage of the income from gold extracted from their lands. While the nomadic Guajá continue to flee from encroachment by farmers in the area of the Greater Carajás Programme, other indians already faced with the fait accompli of the 900-kilometre Carajás railway in their territory, such as the Gaviões, have secured compensation which has been invested in the production and marketing of Brazil-nuts. Still others in Maranhão have been encouraged by State-run schemes to produce non-subsistence crops for sale, giving them access to cash purchases of consumer goods. Whereas the indigenous communities of Acre have joined with rubber-tappers in proposing the creation of 'extractive reserves' to protect the forests upon which they depend, others who live close to the Carajás railway will face the option of selling timber for the charcoal-burning pig-iron smelters now being installed in Marabá and Açailândia. The increased demand for and dependency upon commodities not available within tribal societies is likely to precipitate the penetration of the cash-based economy into those societies, and the attractiveness of the marginally higher levels of income offered by wage employment.

The consequences of such pressure for cultures traditionally based on egalitarian, communal and subsistence forms of social and economic organisation and on sustainable exploitation of forest resources are obvious, and some specific examples will be considered later. However, the crucial issue for any evaluation of the future prospects for the tribal peoples of Brazilian

Amazonia is not simply the degree to which the irrational dynamic that presently determines the development process in the region can be controlled by rationally enlightened State intervention. Nor can discussion usefully be focused on the desirability of the indians' involvement in that process since, as I have already indicated, for many, if not most, communities integration into the regional economy is already a reality.

Rather, the question which both these approaches beg is a political one: what is the relationship of the tribal peoples to the planning, authorising and administration of the development process as it concerns them? What degree of control do they enjoy over the pace and extent of their involvement in that process? It is a question which inevitably raises similar ones regarding other groups in the region and within Brazilian society at large who are directly or indirectly affected by State-led projects and by the wider flux of development in Amazonia. These are addressed elsewhere in this volume and cannot but lead in their turn to the urgent necessity of confronting the structural contradictions which face Brazilian society and economy. However, it is only by beginning to answer that initial question, considering some specific cases not only in terms of the social impact of economic development on tribal peoples, but also in terms of the political, institutional and legal processes which have so consistently excluded them, that we can realistically arrive at proposals which arise out of *their* experience, which express *their* aspirations and which make them the subjects, rather than the objects or victims, of their future.

THE MILITARISATION OF DEVELOPMENT AND INDIGENIST POLICY FOR AMAZONIA: CALHA NORTE AND THE YANOMAMI

Ironically, one of the first concrete steps in the direction of defending the indians' territorial rights arose out of early considerations of what was to become a geopolitical strategy for the occupation and development of the Amazon basin. The Villas-Boas brothers, whose campaign resulted in the creation of the Xingú National Park in 1952, were members of an exploratory expedition to the region organised by the Vargas government with a view to establishing a series of emergency landing strips in order to control the expected wave of European immigration following the Second World War (Branford and Glock, 1985, p. 184).

But with the installation of the military regime in 1964, and its strategy of rapid, export-oriented growth, the contradiction between the Villas-Boas's protectionist, isolationist indian policy and the integrationist needs of regional economic development became acute.

For if the aim was the promotion of large-scale agroindustry, the turning over of huge tracts of land to the production of monocultures or beef, and later the exploitation of mineral reserves, then alternative small-scale and non export-oriented forms of occupation, such as peasant agriculture, and tribal hunter-gathering and subsistence farming, could not be permitted to predominate. The indians represent more than a physical obstacle to the regime's project to push back the frontiers of Amazonia – their cooperative, non-individualistic, non-accumulative, sustainable forms of economic and social organisation pose a political challenge to the capitalist model of development desired for the region. The sharp end of that contradiction is that such indigenous forms of socio-economic organisation have frequently been associated with expressions of active resistance to the regime. Thus the State-led development of Greater Carajás (see Figure 11.1) has partly served the military's need to guard against the possible re-emergence of guerrilla-style activity such as that conducted by Maoists in the Araguaia basin during the 1970s. Perhaps more importantly, through the creation of specialised administrative bodies, such as GETAT (Executive Group for the Lands of the Araguaia–Tocantins), the regime has sought to build a local power base in order to combat the influence of the radical Church in mobilising and supporting the struggles of peasants and indians (Branford and Glock, 1985, pp. 154–5).

Official indigenist policy during this period has increasingly reflected the same strategy of breaking down the indians' economic, social and political resistance to the expansion of the frontier. While successive Constitutions have paid lip-service to the principle of the tribal peoples' inalienable and exclusive rights to the occupation and use of their traditional lands (Carneiro da Cunha, 1987, pp. 53–101) the Civil Code and Statute of the Indian (1973) have assumed the inevitability and desirability of the indians' integration into national society – that is, the elimination of their distinct social and cultural identity (p. 218). Similarly, the chief obligation of the State institution responsible for indian affairs, FUNAI (National Indian Foundation), is the demarcation of tribal lands, or the implementation of the indians' constitutional guarantees of territorial security. Yet FUNAI's brief as laid down in the Statute of the Indian is the 'harmonious and gradual integration of the Indian into the national community'. The economic significance of this process was made crystal-clear by the second President of FUNAI, Costa Cavalcanti, who in 1969 announced: 'We do not want a marginalised indian, what we want is a producing indian, one integrated into the process of national development' (Beltrão, 1977, p. 26).

Thus FUNAI has more often than not stood in the way of the process of indian land demarcation. The deadline of 1978 laid down in the Statute of the Indian for the completion of demarcation of all tribal territories has come and gone; today less than 4 per cent of the land area traditionally occupied by the indians has full and effective legal protection (CEDI, 1987). While this can be partly explained as a function of the underfunding of the agency and its notorious record of corruption (more of which below), the principal

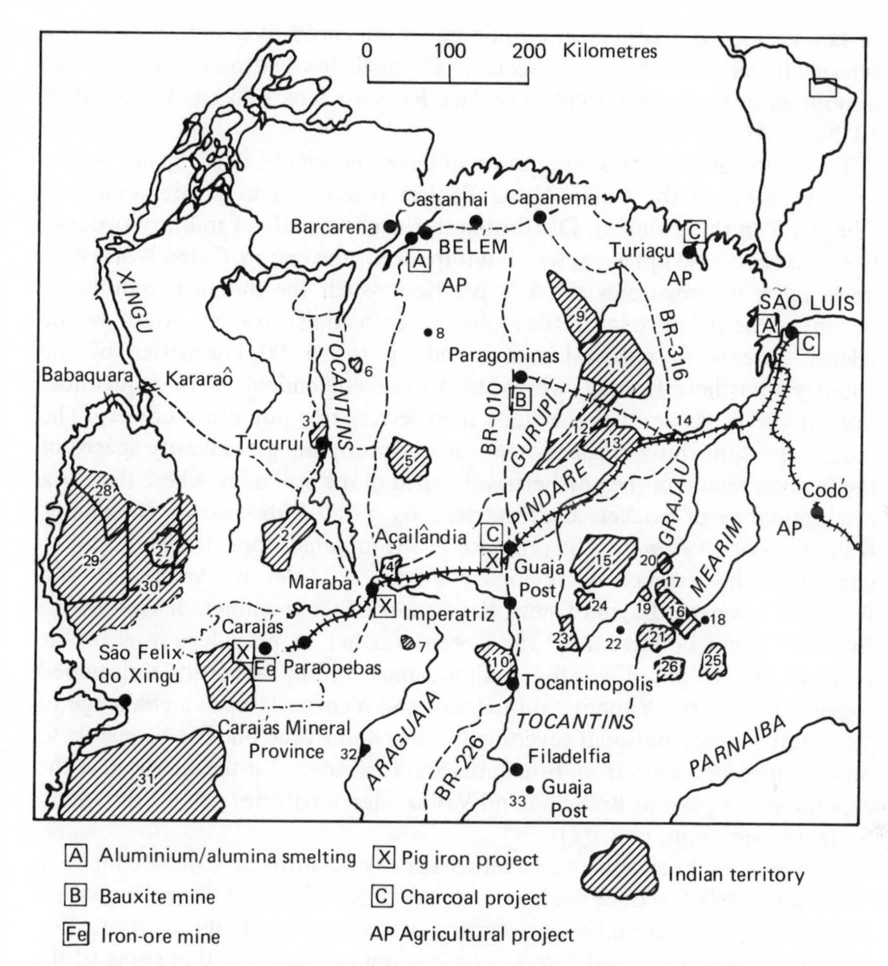

Figure 11.1 The Greater Carajás programme: projects and indian territories

reasons are political. During the course of at least five alterations to FUNAI's statutes, the process of identification, demarcation and regulation of indian lands has become increasingly distanced from even the faintest measure of public accountability or control, let alone that of the indians themselves. Before being signed by the President of the Republic, the relevant documentation must pass through the hands of a Regional Superintendency of FUNAI, the Superintendency for Land Affairs and an Interministerial Working Group including the Minister of the Interior and the Minister for Agrarian Reform and Development. In recent years the Interministerial Group has become progressively dominated by representatives of the National Security Council, who have consistently frozen the demarcation process in key areas. To complete the military stranglehold over indian affairs, a senior official of the National Security Council, Iris Pedro de Oliveira, was appointed in September 1988 to replace Romero Jucá Filho as President of FUNAI (Anon., 1988c).

The culmination of this integration of developmental and indigenist policy for Amazonia is the Calha Norte Project (examined in greater detail in Chapter 7 in this volume). Disturbing in the sheer scale of military occupation and economic upheaval which it implies for the region, Calha Norte also once again demonstrates the key position which the indian communities occupy at the cutting edge of the regime's expansionist strategy. Amongst the alleged threats to national security along the 6,500 kilometres of the country's northern frontier, which the project's defenders cite in its justification, is the presence of the indians themselves. For one army officer, 'The most important thing at the moment is to occupy the great empty spaces of the frontier region of the northern watershed of the Solimões, where there is a predominance of pockets characterised by a total absence of Brazilians, white men or *civilizados*. The presence of the indians, alone, is insufficient to guarantee the defence of regions such as we have in Amazonas and Roraima.' General Bayma Denis, the author of the document leaked to the Brazilian press in December 1986, went further in his allegations of the political threat posed by the indians, making the absurdly unfounded suggestion that the Yanomami of Brazil and Venezuela pose a challenge to the country's very national sovereignty: 'For some time one has been able to observe pressure, both from Brazilians and foreigners, for the creation, at the expense of the present Brazilian and Venezuelan territories – of a Yanomami State' (Anon., n.d., p. 3.1(a)).

However, as the massacre of steelworkers by the army at Volta Redonda in November 1988 has once again confirmed, 'internal security', or the guarantee of unimpeded capitalist exploitation of resources and labour, is of equal concern to Brazil's armed forces. For it is not coincidental that some of the largest tribal communities to be affected by the Calha Norte Project inhabit land containing valuable mineral and fossil fuel reserves. Gold is already being mined on Tukano and Baniwa lands on the Upper Rio Negro, and further south-west petroleum prospecting has led to violent conflicts with the Korubo of the Javari valley. In a key area of northern Roraima, 27 concessions have been granted to companies including BP and Anglo–American for mining operations on the Yanomami territory, which contains gold, diamonds and cassiterite. A further 363 applications have been lodged, representing a direct threat to one third of the indians' land (CEDI, 1988, 11, pp. 57–8).

The central objective of the indigenist policy and Special Projects established under Calha Norte is therefore to ensure unrestricted access to the mineral reserves located on tribal territories and to incorporate the indians themselves into the colonisation and development programmes planned for the region. In order to achieve this objective, one of the Project's stated policies determines that all indigenous communities inhabiting territories within 150 kilometres of the frontier are to be denied their constitutional right to legal and effective protection, or demarcation, of their lands. In their place, FUNAI is being allocated resources for the creation of 'indigenous colonies' in which the indians are to participate in 'Community Development Projects' and agricultural schemes. In addition, missionary and health workers fighting to provide medical protection for the indians against introduced disease, and to monitor and publicise their situation, have been expelled on the orders of FUNAI and the National Security Council. In total, up to 60,000 indians from 83 areas, including the 3,000 inhabitants of the Javari valley, the 18,000 of the Upper Rio Negro, the 18,500 Tikuna of the Upper Solimões and the 8,500 Yanomami of Roraima, face the loss of their right to territorial security, the devastation of their forests and the shattering of their social and cultural systems by the labour and commodity markets.

Several 'indigenous colonies' have already been earmarked for the lands of the 14,000 Macuxi of Roraima, and the indians of the so-called Tukano Triangle in the Upper Rio Negro have recently been struggling against the efforts of the local FUNAI administration and the National Security Council to coopt some of their leaders with promises of education, tractors and air travel (Anon., 1988a). The frontier policy has already clearly made a significant impact on the legal status of many indian territories; such were the delays faced by the Tikuna that they were forced to carry out the physical demarcation of their lands themselves and to remove a timber merchant who was carrying out illegal logging. He subsequently retaliated by organising a massacre of at least twelve Ticuna in March 1988 (Survival International, 1988a). Until late in 1988, a freeze had been imposed on the demarcation of eight territories in the state of Acre, that had been promised legal protection before the completion of paving of the BR-364 highway. More recently, possibly in anticipation of the constitutional changes of November 1988, FUNAI and the National Security Council have changed tack, accelerating the demarcation process in order to push through drastic reductions of the traditional areas recognised and to begin establishing the indian colonies. In a meeting lasting 45 minutes in September 1988, the Interministerial Working Group approved the demarcation of 19 areas and colonies in Acre, refusing to allow any hearing for the communities' representatives, who had travelled to Brasília to demand the recognition of continuous areas.

The most scandalous application of such policies has been the case of the Yanomami, who for more than ten years have been campaigning for the creation of a Park in northern Roraima (see Figure 11.2). Public pressure led FUNAI to take the preliminary step in 1982 of interdicting almost the whole of the Yanomami territory, but despite numerous proposals and intensive national and international lobbying no further progress has been made. The absence of effective legal protection opened the door to invasions by freelance gold-diggers and commercial mining companies, who quickly overran the area after the discovery of gold, diamonds and cassiterite in the late 1960s and mid-1970s. Pressure from miners, businessmen and local politicians for a more systematic, legalised presence on the Yanomami lands culminated in February 1985 when Surucucús, at the heartland of the territory, was invaded by 60 armed and uniformed men. The operation was planned and organised by the President of the Confederation of Trade Unions and Associations of Gold-prospectors of Legal Amazonia, José Altino Machado, who airlifted his men into the area with the cooperation of friends in the local council. Although eventually removed and temporarily arrested, Machado has repeatedly threatened a second, bigger invasion (Survival International, 1985).

However, such plans, and announcements in 1987 and early 1988 of a demarcation decree for the territory, were dramatically overtaken by one of the largest spontaneous movements of people to be witnessed in Amazonia. Within four months from the end of 1987, 20,000 gold-diggers had entered the Yanomami lands in the Couto de Magalhães area, arriving from all over the country at a rate of 100–200 every day by river, on foot and in small planes which were able to take advantage of the new and enlarged airstrips installed under the Calha Norte Project. Precipitated by the announcements of the area's demarcation, the gold rush had brought 40,000 miners into Roraima by the middle of 1988, in the knowledge that FUNAI had no adequate structure to deal with the invasion and that, as has been proved the case, the efforts of the Federal and Military Police to stem the flood would be ineffective. Official estimates of the numbers occupying the Yanomami lands by the end of 1988 were of the order of 100,000.

The latest government responses to this critical situation, far from seeking to restore to the indians their constitutional right to the exclusive occupation and use of their traditional lands, serves on the contrary to legalise the presence of the gold-diggers, to pave the way for the entry of large-scale commercial activities and to accelerate to a genocidal pace the physical, social and cultural disintegration of the Yanomami people. Already, following a meeting in January 1988 with the Governor of Roraima and representatives of the National Security Council, José Altino Machado had warned that: 'There is one thing of which we are certain, the indian areas will never be continuous as before'. Interministerial Decree no. 250, announced on 18 November 1988, gave legal substance to his words, superseding an earlier Decree (no. 160, of September 1988) which already foreshadowed the dispossession of 70 per cent of the indians' traditional lands. According to the earlier Decree, approximately 8 million hectares were to be embraced by the 'Yanomami Indian Land', a concept which has no basis in existing legislation. However, two thirds of this area was to be categorised as forest or national park, and thus subject to laws which do not take indian land rights

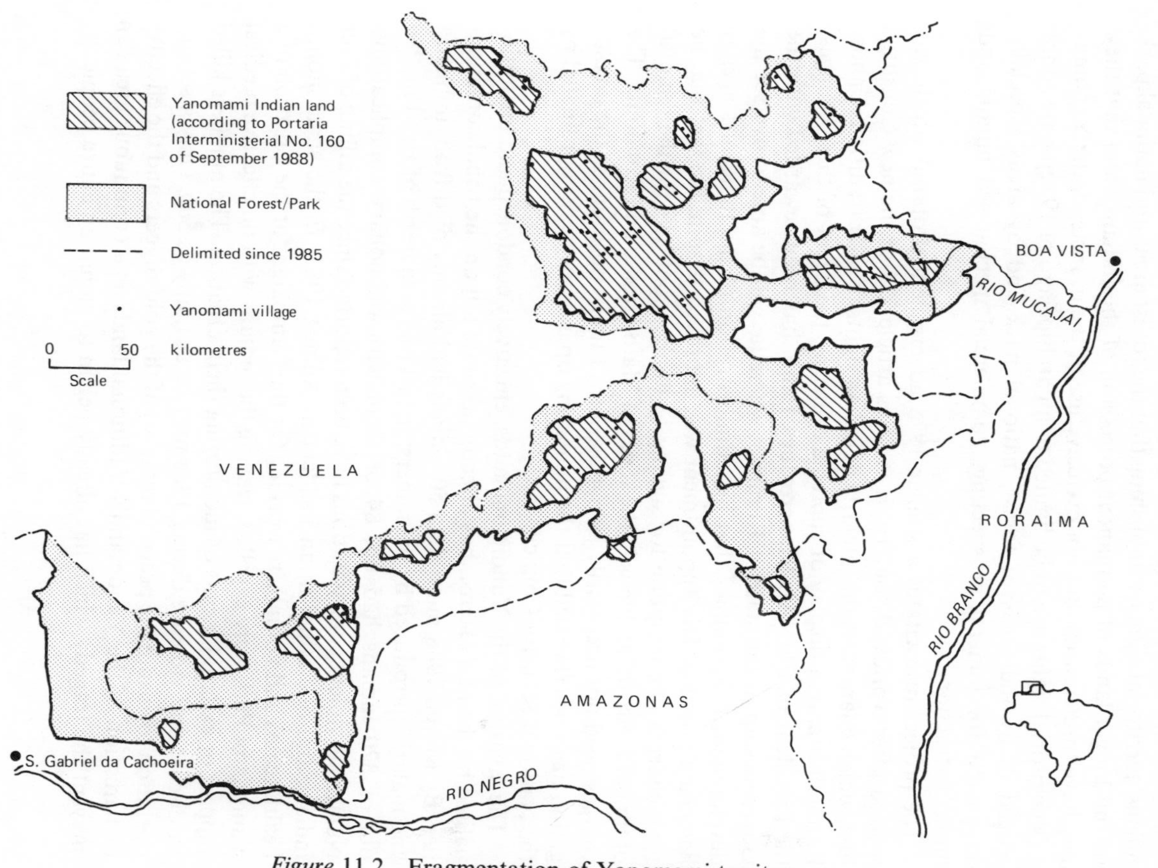

Figure 11.2 Fragmentation of Yanomami territory

into consideration, allowing for economic development under the administration of FUNAI and the Brazilian Institute of Forestry Development (IBDF). Decree no. 250 has removed all ambiguity from the question of the indians' territorial rights, establishing that the forest areas and national park are 'no longer lands of permanent possession of the Indians', and that they may have 'preferential' but not exclusive use of them. Instead, the Yanomami's territorial rights are to be limited to an archipelago of 19 separate zones hedged in by the areas of forest, national park and by areas especially designated for mineral prospecting, commercial mechanised mining and timber extraction.[1]

The project resuscitates a similar 1978 government proposal which was shelved after worldwide outcry. It violates article 234 of the new Constitution, which guarantees the integrity of indian rights to their traditional lands. Furthermore, in a display of supreme cynicism, the agent of these changes, the President of FUNAI, Romero Jucá Filho, has been rewarded with the Governorship of the state of Roraima, where he will be able to oversee the development programmes that his administration of the indigenist agency has made possible. His appointment was approved despite the fact that he faces charges of corruption for signing 13 lumber contracts for the removal of timber from indian lands in the states of Mato Grosso and Rondônia. The charges need not necessarily jeopardise Jucá Filho's position, as the investigation has since revealed that the lumbering operations were coordinated by the National Security Council.

The invasion of the Yanomami lands, effectively condoned and institutionalised by the new Decree, has already taken its toll on the tribal communities. By March 1988 more than 50 indians had already died from influenza and malaria introduced by the miners; in just one area north of Paa-Piu, 280 out of 320 indians had flu, 84 of them with pulmonary complications. Numerous violent confrontations have been reported since the killing of four indians and one miner in an incident in August 1987. In the same month health teams and volunteers working for the Campaign for the Creation of a Yanomami Park were expelled, leaving the indians with no effective medical support or external means of monitoring their situation. The numbers killed by disease and violence during 1988 may be as many as 500. Only a serious operation to ensure the peaceful removal of the gold-diggers and the effective demarcation of the Yanomami's traditional lands as a continuous area can prevent the tragedy which has already begun to engulf an entire people.

INTERNATIONAL FINANCE IN PARTNERSHIP WITH THE BRAZILIAN STATE: TRIBAL PEOPLES AND INDIGENIST POLICY IN THE GRANDE CARAJÁS AND XINGÚ BASIN HYDROELECTRIC PROGRAMMES

In the discussion so far of the recent impact of Amazonian development on the tribal peoples and the relationship of the latter to the implementation of economic and indigenist policies, in particular the Calha Norte Project, the chief protagonists have been Brazil's State institutions, the armed forces and FUNAI. However, two other major projects, the Grande Carajás Programme, which has already brought profound changes to Eastern Amazonia, and the Xingú Basin Hydroelectric Programme, on the threshold of implementation, now oblige us to examine an additional dimension – the role of international governments, private banks and multilateral financial institutions in influencing, for better or for worse, Brazil's internal policies on development, the environment and the tribal peoples of the region. For in both cases, these institutions, and in particular the World Bank and the EEC, have been and are in a position to learn from past experience and to avoid contributing to the further destruction of tribal societies and ways of life. To date, however, somewhat predictably, they have shown little evidence of the political will which could prevent those historical lessons from repeating themselves.

When it agreed a loan of US$ 600 million in September 1982 for a project to extract iron-ore from the Carajás mountains in Pará, and transport it via a 900-kilometre railway to a deep-sea port at São Luís, the European Coal and Steel Community endorsed a small, vaguely worded clause in the contract requiring that the project should respect human rights and environmental conditions in its area of influence. However, foremost in the mind of the Commission, to be sure, were the associated contracts which guaranteed EEC countries cheap, high-quality, iron ore amounting to an average of 13.6 million tons per annum over a period of fifteen years. The Community was eventually required to disburse only US$ 269 million of the loan, but the amounts of ore reaching West Germany, France, Belgium, Luxemburg, Italy, the UK and Spain have been rising steadily towards the agreed figure – estimates for 1988 total almost 11 million tons.[2]

It should not be surprising, then, that any potential conflict with the administrators of the Carajás Iron Ore Project, Companhia Vale do Rio Doce (CVRD), and the Brazilian government over the human rights and environmental issues should have been subordinated to the smooth running of the agreement. Nevertheless, it is worth examining the extent of the Commission's negligence, omission and complacency, and the consequences for the indians of Carajás, not least in the light of calls, as early as April 1982, by the General Assembly of Non-Governmental Organizations (NGOs), for the Commission to reconsider its support for the investment and to impose concrete conditions with regard to the protection of the environment and the tribal peoples of the region.[3]

In none of the six years of the Project's existence, again despite frequent questions and motions from NGOs and Members of the European Parliament, has the EEC carried out its own independent monitoring of the social and environmental impact of Carajás to ensure that it has complied with the loan agreement. Instead, it has relied, in the first place, upon the annual missions of the German banking consortium KfW,[4] which, as well as lacking the technical competence to assess such matters, is highly unlikely to criticise these aspects of the Project, since this might delay financial returns on its own US$ 122.5 million investment in the scheme.

Secondly, the Commission's very non-specific commitment to respect the social and environmental conditions of the loan is tied to the Amerindian Sub-Project which the World Bank attached to its own loan of US$ 304.5 million for the Project. It has accepted the Bank's repeated assurances in its quarterly reports that adequate protective measures were being taken. However, the Bank, in its turn, has not administered the Amerindian project itself, but entrusted it to the overall supervision of CVRD, which has channelled the initial US$ 13.6 million and subsequent funding to FUNAI. The incompetence of FUNAI to guarantee effective protection and land rights to Brazil's tribal peoples, indeed its political collusion in the process of tribal disintegration and assimilation into national society, have been described above. Moreover, by the time the Iron-Ore Project was being implemented, detailed, incontestable evidence of this was available to the World Bank in the form of its own commissioned report assessing FUNAI's programme of aid for the indian communities affected by the POLONOROESTE Project in Rondônia (Price, 1985). In this and another independent report by the human rights organisation Survival International, there was ample evidence of the military domination of FUNAI, its failure to carry out land demarcation and its deficient standards of health-care.

Yet these lessons were ignored and, without informing, let alone consulting, the 15,000 indians affected by the Iron-Ore Project, the scheme of 'Support for the Indigenous Communities' was imposed upon them in January 1982. Its priority has never been the legal and physical integrity of the communities' lands, the prequisite for any serious programme of protection. The initial budget allocated a derisory 1.8 per cent to land demarcation, which was raised to a still paltry 10.5 per cent in July 1986 as a result of external pressure, just a year before the scheme was due to terminate. Although it has now been extended indefinitely, there is no evidence to suggest that any progress is being made in guaranteeing territorial rights. 16 out of the 27 areas in the region still lack the necessary legal and physical demarcation and ratification to ensure their freedom from invasion and the preservation of their resources. Despite claims to the contrary by CVRD, less than 37 per cent of the area traditionally occupied by the tribal communities in the Carajás region has completed the legal process (CEDI, 1987).

In perhaps the most urgent case, that of the 250 or so Guajá of Maranhão, progress has in fact been retrograde. The Guajá are Brazilian Amazonia's last purely nomadic, hunter-gatherer people, ironically forced to abandon their originally sedentary culture by the first waves of colonisation 300 years ago. Now their forests in the Awá territory are under threat of destruction by the charcoal-burning pig-iron furnaces being set up along the Carajás railway in Marabá and Açailândia. In May 1988, a decree was ready for ratification by the President of the Republic, which would have demarcated an area of 147,500 hectares (this little more than half the recommended area) for the Guajá. However, in the following August an injunction suspending the demarcation was granted to 36 cattle ranchers who have claims totalling 111,000 hectares in the indians' traditional territory. The area demarcated is to be correspondingly reduced by a further 60 per cent to meet the farmers' demands (Anon., 1988b).

If FUNAI's ineffectual intervention, its connivance even, against the territorial security of the indians did not already deprive the Carajás Amerindian Project of all credibility, then the Project's administration by CVRD, a state mining company motivated first and foremost by commercial interests, hardly does less to inspire confidence in its success. Embarrassed by the devastating environmental consequences of 25 or so pig-iron furnaces adjacent to the Carajás railway, set to consume 10.3 million cubic metres or 300,000 hectares of forest per annum by 1991, CVRD has sought, like the World Bank and EEC, to disassociate the Iron-Ore Project from the wider developments of the Grande Carajás Programme, including the furnaces themselves. However, as the company's own Environmental Superintendent, Francisco F. de Assis Fonseca, has stated in an internal document, 'CVRD's responsibility is greatly increased by the fact that it holds the monopoly on the mineral and on the rail transport. Any iron and steel project in the northern region can only exist with the support of CVRD (Anon., 1987, p. 4). Historically, there is little doubt that the viability of the Iron-Ore Project derived from its role in providing the economic and transport infrastructure, the so-called Carajás Export Corridor, which was necessary to attract the other mining, metallurgical, timber and food-processing companies into the region.

Indeed, CVRD has not hesitated in taking advantage of this infrastructure. Its Docego subsidiary is one of several mining companies which invaded the lands of the Guajá in 1985. Having discovered a deposit of bauxite in the Awá area, CVRD refused to hand over the funds awaited for the demarcation of the territory, in an attempt to force FUNAI to remove the indians. Meanwhile, the Catete Indian Area, adjacent to the Carajás Mineral Reserve itself, has been carved up by Docego and CVRD's other 17 subsidiaries in the form of claims for mineral exploitation. This has attracted freelance gold-panners onto the lands of the Xikrín, leading to the pollution of their rivers with mercury and the exposure of the indians to malaria and other lethal diseases.

The 'Support for the Indigenous Communities Scheme' is concerned, not with safeguarding the indians' interests and rights, but with subordinating those interests and rights to the developmentalist plans for Carajás, incorporating the tribal peoples into the new regional economy. As Nester Jost, the Executive-Secretary of the Interministerial Council which oversees the Greater Carajás Programme, put it: 'The Indians will reach a degree of acculturation sufficient for them to be assimilated as workers on the Project' (Anon., 1982). In addition to using the scheme's resources to meet its own budgetary deficiencies, hiring unqualified personnel and making extravagant purchases of equipment, FUNAI has promoted ambitious agricultural projects. In the case of the Krikati of southern Maranhão, FUNAI's urge to maximise production has even meant contracting additional outside labour, reinforcing the presence of non-indians on their territory. Such has been the investment in the marketing of the produce that the Krikati have been left with little money to buy food before the annual harvest. For the Apinayé of northern Goiás, meanwhile, FUNAI's discriminatory allocation of resources has fomented a divisive rivalry between the communities' two main villages. While the rice-growing projects supported in the Mariazinha village have served to increase its financial dependency and its ties to local landowners, the less favoured São José village, which has led the campaign for demarcation of the Apinayé lands, has been identified as lazy and aggressive.

Such policies have therefore precipitated, rather than checked, the pressures towards economic integration and socio-cultural disintegration brought by the major disruptions of the Iron-Ore Project itself. 12 of the communities in the region have roads, the railway, electrical transmission lines or a combination of these passing through or close to their territories. In one location the railway, with its daily traffic of 12 trains, each with 160 wagons stretching 2 kilometres, comes to within 3 kilometres of a Gavião village. It also intersects the Rio Pindaré reserve of the Guajajara and runs adjacent to the Carú territory, acting as a flood-gate for thousands of landless settlers to enter the indians' lands and set up claims. The indians are thus forced to compete with the settlers for an ever-decreasing supply of game and farmland, heightening the atmosphere of aggression and violence.

The loss of the forest resources essential to their traditional economy, in its turn, is accelerating the cycle of integration into the regional economy. The Asurini of Pará, for instance, are suffering an increasing scarcity of fish, their chief source of protein, and have been encouraged to produce crops for the market on community farms, so neglecting their other subsistence activities and becoming ever more dependent on supplies purchased in the town. Similar pressures on the lands of the Suruí and Gaviões of Mãe Maria have created a dependency on their external sale of Brazil-nuts, which has been financed by resources from the 'Support Programme' and by compensation for the presence of electrical transmission lines on their lands.

These, then, are the consequences of the collaboration of international institutions such as the World Bank and EEC in a development process which they are content to see administered by State agencies whose proven record is one of collusion and active involvement in the violation of tribal people's rights and the destruction of their identities and social systems. Worse still, despite the principles endorsed in the Bank's published policy guidelines, that the tribal communities affected by Bank-supported projects must be consulted and their agreement obtained in advance of any project work, the indians have been systematically excluded from any participation in the planning and administration of projects and from any control over the application of resources supposedly intended for their benefit. The same non-accountable, authoritarian administration policy has been extended even to those non-governmental organisations with effective expertise in the field. The anthropologists appointed by CVRD and nominated by the Brazilian Association of Anthropology to accompany and advise on the work of FUNAI in the Carajás area found their work obstructed and their conclusions and recommendations overruled by the rigid budgetary priorities established in advance by the agency. As the anthropologists themselves have stated, it is vital 'to choose the appropriate institutional channels through which the financial funds intended for the protection of the indians will actually reach those communities ... the administration of funds through institutions subject to pressures contrary to the interests of the indians is a permanent risk. In this sense, the planning and allocation of resources should be made through agencies less subject to economic pressures, regulated by scientific bodies and universities and not just by private consultancies' (Vidal *et al.*, n.d., pp. 4–5).

Has this advice been absorbed by the World Bank, or is the experience just described to be repeated in the lives of the 4,000 indians who inhabit the Xingú river basin? The question is raised by the awaited decision (delayed until late February 1989, at the time of writing) of the World Bank over its proposed second loan of US$ 500 million to Brazil's Power Sector. If agreed, the loan will indirectly contribute to the financing of a series of hydroelectric dams on the Xingú, most immediately the so-called 'Altamira Complex', which is just one element in a much more ambitious energy strategy for the country, the 2010 Plan, which envisages the construction of 297 new hydroelectric plants. In Altamira itself, 7,200 square kilometres will be flooded, dispossessing 70,000 people including 1,400 indians from ten communities – the Xikrín, Asurini, Parakanã, Kararaô, Arara, Araweté, Juruna, Xipaia, Xipaia–Kuruaia and Kayapó. The impact will not be confined to the environmental consequences of the flooding; 100,000 people are expected to be attracted into the area by the construction work, bringing all of the socio-economic upheaval of the kind already experienced in the

Grande Carajás region (Comissão Pró-indio de São Paulo, 1988, pp. 6–8).

The response of the Brazilian state to efforts by two Kayapó leaders, Paiakã and Kubei, to publicise the issue, has set the tone for the likely administration of the project if it goes ahead. In July 1988 the two indians travelled to Washington to present their concerns to the World Bank and US Congress. On their return they found themselves and the US anthropologist who had accompanied them, charged under Brazil's 'Law of Foreigners' for having 'denigrated the image of Brazil abroad' (Survival International, 1988c).

Yet, for the World Bank, the Brazilian Power Sector loan represents a test case, the proof of the major reforms which it announced in 1987 to improve the social and environmental quality of its lending operations. In its note to the Development Committee Meeting of 13 April 1988, titled 'Environment and Development: implementing the World Bank's New Policies', the Bank specifically mentions its loan to the Brazilian Power Sector as an example of a loan 'which contains certain conditions concerning resettlement of affected populations in power project areas, staffing and equipping environmental protection agencies, and use and designation of wetlands'. Indeed, a requirement of the loan, imposed by the Bank on the first tranche approved in 1986, was that the Brazilian government would draw up an Environmental Master Plan before the disbursement of the second tranche.

However, not only is this Master Plan still in the preparatory stage, it provides for no specific protective measures and has been developed with almost no consultation with non-governmental institutions in Brazil, let alone with the peoples likely to be affected by specific projects. The Plan's authors, the state power company ELETROBRAS, make no evaluation or reference to any evaluation of past projects in order to identify the institutional, legal, policy, planning and funding problems which have prevented sound social and environmental practice in the execution of previous power projects. The Plan also fails to make clear by what structure of authority the executing agencies would be obliged to apply the provisions set out in the Plan. For example, at present it appears that ELETROBRAS does not have the authority to insist that certain procedures are enforced in projects carried out by its regional subsidiary, ELETRONORTE. This raises the question as to how any conditions imposed on the Power Sector Loan in the loan agreement with the World Bank would actually be enforced.

The two most recent examples of projects implemented by ELETRONORTE in Amazonia give historical substance to these concerns regarding the lack of the institutional safeguards and structures necessary for any programme which seriously seeks to guarantee the rights of local communities. 20,000 people were dispossessed when a quarter of a million hectares of forest were flooded by the Tucuruí dam, a scheme which supplies power to major industries in the Grande Carajás region. Amongst them were the Parakanã indians, who were subjected to a total of eleven moves during the official contacting and 'relocation' process. Such was the trauma of this upheaval and the impact of epidemics of malaria, venereal disease, influenza and dysentery introduced by FUNAI teams and construction crews, that for some time the indians gave up their traditional singing and dancing. Only in 1985, eight years after ELETRONORTE announced that the indians' lands were to be flooded, did the Parakanã receive the demarcation of a territory, and at that an area excluding valuable farmland, and occupied by settlers.

Similarly, the Waimiri–Atroari of Amazonas state had their lands demarcated only after the filling of the notorious Balbina dam had been started. Between 1975 and 1981, FUNAI carried out a series of relocations in order to clear the proposed Balbina catchment of indians, and to free ELETRONORTE of responsibility for indemnifications. The resettlement, hurried and virtually unplanned, has resulted in serious social conflicts amongst the Waimiri–Atroari, whose numbers have fallen from perhaps 1,000 in 1975 to no more than 400 today.

The World Bank's standards on tribal peoples are set out in its internal Operational Manual Statement 2.34, titled 'Tribal Peoples in Bank-financed projects'. Included in the Statement is the provision that adequate safeguards must be guaranteed before the Bank assists development projects encroaching on traditional tribal lands; that the Bank will support projects only when satisfied that the borrower or relevant government agency can implement measures to effectively safeguard the integrity and well-being of the tribal people; that projects must include a parallel programme providing for the recognition, demarcation and protection of tribal lands and resources, appropriate health and social services, the maintenance of the tribe's cultural integrity to the extent desired by the community itself, and the creation of a forum for the participation of the tribal people in decisions affecting them, and providing for adjudication and redress of grievances.

Yet nothing contained in ELETRONORTE's Environmental Master Plan so far indicates that these standards can possibly be adhered to under the present conditions envisaged for the implementation of the Xingú Basin hydroelectric programme. The administration of existing projects, such as the Grande Carajás Programme, whether with or without the participation of the World Bank, has demonstrated the total incompatibility of the interests of handling agencies such as CVRD, ELETRONORTE or FUNAI with the defence of tribal rights. The military domination of indigenist policy in Brazil as revealed by the Calha Norte Project has made nonsense of legal and institutional provisions for the protection of tribal territories and the participation of indigenous communities in the planning of regional development. It is therefore clear that if they support development projects in Brazilian Amazonia, foreign governments and international financial institutions will be collaborating, knowingly or by omission, in the dispossession of the tribal peoples' right to pursue their traditional way of life in the lands they have occupied for centuries, and of their right to determine for themselves the nature and extent of their involvement in the process of capitalist development in Amazonia.

THE PROSPECTS FOR THE TRIBAL PEOPLES OF AMAZONIA IN THE 1990s

The events which marked the preparation of Brazil's new Constitution are a reminder of the critical political position which the tribal peoples occupy in the struggle for control in the Amazonian economy. Not only that, as so often in the past, the situation of the indians is a touchstone of the wider struggles within Brazilian society for some genuine control by the majority over their lives.

In August 1987, the Constituent Assembly appeared to be on the verge of endorsing considerable improvements in the status and rights of the indians. That all changed dramatically when a leading daily newspaper, *O Estado de São Paulo*, mounted a week-long libellous campaign of front-page articles denouncing the entire pro-indian movement, in particular the Catholic-based Indian Missionary Council (CIMI), of participating in an international conspiracy. By defending the territorial rights of the indians, it was alleged, this conspiracy, supported by the mining interests of Europe and the United States, sought to undermine Brazilian sovereignty over its Amazon region and to prevent Brazilian minerals from entering onto the international market. Although subsequently exposed as the fraudulent fabrication of a major private mining company, Paranapanema, the campaign was temporarily successful in reversing the gains achieved for the indians thus far.

It is therefore all the more extraordinary an achievement that the intensive lobbying of tribal representatives and non-governmental organisations was able both to recover those gains and to produce a Constitution in which indian rights are significantly strengthened. According to the Constitution finally approved in November 1988:

> the integrationist thrust of existing indigenist legislation has been replaced by the affirmation of the Indians' right to cultural diversity;
> Indian land rights have been recognised as original, preceding the establishment of the state itself, and 'Indian lands' have been given a broader, more precise definition;
> affected Indian communities must be heard before Congress can approve requests for mining or hydro-power exploitation proposed for Indian lands;
> the temporary removal of Indian communities from their lands due to epidemics or reasons of national sovereignty must be approved by Congress;
> any claims on Indian lands by non-Indians are null and void, any

exceptions to this principle being subjected to complementary legislation still to be approved;

the legal competence of Indian individuals, their communities and organisations to initiate lawsuits in defence of their rights and interests is recognised, it being the obligation of the Public Ministry to intervene in the Indians' defence;

cases concerning Indian rights are the jurisdiction of the Federal Justice;

the State must complete the demarcation of all Indian lands within five years of the promulgation of the Constitution.

The Constitution represents a number of major advances, then, in the legal status of the indians, clarifying their right to the protection of their traditional lands; transferring from the Executive to the Legislative body the power to authorise industrial activities on indian lands, and ending the indians' 'relative incapacity' as minors subject to the legal and administrative 'tutorship' of the state agency, FUNAI. A potential space has therefore been opened up for the tribal peoples and their supporters to intervene directly in the political and juridical arenas in defence of their rights.

The capacity of the indians to take up this challenge has been proved by their participation in the lobbying of the Constituent Assembly, which received a daily presence of more than 200 representatives from over 30 indian nations. Since 1980 the tribal communities have had a national autonomous organisation, the Union of Indian Nations (UNI), which has supported individual campaigns over land demarcation and other issues, as well as representing the country's 220,000 indians at Congress and in negotiations with Government departments. Increasing numbers of indians have stood as candidates in municipal, state and federal elections, enabling them to develop their experience and understanding of the political process in Brazil.

A national and international network of non-governmental organisations now exists to advise, support and amplify the demands posed by the tribal communities themselves. The documented experience of developmental and indigenist policy and practice in Amazonia is more than sufficient for the deficiencies of this policy and practice to be identified and for demands for structural reform to be placed before the Brazilian government and those international institutions which support development projects. These demands might include the following:

1. The full demarcation and protection of all traditional indian lands as defined in the new Constitution must be the pre-condition for the approval of any new development projects in Amazonia, and must precede the beginning of any work on new projects.
2. All indian lands in areas of existing projects must be completely demarcated and any non-indians occupying them resettled and compensated where appropriate.

3. The planning of new projects must be opened to public debate in Brazil and must take into account the social and environmental consequences of previous projects as well as the foreseeable impact of the specific project under consideration.
4. Any official body responsible for administering indigenous affairs must be financially and politically independent from control by bother government departments or the armed forces, and must include the effective participation of tribal representatives and representatives of non-governmental organisations with expertise in the field.
5. The administration of development projects affecting tribal communities must not be entrusted to institutions whose economic or political interests are in conflict with those of the tribal communities themselves; the administering institution must be fully open to public accountability for its actions.
6. The tribal communities, through their own chosen representatives, must play a full and effective part in the process of Indian land demarcation and in the planning and administration of projects affecting them, and their authorisation must be obtained before any such project proceeds.

Clearly, a series of demands such as this, offering a minimal starting-point for the democratisation of the development process in Amazonia as far as the indians are concerned, proposes much more than is already available to the vast majority of Brazilian society in the way of political rights. As such, it assumes some fundamental changes in the distribution of power in Brazil and, where foreign participation in projects is concerned, in the control of wealth by the societies of the developed world. As is evident from the examples discussed above, Constitutional rights or World Bank guidelines are not of themselves any guarantee of the integrity of indian lands and societies. Indeed, the ban on the land demarcation in the Calha Norte region and the latest proposal to fragment the Yanomami lands are in blatant violation of the new Constitution.

In the short term, effective gains in terms of territorial protection and economic compensation have most often been those won through the organised direct action of the indians themselves. The Txukarramãe, for instance, who were isolated from the other tribes of the Xingú Park by the construction of a major highway in 1971, had conducted their campaign through legal channels for many years. In 1984, exasperated by the lack of official action, they blocked the highway, confiscated the ferry linking it across the Xingú river and took three FUNAI officials hostage, with the result that most of their demands were granted. The Kayapó of Gorotire brought goldmining activities on their lands to a halt by immobilising the aircraft used to transport workers in and out of the area, and so secured a percentage of all takings from the goldmining operations. At the time of writing, the Kayapó have planned a major gathering of tribal representatives from the Xingú basin in Altamira during late February 1989, for the creation of a permanent settlement in protest at the proposed hydroelectric programme for the region.

However, despite these successes and the unarguable courage of the tribal peoples, they are few and scattered in comparison with the concentrated manpower and resources at the disposal of landowners, mining companies and the State. Whatever the individual victories of isolated communities, Brazil's distorted structure of land ownership continues to exercise increasing pressure for the occupation of Amazonia by ranching and settlement; the burden of foreign debt and the irrational dynamic of capitalism are hurrying on the industrialisation of the region at a dizzying pace. Ultimately, any real advance in the control of the indians over the process of development in their lands depends upon the balance of political forces in society at large, upon the ability of workers to wrest control of the heart of Brazil's economic power, in the industrial centres of Belém, Recife, Belo Horizonte, Rio de Janeiro and especially São Paulo – and to return it to those whom it exploits most and benefits least.

The coincidence of two recent events – the defeat of the ruling PMDB party and the election of socialist mayors to 37 municipal councils, including São Paulo, following the massacre of workers at the strike-bound Volta Redonda steelworks in November 1988, and the murder of rubber-tappers' leader Chico Mendes in late December 1988, which provoked an unprecedented national outcry – has served to underline the identity between the urban and rural struggles. The Amazonian Alliance of the Peoples of the Forest, created in Acre in 1986 to unite the National Council of Rubber-Tappers and the Union of Indian Nations in their fight to defend their forests and way of life, has demonstrated how the realisation that all forest peoples face a common enemy can forge a common struggle (Survival International, 1987). In September 1988, the Rural Workers Union of Xapuri–Acre took that message to the 3rd National Congress of the free trade union confederation, CUT, calling on the Congress to defend the self-determination of the indian nations and to strengthen the ties of solidarity between the forest peoples of Amazonia and workers all over Brazil, in order to prevent the fragmentation of struggles against projects such as Calha Norte (CUT, 1988, p. 73).

The task ahead must be to build upon the incipient links between these individual struggles so as to challenge the impunity with which mining companies, government agencies and the armed forces have violated the rights of tribal peoples. Only the advance of this movement can guarantee the indians, as well as workers and peasants, the possibility of genuine control over their lives, not by waiting for it to be handed down by a benevolent state, but by taking it for themselves.

Notes

1. Survival International (1988a and 1988b) and communications from Centro Ecumênico de Documentação e Informação and Conselho Indigenista Missionário of 24 August 1988 and 30 November 1988 respectively.
2. Information from Stichtung International Union for Conservation of Nature and Natural Resources, Amsterdam.
3. Treece (1987) except where otherwise indicated, further information on the Grand Carajás Programme is taken from this source.
4. Information from meeting between Survival International and E. Cervino, Director of Credit and Investments, European Community, Luxembourg, on 25th February 1988.

References

Anon. (n.d.), *Projeto Calha Norte – Desenvolvimento e Segurança na Região ao Norte das Calhas ds Rios Solimões e Amazonas.*

Anon. (1982), 'Projeto Carajás: Nove Povos Indígenas na Rota do Extermínio', *Porantim*, 43, p. 9.

Anon. (1987), *Centrais de aço ao longo da E. F. Carajás – Estudo de viabilidade da KTS*, SUMEI/SUPES-186/87 (7 April 1987).

Anon. (1988a), 'Lideranças não aceitam colônias', *Porantim*, 110 (July/August) p. 3.

Anon. (1988b), 'Negociados 65% da terra Guajá', *Porantim*, 112 (October) p. 4.

Anon. (1988c), 'Militares conduzem a política de ocupação', *Porantim*, 113 (November) p. 3.

Beltrão, Luiz (1977), *O índio, um mito brasileiro*, Rio de Janeiro: Vozes.

Branford, S. and Glock, O. (1985), *The Last Frontier: Fighting Over Land in the Amazon*, London: Zed Books.

Carneiro da Cunha, Manuela (1987), *Os Direitos do índio: ensaios e documentos*, São Paulo: Brasiliense.

CEDI – Centro Ecumênico de Documentação e Informação (1987), *Terras Indígenas no Brasil*, São Paulo: CEDI/Museu Nacional.

CEDI (1988), *Empresas de mineração e terras indígenas na Amazônia*, São Paulo: CEDI/CONAGE.

Comissão Pró-índio de São Paulo (1988), *Campanha as hidrelétricas do Xingú e os povos indígenas: dossiê para coletiva de imprensa*, São Paulo: Comissão Pró-índio.

CUT – Central Única dos Trabalhadores (1988), *Boletim Nacional da CUT, edição especial.*

Frank, André Gunder (1968), *Capitalism and Underdevelopment in Latin America: historical studies of Chile and Brazil*, New York and London: Monthly Review Press.

Price, David (1985), The World Bank vs Native Peoples: A Consultant's View, *The Ecologist*, 15(1/2).

Quintanilla, Antonio Paulet (n.d.), La reforma agraria y las comunidades de indígenas, *La reforma agraria en el Perú*, Lima: Comisión para la Reforma y la Vivienda, cited in Frank (1968), p. 142.

Stavenhagen, Rodolfo (1963), Clases, colonialismo y aculturación: Ensayo sobre un sistema de relaciones interétnicas en Mesoamérica, *América Latina*, 6(4) (October–December) cited in Frank (1968) p. 135.

Survival International (1985), *Urgent Action Bulletin*, BRZ/1f/SEP/85.

Survival International (1987), Amazon Alliance, *Survival International News*, 17, p. 4.

Survival International (1988a), *Urgent Action Bulletin*, BRZ/1I/APR/88.
Survival International (1988b), *Urgent Action Bulletin*, BRZ/1j/SEP/88.
Survival International (1988c), *Urgent Action Bulletin*, BRZ/14/NOV/88.
Survival International (1988d), Brazilian Amazon: loggers murder Indians, *Survival International News*, 21, p. 1.
Treece, Dave (1987), *Bound in Misery and Iron: the impact of the Grande Carajás Programme on the Indians of Brazil*, London: Survival International.
Vidal, Lux, *et al.* (n.d.), Sobre a participação de antropólogos na assessoria de órgãos públicos ou de projetos de desenvolvimento regional (mimeo).

12 Small-farmer Protest in the Greater Carajás Programme

João Hébette and José Alberto Colares

In recent years Brazilian Amazonia has been the subject of many journalistic articles and conferences. The representatives of large capital and of the Brazilian State have prioritised the modernisation of this 'backward' region; large and ambitious projects include extensive highways, hydroelectric schemes, iron-ore, bauxite and gold mining, ports and industrial plant. Indigenous names are cited in newspaper and magazine associated with huge construction works, bulldozers, mechanical diggers, sophisticated turbines, cranes, floating factories: Carajás, Tucuruí, Oriximiná, Itaquí, Jarí. Modern buildings on ancient foundations, the future born from the past! From the crushing of the 'primitive' indian's past, from the 'backward' peasant, the 'illiterate' small farmer. The voice of the indian, the suffering of the peasant, the protest of the coloniser has been reported, at best, together with news about crime and rape. In practice it has been disseminated almost clandestinely and surreptitiously from one end of the country to the other, via the proliferation of community news-sheets, pamphlets and popular ballads so common amongst the poor, or through the short-lived underground press.

This is the muffled voice, the sorrow, the indignation, the protest, the curse, the testimony collected by chance during our researches on small farmers (*posseiros*). Expelled from the Centre, South and South-East of the country, they came in search of 'free lands' in Amazonia, a small plot to cultivate. We found them along the highways built during the past 30 years in the area known today as the Greater Carajás Programme. The focal point of this region today is the town of Marabá, isolated during the 1950s with its 57,000 inhabitants, bubbling today with trucks and people, its population grown to 150–200,000. Several highways meet here; the Transamazon (over 2,000 kilometres), the northern and southern sections of the state highway PA-150 (over 1,000 kilometres) and the old PA-70 which is today a federal road, the BR-222, linking the first two to the Belém–Brasília (almost 2,000 kilometres). During the last three decades the population has almost doubled, penetrating the forest and settling particularly on the outskirts of towns such as Marabá and Imperatriz, at construction sites, at the mines and gold-panning operations. From an average of 1 inhabitant per 10 square kilometres, the population density has risen to 1 person per square kilometre.

The evidence collected here was given by rural workers from the municipalities of Vila Rondon, São João do Araguaia, Jacundá, Marabá and Xinguara. They are but a stifled echo, timid and distant, of a cry of pain and indignant protest against those who monopolise the land. Their words have been reproduced literally, maintaining their spontaneity, respecting the authenticity of the spoken voice. Because of this, they demonstrate that it is often more important that the researcher listens rather than asking a multiplicity of questions. The testimonies have been grouped around different situations and phases in the settlement process; the PA-70, the PA-150, the 'Brazil-nut Polygon' (Marabá and São João do Araguaia), and Xinguara.

THE PA-70: THE APPEARANCE OF VIOLENCE AND THE APPRENTICESHIP OF PROTEST

The settlement history of the area which lies within the curve of the River Tocantins on its course between Imperatriz, Marabá and Tucuruí, west of the Belém–Brasília highway, is extremely complex and tragic. At the end of the 1950s, precisely when the highway was opened, speculators from the South appeared in order to take possession of large tracts of 'free land', covered by virgin forests rich in valuable hardwoods. Until then this land had been under the domain of various indigenous groups which lived there in relative peace. Not even the Portuguese colonisers during two centuries, nor Brazilians in over a hundred years, had penetrated the traditional habitat with its primordial owners. Within the space of a few years, however, this situation changed dramatically.

In 1959 a São Paulo businessman acquired an area of no less than one million hectares, with help from the state of Pará, in the municipality of São Domingos do Capim. In 1960 other *paulistas* took possession of two more areas, of 200,000 and 500,000 hectares respectively, at a place called Frades, in the municipality of Imperatriz. In 1967 an American set up home here with his family, supposedly in order to develop an agrolivestock project over 250,000 hectares. However, at the same time a large number of small farmers also arrived, expelled from other regions by the expansion of large estates and the greed of big landowners. The indians who lived in the area were then driven out by the *latifundiários*, with assistance from the federal agency charged with their protection, the National Indian Foundation (FUNAI). They managed to play off groups against each other; for example, small farmers against the Gavião indians, who ended up confined to an area of 60,000 hectares, the Mãe Maria reserve, but not without having risen up in protest, burning peasants' houses and crops, and killing a few in the process. Today the small reserve has been totally disfigured by a highway, power transmission line and railway which bisect the area in different directions.

Innumerable conflicts arose at this time between the estate-owners and newly-arrived peasant farmers. One such colonist thus describes his odyssey:

> I left [the state of] Espírito Santo and came to Pará. I arrived in Pará; I carried on working on fences, corrals and construction. But then I decided that this was not right and wanted to farm . . . I managed to get the use of some land from an estate owner in Vila Rondon and went to farm on the edge of the forest, which was only possible using a machete. I carried on farming this land . . . In forty days I cleared over one *alqueire* . . . It was a very good, very special plot.

It is revealing to note the system used by estate-owners in the region for converting forest to pasture at no cost to themselves, or even at a profit. They concede an area of forest to a landless farmer, nominally at no charge, so that he may clear it, plant his crops and, after the harvest, give back the land already sown to grass, sometimes converted to pasture, and even handing over half of the crops, as in this case. The colonist cannot understand the estate-owner who thinks he owns the produce of the land:

> The landowner arrived . . . the rice and corn were standing higher than a man. I was angry at seeing him on the plot with his wife and cattle, breaking the corn. So I thought, I will go over and ask him . . . because this isn't right. I suffered a lot and I am still suffering. But I will ask him in a spirit of love . . . I asked him, 'Mr Edward, why have you put your animals among my crops, this is a great shame . . . This land is yours but I want to be compensated for the work I have done so that I may leave'. He replied, 'Is this land yours?'. 'No, this land isn't mine, otherwise I wouldn't have asked your permission to cultivate a plot here'. 'Listen, I'm telling you, get out of here while I'm still in a good mood'. 'Listen friend, I am leaving but I want to sell what is mine because I can't stand this anymore. This isn't the first but the third time that this has happened to me.' So then he said, 'Listen, you shamefaced buzzard, disappear before I cut you up'. 'As you wish, I'm not quarrelling, I don't want a fight, I'm leaving'.

The smallholder was unarmed, the landowner was showing his machete and had a concealed revolver and said:

> 'Wait here, friend. I'll show you what I'm going to do for you'. He got off his horse and walked towards me so I watched him. When he got near . . . his wife shouted at me to run, asking my father-in-law to get me out of the house otherwise Edward would kill me. When he got close with his knife I decided he was too threatening and picked up a lump of wood . . . When he got near I let him have it and he backed off. I went to get another piece of wood to defend myself and said to myself that I would stand up to him

to the death if necessary. He then took out his gun and, from about ten metres, took aim at me. His wife grabbed him by the waist but he took a shot at me; the bullet missed me by about 70 centimetres. Then my father-in-law who was nearby came with his rifle to investigate; if he had found him, with me dead, he would have killed the landowner'.

One of the small farmers still remembers the killings that occurred in the region ten years previously:

'In 1975 a sad, terrible event took place. I can never forget what happened that year on Pedro Alves's land. The Reis family, real family people from Bahia, Minas Gerais and Espírito Santo and even some people from here; they were nearly all killed, one by one. Even to talk about it makes my hair stand on end. Only God can help us win this battle'.

Pedro Alves dos Santos had arrived from Bahia at Vila Rondon in 1969, where he bought a plot of 100 hectares. However, he started to expand his property on land from which the Gavião indians had been expelled, 'the people from here', as they were popularly known. He aspired to the ownership of an area of 35,000 hectares, on which families of small farmers had settled, including that of António dos Reis. Pedro Alves spread terror among the farmers who had, in the meantime, acquired from the land agency, INCRA, confirmation of their right of occupation. In spite of this, Pedro Alves managed to get police help to 'clear the area'; he threatened to have the farmers' tools confiscated and have them prohibited from working the land for 2 years, he had them arrested and imprisoned.

At the end of 1975 Pedro Alves sold the land to one Josélio who set up a ranch called Santa Fé which, to this day, stretches along 5 kilometres of the highway. The massacres continued, in the words of our informant, 'Pedro Alves had 86 widows on his conscience, Josélio six more'. In August 1976, on the pretext of having been summoned by INCRA, two rural workers were led into an ambush, António dos Reis and Onório. This is how, today, a farmer tells the story, much as it was reported by newspapers at the time:

They were the men who fought with us in our own defence. But they had a little more courage than us, poor devils. Onório was killed by the first bullet; they emptied six revolvers into António dos Reis. The more they fired the more he fought back.

They were led into a trap:

He was tricked when they said to him, 'Look, we have come to fetch you because you will get a land title. We told you about that institution in Belém that would settle all your land problems . . . Didn't we say we would

> come to get you? Quickly, hurry to receive your documents. The official is in Vila Rondon and is returning soon to Belém'.

Unfortunately, the peasants were duped by this smooth talking:

> The farmers went with them. When they got to the middle of the forest there were about twenty [gunmen] and they caught these wretches, they shot and shot but António dos Reis just wouldn't die. I don't know why but they got a motorised saw and used it on him. He was very wicked and cut António's neck to kill him ... He cut his head off and António still jumped like a chicken with a broken neck. So what did he do then? He wrapped the body in some cloth, together with that of Onório, tied them up well, placed the bundle in the back of a Ford F100 and made tracks towards Belém. In Paragominas he threw the bodies into a river ... A fisherwoman saw them but she would not speak out, afraid they might also kill her and dump her in the river.

Pedro Alves dos Santos and Josélio de Barroa were caught and detained for ten months, after which they were released.

THE PA-150: ORGANISED RESISTANCE

The southern part of state highway PA-150, to the north of the PA-70, was opened in 1977, in an area which had until recently been under the control of indigenous groups. Its purpose was to facilitate haulage of heavy equipment for the construction of the Tucuruí hydroelectric scheme, the fourth largest in the world (8,000 MW). The opening of the road encouraged the arrival of peasants in search of land, also attracting land speculators and land-grabbers. Some of the latter came from other regions where they had already been practising violence, such as António Abreu, accused of murdering a rural worker who served as bodyguard for the President of the Rural Union, in the town of Bom Jardim, Maranhão. Hundreds of small farmers peacefully settled on unoccupied land near the PA-150 were evicted or threatened; hundreds of houses and barns were burned down and workers beaten by hired thugs.

The president of the Rural Workers' Union of Nova Jacundá, a young peasant woman from Maranhão, recounted the following:

> I am from Maranhão, a peasant's daughter. We came to Pará in 1981. I am now 22 years old ... On the day I arrived, 4 January, I was immediately shocked. Only a few days before they had murdered Zé Piau.

Jose Manuel de Lima, known as 'Zé Piau', was a militant activist of the

Association for the Defence of United Workers in Nova Jacundá, a professional organisation. He had a small plot of land within an area of 18,000 hectares where the National Institute for Colonisation and Land Reform (INCRA) had authorised the resettlement of some 100 small farmers. The area was claimed by a large landowner who was protected by an army of gunmen. Zé Piau, defender of these harassed farmers, was killed in front of his own house by six gunmen who had, the year before, already murdered a rural worker by the name of Lourival Marques and wounded another, paralysing him. The president commented:

> I thought it strange when I heard about it. A rural worker being murdered . . . because there, back in my native land . . . I had really never heard of a farmer being killed by a gunman in a land dispute.

Along the PA-150 a climate of terror had been created which affected entire families:

> It was very difficult for me to face up to the situation here in Pará, you know. I saw the suffering of women being threatened such as in the Gleba Pitanga, in Madeira and others where the direct threats of gunmen and landowners were difficult to accept. We saw the anguish of women who had to put up with that repression.

People lived then, as they do today in many places, in a climate of continuous tension:

> Sometimes the husband had to leave home in order to hide because someone wanted to kill him; and the wife had to stay at home with the children, looking after everything.

Even then, however, they did not leave the wife alone:

> Sometimes the gunman would arrive and ask where he [the husband] was. I would say, 'He is away, he is in such-and-such a place, and I don't know when he will be back. At times he was actually hiding inside the house. The wife had to put up with all of this, knowing that she too could also be killed; just as they were sometimes tortured and had their houses burned down'.

The strength of the small farmers lay in their cohesion; the unity of the married couple, the solidarity of friends, the support of religious workers:

> The women reacted very well at that time and were a very important source of strength. When groups of rural workers occupied land, the

women were courageous enough to get together and accompany them, not allowing their husbands to enter alone. She also stayed there, risking her own life and those of her children.

The police often cooperated with the gunmen in practising violence; they travelled openly in landgrabbers' vehicles; they drank in bars and clubs with them:

> They [the policemen] often threatened rural workers; they even went after Father Paul and Sister Dorothy. The number of times they set up ambushes for the Father and Sister, who gave us so much support in this area.

They even killed to steal, as happened in Goianêsia with a couple and their two little daughters:

> In 1980 two little children were martyred, Saints Elineuza and Elizabeth, and every year we go on a pilgrimage to the chapel, known as Saint Elineuza, were they are buried; one was two-and-a-half and the other was four years of age. The mother was stabbed 17 times by a policeman but survived; the two children were burned to death. Several cases like this have occurred.

This woman was travelling along the PA-70 with her husband, his brother and the two daughters:

> They were travelling by cart selling clothes and suchlike. And they [the policemen] hitched a ride; they killed the husband and the brother, then killed the two children and attacked the woman. They thought she was dead; they are perverted because they care nothing for the lives of workers.

According to the union president, this was no isolated example and she maintained that several similar incidents had taken place. Young and sensitive to the situation of her companions, the female president emphasised how hard is the peasant woman's life:

> The rural woman suffers a lot. Firstly because she has a double work-load through having to look after two things, three, four. In farming she sows, harvests, helps to prepare the land. Sometimes she clears it as well, she helps burn the vegetation and collect firewood. She does the same work a man does. The only thing she doesn't do in this region is chop down trees. Even so, in this municipality, there are cases of women who also do this when necessary. Apart from this work in the fields, she is also a mother

and must take care of the home; wash clothes, cook the food, look after the children.

The burden of domestic work is greater still in the case of women living on their own, whether divorced, a single mother or abandoned, or the wife of a man who is sick, injured or who has simply disappeared.

> Sometimes she is alone, she is a widow; she has a house full of small children, is without money and has no choice but to take up the machete and clear the bush. She alone prepares and works her land, right up to the harvest.

Even when she has a man, the woman is responsible for most domestic work, a fact which is looked upon critically by the president:

> Our workers, our people, are too complacent about work in the home. They allow the woman to take charge of several duties which are also the man's responsibility, but they place a double burden upon the woman's shoulders.

The excessive demands made upon women in the home are matched by a lack of confidence in her ability to perform a public role although, according to the president, this is changing. As secretary of the union, she had to replace the deposed ex-president:

> When I took over the presidency, the fact that I was a woman caused great concern. Some people said that I would not be able to cope. I was what? Twenty years old. Even the women themselves didn't believe that I, as a woman, was up to the job. Of 14 members on the committee, some resigned on the grounds that they would not work under a woman boss.

This case demonstrates that a woman often shows more fighting spirit in support of peasants' rights:

> When we make our demands, for example a road or school, committees are formed and all you can see are women. There are some men present but many believe that women have greater strength to fight the mayor or the authorities to press home these demands.

The president highlighted the participation of a religious worker in this battle:

> To this day Sister Dorothy is much-admired. She was a great source of strength here in the region. At the beginning, which was a time of much

repression by police and gunmen, she was a tremendous support. She was persecuted several times, she was shot at but never injured. She took part in meetings with peasant farmers and went with them to the government, to the authorities, to pursue these causes. A Sister who understands that we must change this society, with all its inequalities, is a source of great strength.

THE BRAZIL-NUT POLYGON: FROM RESISTANCE TO DIRECT ACTION

Until the 1970s the municipalities of Marabá and São João do Araguaia, situated on the west bank of the rivers Tocantins and Araguaia, were still totally covered by forest. The population was very sparse and lived on seasonal work: diamond prospecting in the Tocantins river, hunting of wild animals and the sale of skins, and Brazil-nut gathering. The nut-bearing forests were gradually brought under the control of a mercantile oligarchy, which was favoured by the Pará state government with leases for areas of 3,000 and 4,000 hectares that were eventually claimed as their own private property. Furthermore, they illegally expanded these areas to include 'excess' or 'surplus' land and in this way they hoped to become the exclusive owners of the 'Brazil-Nut Polygon'.

A rural union president explained the background to the peasants' decision to carry out their own 'agrarian reform':

> The surplus is accounted for by the fact that someone may have a land title for 10,000 hectares, but the actual property boundary may cover 20,000, 15,000 or 12,000 hectares; sometimes there are 6,000, 5,000 or 3,000 hectares more than the figure officially declared in the land registry.

They maintained a strict watch over the estates by means of private guards or *fiscais*:

> The *fiscal* guarded the area on behalf of the self-appointed owner. What for? To stop anyone from hunting inside, to prevent them from cutting down the trees, to keep everyone out except the Brazil-nut collectors (*castanheiros*). The guards patrolled the area permanently and were armed. If they found anyone inside they might even kill them. These were their orders.

The Brazil-nut gatherers spent most of their lives in the forest:

> In the largest Brazil-nut regions there were many collectors who would gather nuts in the winter and, in the summer months, clear shrubland,

> making fences and rope, stripping more Brazil-nut trees for the owner. So these people used to spend all their time in the woods.

At the end of the 1950s and start of the 1960s there was much talk in Latin America of agrarian reform. Brazil even signed an agreement at Punta del Este to carry this out:

> In the early1960s I was a nut-collector in Pará, mainly in the Marabá region. On the radio I listened to politics and heard about agrarian reform, which they said they were going to do.

This was one of the reasons behind the military coup of 1964, so that instead of agrarian reform people afterwards talked about colonising Amazonia. 'People were very enthusiastic and flocked to Pará. When they arrived, there was no land left'.

After the military regime left office, there was renewed talk of agrarian reform; a Ministry of Agrarian Reform was even created. Many hopes were raised, soon to be dashed. The acquisition of a small plot of land was only possible through the actions of the workers. The Brazil-nut collectors already occupied these lands, whose boundaries and 'surpluses' they were familiar with:

> These people knew exactly what was happening, and when they found out that agrarian reform no longer existed, many who already worked the land said to themselves, 'I shall not leave here because this land has no owner', and they started to occupy it . . . The beginning of the occupation movement, in 1981–82, came immediately after the peak of the military regime, which was very repressive. When small farmers started to occupy the land in a disorganised way they were beaten by the landlords, who called in the police and gave us a very bad time.

The lessons of occupation without organisation were very hard:

> During this period, from 1980–82, we were not at all active in the area. It consisted simply of landless farmers (*posseiros*) trying to carry out a spontaneous occupation on their own, with no assistance from any organisation. On the first occasions the repression was fierce. They were arrested and beaten. But they returned to the area . . . As a community leader I had various meetings with the local population in areas undergoing land invasions. Reading the Bible, we saw that the first Christians suffered a similar fate. And they had to find a way of organising themselves . . . Nobody can do anything on their own; if anything has to be done it must be in an organised way, because the strength of the *latifúndio* is great indeed.

Police repression was considerable:

> Often in the city it was impossible to meet to discuss our more serious problems because persecution still existed . . . The National Security Law, SNI (National Intelligence Service) informers who lived in the area accused us of holding terrorist meetings . . . It was often difficult to get together in the town, for as soon as three of four people met the police arrived and beat us. So we decided to hold our meetings on the lands where the problem itself had arisen . . . If it was a question of taking legal action we set out from there with the decision taken.

Due to the severity of landowners' and police repression, the land occupation process would stand for no weakness or cowardice:

> Whoever took part in an occupation had to be prepared to stay or be arrested and suffer, but never to give in. Whoever gave up and left lost any rights . . . People went in, they battled and put up a fight; then someone who had given up would arrive and claim the land for his own. So it was decided that these people had no further right to return to the area. He ran . . . so why come back here? You must be joking! We are here putting our lives on the line. If a colleague runs away at the most difficult time, why should he be allowed back?

This was no joke, no game. During one such occupation the farmers were on a war footing; they cut off access to the area to prevent the entry of gunmen and they dug themselves in . . . A peasant woman tells of their suffering:

> We've been through a lot here. We fought to survive, to get our little piece of land . . . One evening, at six o'clock when we were having supper, we heard that gunmen were on their way to attack us. So we ran to gather our children together, six altogether, and prepare ourselves. We went to the next house. There were ten of us women in the neighbour's house and two men, while others kept watch on the road from trenches so that if the gunmen did come they could protect us. Some of us had rifles, sickles, hoes and ploughs. My husband had no gun but took a sickle. Others had no proper weapon but had a machete instead. We suffered a lot in all of this confusion. The children cried; they were afraid they might die. A twelve-year-old daughter of mine could not sleep at night but only lay down alongside us, and she said, 'Can you hear, mum? They are circling the house. Can you hear?' I was also very nervous and called my husband, 'Tonio, I'm not staying here anymore, I'm too afraid; there are gunmen on the way'. We would hear footsteps outside and thought that it was the gunman. We all became very afraid . . . So my husband took the most

scared daughter to São Domingos and we stayed there a while. But after three days she said to us, 'If you are going to be killed by these bandits I am going to die with you; I'm not staying here'. We suffered an awful lot, you understand . . . we fled barefoot even because we had no time to pick up our shoes and those gunmen were heavily armed, and all we and the children had were sickles. It was cowardly of us to flee and abandon our plates of food, running away in fear of the thugs, hiding among the cassava plants, behind the leaves. We were cowards. What were we to do for food? We worked the land, grated our cassava, made flour for drying in the pan and made tapioca to feed ourselves. We planted yams, okra, tomatoes. On troubled days tapioca was the easiest; there was no time to cook beans, not time even to light the fire because we were so worried. For the same reason we couldn't slaughter a chicken nor enjoy eating it in case the gunmen were about to descend upon us. So this was our struggle. Sometimes my husband remained and was more courageous. He dehusked the rice, put it into sacks and took it to the neighbour's house where the men were on guard. The women cooked white rice because there was no time to make beans, kill a chicken or prepare proper food.

They practised military discipline:

When there was an attack everyone assembled in the village. The women took care of the children and the food for the men, while we manned the barricades. Quite often one of us would run away, pretending to be ill, with his children, cattle and animals, losing chickens all over the road and abandoning the ditches. But I stayed in the trenches and went from one to another, organising. So then they started using a pathway from Pedra de Amolar and started to use that. So when we discovered what they were up to we stopped them and made them go back. 'I'm not well', they would say. 'But you will not die, your case is not fatal', we would reply. So we turned them back and made them man the defences. Those who were too afraid had to help the women in the village, carry water and prepare food for the others. This is what we used to do.

Those who were not organised, who did not take care to defend themselves, became easy targets, as recounted by one villager who barely escaped:

The first land conflict I experienced here in Pará was on the Dois Irmãos estate of Almir Morães. I went to visit my father's place; when I got there I went to hunt in the forest with my twelve-year-old son and pick some *cupú* fruit to bring home.

This innocent activity almost cost him his life:

> When I got there I was grabbed by a band of 20 gunmen, me and three friends. And do you know how I managed to escape from the grasp of these 20 gunmen? . . . by telling the truth, because I could not tell a lie. They wanted me to take them to those that had invaded the land but I didn't want to implicate my friends. So I told them if they wanted to kill me they could do so, but I didn't know these people, I didn't know anyone around there. So they replied, 'O.K., we shall kill your three companions but as you don't know the names of these people we'll set you free. Then they accompanied me to my father's house'.

The violence and destruction started there:

> When they got there they burned down 17 houses, completely. My father had a small shop and they stole all the merchandise. At that time I had nothing to do with land conflicts and didn't live there; I had only gone to visit.

Not satisfied with these acts of vandalism, they also killed several farmers:

> They took my friends and killed them. They cut them into pieces, little pieces, threw them into sacks and dumped them into the river. I didn't see it but I saw the bones of all three. They were displayed in the church for all to see.

They went on to have another, similarly tragic, adventure:

> As a result of this problem we left another land occupation in the area owned by the Bamerindus bank. We were there for a year and five months, more or less. All we had was trouble; gunmen, soldiers. In the end 92 soldiers came and took everyone inside the Monte Santo estate; we were arrested, beaten and herded together on the back of a truck, only one truck. It carried the soldiers and about 70 farmers, so it was very cramped. So all we got for our pains was prison, and we were forced by the landowner and police to give up our struggle to stay on the land. So we were homeless and went to look for another area to occupy.

This happened in February 1987 during an operation by the Ministry of Justice to disarm the countryside, during which 32 people were tortured and two women raped. Violence was built into normal police tactics:

> When people manned the defences, where were the police then? They never dared to enter. They grab people on the road to humiliate and intimidate them. They make the poor farmer taste bullets until he gives up and leaves the area. They beat one poor old man too much . . . Another time they knocked a young lad over a bridge and he broke his ribs.

The pioneers' struggle has made it easier for the more recently-arrived migrants with no resources, as explained by one person from the neighbouring state of Maranhão:

> I came from Maranhão with much sacrifice ... with a family ... producing more children. I came to São Domingos do Araguaia and went to work for Pedro Borba, a landowner. He put me on a small piece of land and I prepared a plot. We had a good harvest because we know how to farm. The following year, when I wanted another plot, he wouldn't give it to me and I was left without any land or anywhere to live. I came to live in the village of São Domingos and worked as a day labourer here and there. The family was getting larger, as it still is today. So then the chance to work this land came up and they said I could help them in the struggle, together with the others. I thought it was a good idea. I have already been here four years. My children have never suffered like that again or gone hungry, or been without rice or beans to eat.

1985 saw the most tragic massacre known in this region, on the Castanhal Ubá estate, which belonged to an important Marabá family. This family of merchants used to gather Brazil-nuts on a 4,000-hectare area, but had latterly ceased such activities. Some 20 nut-collectors had settled on the estate and farmed the land. In April 1985 another ten families arrived, in search of land to cultivate. They started to clear the scrub ready for chopping down the trees and farming:

> I think it was on 18 May that he [the landowner] fixed a meeting with the farmers to say that he would come to some arrangement with them over the area. This was only way to identify those on the property.

With this information in hand, the owner planned the massacre:

> After this meeting, this identification, he went back on his word and hired some gunmen. Without any further warning, the small farmers were attacked.

The landowner himself took charge of the slaughter.

> On 18 July they were attacked by Sebastião da Terezona and his band of gunmen, who murdered seven people there.

Sebastião da Terezona was the most feared gunman in the region:

> They were taken by surprise because they were all preparing the land, scattered over the forest. The gunman, accompanied by the landowner in a Corcel car, entered the forest and found them at their work. And they

went in firing, attacking people in their groups of two, three, four or even five people. The five bandits managed to kill a few, while others fled.

The brother of one of the settlers was a member of the gang:

I was warned by my own brother to leave the area immediately unless I wanted to die, that I should disappear because he and his friends were going to attack ... They managed to kill seven, including a woman, a pregnant woman, the young wife of one farmer, who was busy cooking.

The bodies were left there.

The police went there, identified the bodies and managed to retrieve a few ... three bodies remained until the day we were able to bury only the bones ... We took the lawyer and he confirmed the existence of the bodies but we could not take them with us because they had decomposed.

The bandits were not satisfied with this massacre. Payment for the job may have depended also on the head of the community leader, Zé Pretinho, who was a natural leader of people, and had lived in the region a long time. He had taken part in the occupation of the Brazil-nut estate but was not there at the time of the tragedy. 'On 18 June they were at his house at kilometre 40 on the Transamazon highway. They killed him inside his own house; him and his nephew called Valdemar'. It was in broad daylight, in the morning. Zé Pretinho's widow, mother of five children, finds it hard to tell the story:

They committed the first murders on 13th. They killed a woman who was three months pregnant. On 18th they came to our house. When they arrived I was sitting down like this, with my back to them. And they asked, 'Where is Zé Pretinho?' ... There were three of them, but I can only remember the name of one, Sebastião. All they could ask was, 'Where is Zé Pretinho?' So I got up to warn him that the gunmen had arrived. As I went into the kitchen they passed by me and said he was under arrest. Then I heard him ask. 'Why am I under arrest?' And the gunman said, 'No talking!' and told him to prepare his clothes. As I turned around the revolver was pointed at my head and I didn't know what to do. So I went out of the kitchen door and thought of returning to collect the children who were all inside ... Then they started shooting and I couldn't enter anymore. After that I can't tell you what happened.

All she can remember is what others told her afterwards, 'They shot him five times, even in his navel. When he came out of the house he raised his hand, like this ... and the boy was saying, 'Daddy, get up, daddy'. She added, 'I was eight months pregnant and lost the child'. A neighbour told us how she felt after these barbaric acts:

> It was heartbreaking and I felt really sorry for her ... I thought that in future there would be less killing. But you turn one way, and I don't know how many people were killed; you turn the other way, even more deaths. They murdered some at Cuxiú, and who knows where else. My God, what kind of assault upon us is this? My God, how terrible, dear Father in Heaven. I often stopped my husband Joaquim from going to parties, to certain gatherings, certain meetings, in fear ... At this time a story was even circulating that a landowner in Marabá kept 40 gunmen in an apartment for the sole purpose of killing people. That's when I got really, really frightened.

The municipality of Xinguara, which includes the southern part of the PA-150, has in recent years been one of the strongest focii of violence inspired by the *latifúndio*. At the beginning of the 1960s, when it still formed part of the municipality of Conceição do Araguaia, the area now known as Xinguara was sub-divided into dozens of 4,356-hectare quadrants (equal to half a 'league', the norm for a medium-sized estate) which the state of Pará then distributed to speculators from southern Brazil, who rarely worked the land. There were so many speculative transactions that today nobody really knows who are the rightful owners. In any case, these lands were settled peacefully by small farmers anxious to support their families who, by law, acquired the right of usufruct.

Such was the case of Maria Aparecida : 'We came from Ourilândia. My husband got some land in the Santa Paula area ... He settled there because many others were doing the same thing, and we have no money to buy land'. They started to work the land peacefully:

> When he had already set himself up, his plot prepared and everything, the people here were called to the INCRA office here in Marabá. They went and the judge told them that they could stay on the land, that there would be no problems, gunmen would not trouble them and the police would not bother anyone.

Within the government everyone gives orders but nobody takes them, so one person's word is not binding on another; it is the police who have the guns, however:

> We were in the forest when suddenly there were police everywhere throwing everybody out and confiscating our weapons; and some very ugly threats were made ... They were very cowardly because they burned

> down our houses, they burned all our possessions . . . They told everyone to leave then arrested them and brought them to Marabá.

Maria Aparecida's husband had avoided being arrested, but at the time she did not realise:

> My husband wasn't arrested. He wasn't caught because he was walking in the forest with some others and he was late getting back. The police arrived and took those who were in the hut, 18 men they took. Some were left behind and said they should follow those who had been arrested to find out what was happening.

These events were repeated dozens of times in Pará. Hundreds of women were left on their own to face a hard life, as if in wartime:

> I stayed there in the forest. I am a woman surrounded by children, I have five. So I stayed there, unable to leave and I didn't even know what had happened to the others. So I left my children with friends and came to find them. When I got there I discovered what had happened.

What solution could this woman find? Take her family to the periphery of a large town, to live in a slum with no guarantee of a job, ceasing to grow the food that so many families need?

> Take a woman such as myself, surrounded by children, I can't live in a city because that life is not for everyone. Because if everybody went to live in the city, I believe that all Brazilians would die. Some must work on the land and others in town.

Rural workers suffer all manner of humiliation, aggression and violence. They cannot get land to cultivate, nor employment in the city. Nonetheless, they are conscious of the fact that it is they who farm to feed the nation:

> I have never seen an estate-owner carrying a sickle. Only poor people . . . I have never seen a landowner with a hoe or a rice-planter or suchlike. He has the money . . . He can buy any machine he wants but without arms he can produce nothing . . . We feed Brazil and even overseas as well. When our needs are satisfied, the rest can be exported. So they could have some consideration for our class of people, and not slaughter us as they are doing at the moment, fathers of families with 3, 4, 5, 6, 10 children. Or kill all the 10 children together, as happened here at Surubim. Is it only the landowners who deserve to be treated well? Are they the only Brazilians? Don't we need to survive also? And what if we all die, will they survive without us?

Aparecida is no longer under any illusions regarding the agrarian reform announced by the government. The farmers must themselves acquire the land:

> The only way we can get some land is through this struggle. Agrarian reform is an impossibility because the men that could bring it about choose not to, so that it fails. So the only way we can get a small piece of land is by sacrificing the lives of all our friends, that's the situation.

Part III
Towards Sustainable Development

13 The Nature and Sustainability of Brazilian Amazon Soils

Peter A. Furley

Around 45 per cent of the world's humid tropical forests and savannas are found in South America and three-quarters of these are situated in the Amazon Basin. Taking the definition of the humid tropics to be those regions where there is less than a 5 degrees C variation in the mean monthly average temperature between the three coldest and the three warmest months, and with four months where potential evapotranspiration is greater than precipitation, the total world area is in the order of 1,500 million hectares with about a third or 500 million hectares in the Amazon (Nicholaides *et al.*, 1984). According to a recent agronomic point of view, at least half of this area is considered to be potentially arable land or suitable for grazing (Sanchez *et al.*, 1982a). A radically different point of view has been presented by a number of ecologists, such as Sioli (1980), Goodland and Irwin (1975) and Goodland (1980), who do not see either sufficient evidence that an agricultural technology has been developed to overcome the severe constraints of the environment, or that there is sufficient justification for removing one of the world's most diverse ecosystems and replacing it by near-monocultural forms of land-use. Whichever point of view or compromise prevails, it is readily apparent that a sound understanding of the environment is essential and that the nature and dynamic properties of the soils forms a crucial part.

Whatever the outcome of the controversy over the possible use of the Brazilian Amazon soils, deforestation and pressures on the land are increasing relentlessly and pose demanding questions concerning the direction and intensity of development (Hecht, 1981). Increasing information is also being made available on the effects of the longer term climatic changes on the Pleistocene and Holocene vegetation (Whitmore and Prance, 1987). Both long-term and short-term changes have been shown to have major impacts on the environment and indicate that development of the forest areas should be approached with caution. At the same time a number of government and external authorities regard the 'development' of the Amazon Basin as not only inevitable but desirable. For example, the National Research Council (1982) considered that the humid tropic system represents 'a very important, under-exploited resource for tropical countries ... as population pressures increase ... rapid and extensive development will and must take place if even the currently inadequate standard of living in most countries is to be maintained'. This will not be achieved if earlier forms of soil misuse prevail and inappropriate soil management techniques continue to be employed.

This chapter will examine the types of soil underlying the forest and savanna areas of the Brazilian Amazon and assess the uses of this resource for different forms of development. Such an assessment involves a closer look at the mechanisms which support the present natural systems and how soil limitations constrain possible changes in land use. Finally, an attempt will be made to evaluate, from the point of view of the soil resources, some of the alternative management strategies that have been proposed.

NATURE AND PROPERTIES OF AMAZON SOILS

The current state of knowledge is based upon a very limited number of detailed surveys, at scales of 1:200,000 or larger, and therefore depends for the greater part of its evidence, on the Projeto RADAMBRASIL reconnaissance surveys taken in the 1970s with published map scales of 1:1 million reduced from drafts at 1:250,000 (Projeto RADAMBRASIL 1972–78; Furley, 1986). A synthesis of the land resources for the whole of the Basin has been undertaken at CIAT in Colombia (Centro Internacional de Agricultura Tropical), which provides a valuable overview of the climatic, vegetational, pedological and geomorphological distributions (Cochrane and Sanchez, 1982; Cochrane, 1984; Cochrane *et al.*, 1985). Detailed studies (at scales of 1:100,000 or better) have been confined to specific surveys, such as those undertaken by CEPLAC (Centro de Pesquisas do Cacau) in 1973–7, to the local areas around EMBRAPA research stations in each State or Territory, especially work instigated from CPATU in Belém (Centro de Pesquisa Agropecuária do Trópico Úmido) and finally, to specific research areas such as the Reserva Ducke, close to Manaus. Extremely valuable work on soils for sustained cropping can also be extrapolated from the research at Yurimaguas in Peru conducted under the direction of the NCSU (North Carolina State University) and from some of the international research carried out at CIAT. In addition there have been localised and specific investigations on nutrient dynamics both in Brazil and in Venezuela (e.g., Klinge, 1971; Jordan, 1982a). Other useful reviews include Sombroek, 1966, 1979; Beek and Bramão, 1969; Falesi, 1972; van Wambeke, 1978; Correa, 1984; Jordan 1985a.

By far the most extensive groups of soils occupying around 75 per cent of the area, consist of nutrient poor, acid Oxisols and Ultisols (Soil Survey Staff, 1975) (see Figure 13.1, Table 13.1). Poorly drained soils typical of alluvial, flood plain and palm swamp areas cover a further 14 per cent. Moderately fertile and well-drained soils, belonging to a variety of Sub-Orders (especially Udalfs, Fluvents, and Tropepts) occupy around 8 per cent

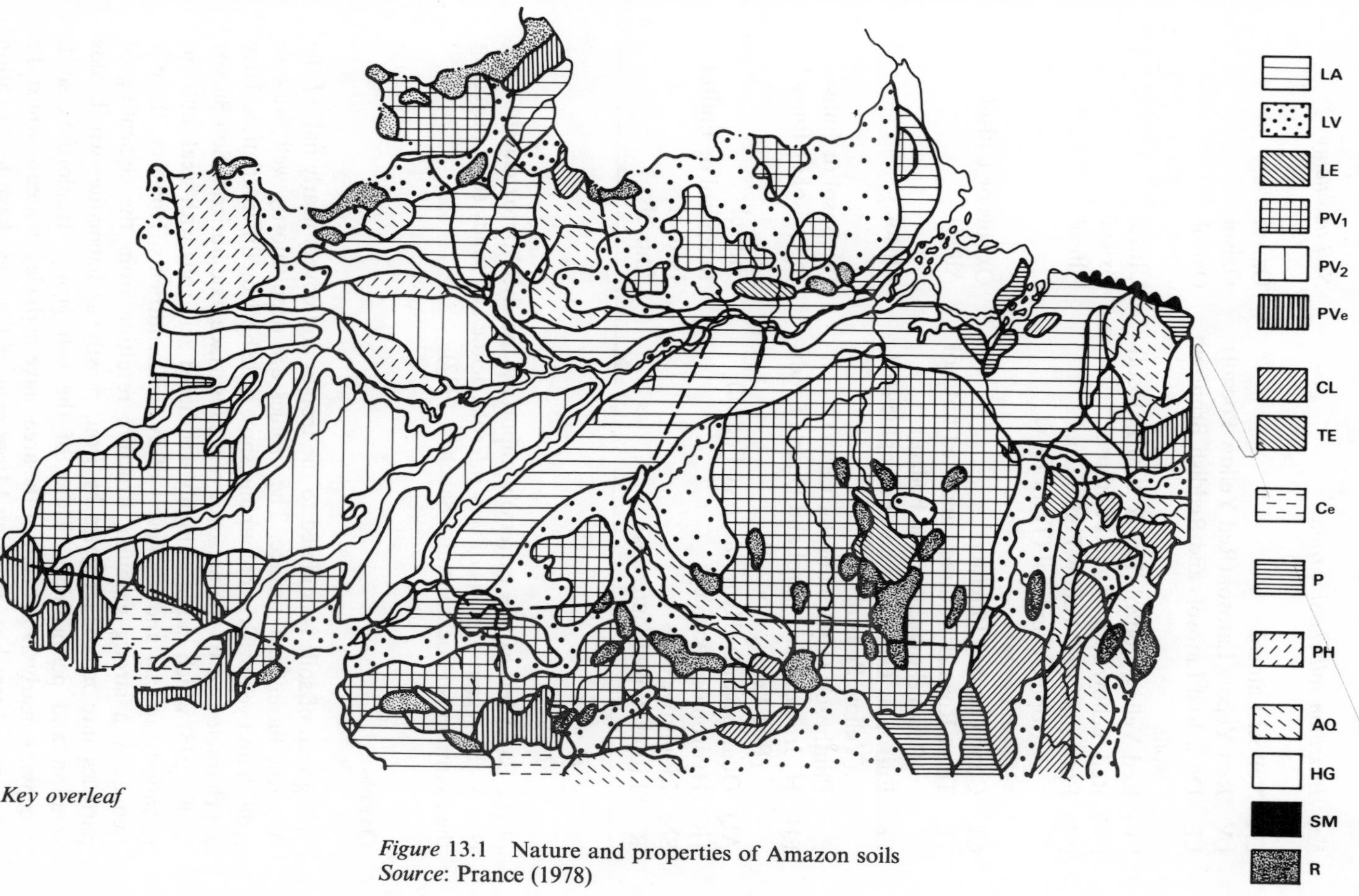

Key overleaf

Figure 13.1 Nature and properties of Amazon soils
Source: Prance (1978)

Key	*Brazilian soil classification*	*Soil taxonomy*
LA	Yellow Latosol	Oxisol
LV	Red Yellow Latosol (Red Yellow Podzol)	Oxisol
LE	Dark Red Latosols and Reddish Brunizem soils	Oxisol
PV1	Red Yellow Podzol (Red Yellow Latosol)	Ultisol
PV2	Red Yellow Podzol and Plinthosol	Oxisol
PVe	Eutrophic Red Yellow Podsol, and Eutrophic Cambisol (Terra Roxa Estruturada)	Alfisol
CL	Concretionary Laterite	Oxisols or Ultisol
TE	Terra Roxa Estruturada and Red Latosol (Red Yellow Eutrophic Podzol)	Alfisol
Ce	Eutrophic Cambisols and Eutrophic Red (Yellow Podzol)	Alfisol
P	Plinthosols (Hydromorphic Gleys)	Inceptisol or Entisol
PH	Hydromorphic Podzol and Quartz Sands (Red Yellow Podzol)	Inceptisol or Entisol
AQ	Quartz Sands (Hydromorphic Podzol)	Entisol
HG	Hydromorphic Gleys	Inceptisol or Entisol
SM	Mangrove Soils	Entisol
R	Lithosol (rock outcrops)	Entisols

of the Basin but can be extremely important locally. Approximately 3 per cent of the region consists of very infertile, acidic and sandy soils, including the Psamments (Cochrane and Sanchez, 1982).

Oxisols

This group of soils is confined to the tropics and covers nearly half of the Amazon Basin land surface. The profiles are deep, mostly well drained, reddish and yellowish in colour and show little horizon differentiation. They are characterised by the presence of a subsurface oxic horizon (Soil Survey Staff, 1975) which indicates highly weathered acid soils with weak nutrient retention capacity and low levels of available nutrients. However, there is frequently a strong granular structure resulting from the cementing of particles into microaggregates by iron, mostly as haematite, which aids aeration and permeability. Many of the sub group, Haplorthox soils, occupying nearly 29 per cent of the area, have high clay contents, whilst the sub-group Acrorthox, covering 14 per cent of the area, have lower cation

Table 13.1 Soil distribution of the Amazon region at the greater groups level, tentative classification

Order	*Sub-order*	*Great group*	*Million (ha)*	*% of Amazon*
OXISOLS	Orthox	Haplorthox		28.5
		Acrorthox		14.0
		Eutrorthox	37.8	0.1
			67.5	
			0.3	
	Ustox	Acrustox	6.6	1.4
		Haplustox	4.8	1.0
		Eutrustox	2.0	0.4
	Aquox	Plinthaquox	0.9	0.2
	Total Osixols:		219.9	45.5
ULTISOLS	Udults	Tropudults	83.6	17.3
		Paleudults	29.9	6.2
		Plinthudults	7.6	1.6
	Aquults	Plinthaquults	12.2	2.5
		Tropaquults	7.1	1.5
		Paleaquults	0.7	0.1
		Albaquults	0.1	0.1
	Ustults	Rhodustults	0.5	0.1
	Total Ultisols:		141.7	29.4
ENTISOLS	Aquents	Fluvaquents	44.8	9.3
		Tropaquents	6.7	1.4
		Psammaquents	2.8	0.6
		Hydraquents	0.6	0.1
	Orthents	Troporthents	6.9	1.4
	Psamments	Quartzipamments	5.5	1.1
	Fluvents	Tropofluvents	4.7	1.0
	Total Entisols:		72.0	14.9
ALFISOLS	Udalfs	Tropudalfs	16.5	3.4
	Aqualfs	Tropaqualfs	3.3	0.7
	Total Alfisols:		19.8	4.1
INCEPTISOLS	Aquepts	Tropaquepts	10.6	2.2
		Humaquepts	0.5	0.1
	Tropepts	Eutropepts	4.3	0.9
		Dystropepts	0.6	0.1
	Total Indeptisols:		16.0	3.3
SPODOSOLS	Aquods	Tropaquods	10.5	2.2
MOLLISOLS	Udolls	Argiudolls	2.8	0.6
	Aquolls	Haplaquolls	0.9	0.2
	Total Mollisols:		3.5	0.8
VERTISOLS	Uderts	Chromuderts	0.5	0.1
Total orders			484.0	100.00

Source: Based on Nicholaides *et al.* (1984).

exchange capacities (CeC). In summary, they are soils which are chemically poor but possess physical properties which are believed to give a number of them considerable development potential for agriculture.

Ultisols

These soils represent a Soil Order found within the tropics but also in the sub-tropics and warm temperate areas. They underlie around 30 per cent of the surface area of the Basin. The profiles are also deep and acid but differ from Oxisols in having greater clay contents with depth and higher levels of weatherable minerals. The physical properties are less advantageous; they are often found on steeper slopes and are more susceptible to erosion as well as being susceptible to compaction where surface clay levels are low. They are found mostly to the west of Manaus in well drained areas of the Basin which have not been affected by deposition from the ancient geological shields.

From the point of view of development for continuous agriculture, these two major groups of soils possess significant limitations (Irion, 1978). They both have high levels of acidity and often possess toxic levels of aluminium. Phosphorous is usually critically deficient and becomes fixed or unavailable in the presence of clay. Several other major plant nutrients such as potassium, magnesium, calcium and sulphur are also deficient whilst there are limitations in the quantities of zinc and other micronutrients. Both groups of soils have low effective CeC levels which are associated with high leaching potentials. (Sanchez, 1976, 1981), together with low nutrient retention capabilities in the kaolinitic, gibbsitic and goethite-rich clay minerals. The decomposing organic matter also has a low nutrient retention capacity since iron and aluminium may partly occupy the exchange sites (Jordan, 1985a). The soils do not tend to form ironstone (laterite, *sensu strictu*) as often believed (McNeil, 1964) and only about 4 per cent of the Basin possesses plinthite within the profile (non-hardened iron accumulation in the subsoil). The plinthite is capable of hardening with exposure to drying and oxidation, if erosion were able to remove the protective cover of the surface soil (Sanchez and Buol, 1985, Moorman and van Wambeke, 1978, Buol and Sanchez, 1978). Erosion has been dismissed as a serious problem by a number of agronomists in view of the low gradients. It should be stressed, however, that erosion can be effective on very gentle slopes and that there has been insufficient monitoring over a long enough period of time to be certain of the fate of cleared land.

Entisols

Although these soils cover less than 10 per cent of the Amazon, they have a very important role in food production within the region. They are principally Fluvaquents, in other words periodically flooded (alluvial) soils of varying fertility depending upon the provenance of the sediment. Alluvial sediments derived from geologically youthful source areas such as the Andes are far more mineral rich than those sediments eroded from the ancient weathered shields of the Guyana and Brazilian plateaus. In well drained positions, a combination of good water supply, and naturally replenished soil fertility provides some of the most promising soils for agriculture. Indeed, most of the accessible and largest stretches of *várzea* are already developed.

Other Soil Orders Occupying Less than 5 per cent of the Basin

The most important soils, from the point of view of possible agricultural development, are found in the Alfisol group (Tropaldalfs and and Paleustalfs, see Table 13.1). Together with small patches of other high base status soils, notably some of the Inceptisols (Eutropepts), Mollisols (Argiudolls) and Vertisols (Chromoderts), such profiles have high levels of natural fertility and often a good physical structure. However, they are limited in extent and, because they are frequently found in more accidented relief, they are often more susceptible to erosion. A good example illustrating these soils, is the group of Paleustalfs found in stretches near Altamira (Pa), Ouro Prêto (Ro) and Rio Branco (Acre) resulting from more favourable igneous parent materials (van Wambeke, 1978; Furley, 1980).

At the other extreme, there are several exceptionally poor soils whose properties exhibit so many limitations to development that they have not been utilised up to the present and furthermore, are unlikely to be developed in the future without expensive engineering and capital-intensive reclamation programmes. Such soils include the coarse sandy soils or Psamments (Klinge and Herrera, 1978; Klinge and Medina, 1979) which may be excessively wet during the rainy season and drought-stressed during dry periods. They tend to be very organic at the surface, extremely acid and infertile and are highly susceptible to erosion. They are often identified with the 'campinarana' and 'Amazon caatinga' type of vegetation. They have been well described in several publications because of the interest in their extreme oligotrophic character (for example, Klinge, 1965, 1975; Stark, 1978; Sombroek, 1979).

Despite the relatively small proportion of naturally fertile soils, the total area of land potentially suitable for agropastoral activities is still in the order of 31 million hectares (Cochrane and Sanchez, 1982) and this does provide a number of localities where permanent agriculture has a better chance of success. Further, it has been argued in a number of publications by Sanchez and his associates (e.g., Sanchez *et al.*, 1982a, Nicholaides *et al.*, 1985) that many of the extensive, well-drained acid soils could be successfully farmed with appropriate technology at various levels of management and capital input. Half of the Amazon Basin has well drained soils with slopes less than 8 per cent which, it is suggested, provide sites suitable for agricultural development (Table 13.2).

NATURAL CYCLING MECHANISMS

The undisturbed vegetation is well adapted to its climatic and soil environment. Natural cycling processes retain a large proportion of the nutrient stock within the '*tessera*', the 3-dimensional unit including vegetation, ground surface and biologically active part of the soil profile. It is therefore helpful to consider the mechanisms responsible for the maintenance of the vegetation in its natural state. This allows an assessment of the effects of disturbance on the plant communities and permits an estimation of the prospective loss of nutrients or favourable soil properties occasioned by the damage to such natural mechanisms.

Detailed studies of nutrient cycling have only been carried out at a limited number of sites, although the dependence of low fertility tropical forests on

Table 13.2 Topographic classes of the major Amazon Basin soils

	Million ha					
	Poorly drained	Well drained				
Soil grouping	level	Slope 0–8%	Slope 8–30%	Slope >30%	Total	(%)
Acid, infertile soils (Oxisols and Ultisols)	43	207	88	23	361	(75)
Poorly-drained alluvial soils (Aquepts, Aquents)	56	13	1	—	70	(14)
Moderately fertile, well-drained soils (Alfisols), Mollisols, Vertisols, Tropepts, Fluvents)	0	17	13	7	37	(3)
Very infertile, sandy soils (Spodosols, Psamments)	10	5	1	—	16	(3)
Total	109	242	103	30	484	

Source: Nicholaides *et al.* (1984) based on Sanchez *et al.* (1982).

effective recycling has long been suggested and partly demonstrated (Nye and Greenland, 1960). Nutrient cycling involves the distribution of mineral nutrients or more accurately mineral elements, within the different biological compartments of the vegetation, and the amounts of elements moving into and out of these compartments over a given period of time. Some of the most detailed studies in the Amazon region have been on Terra Firme near Manaus (Klinge and Rodriguez, 1968) and the Rio Negro in Venezuela (Jordan and Herrera, 1981; Herrera *et al.*, 1978) or in areas of Igapo. The factors which control nutrient losses and their cycling mechanisms have been reviewed by Jordan (1985b), who divided them into four groups:

1. *Temperature*: The constant high temperature permits continuous primary and secondary organic production which is matched by an equally high rate of decomposition. There is therefore, a high potential for nutrient movement through food chains.
2. *Moisture*: Together with temperature, these two factors form the dominant controls over nutrient cycling. Although mean annual rainfall figures are high, a large part of the Amazonian forest experiences some degree of dryness (Figure 13.2), during which time there is a slower rate of decomposition unless fire (artificial or less likely natural causes) activates nutrient release.
3. *Biotic interactions*: The interest here is the balance between the rates of primary production and decomposition. It is generally accepted that there are high decomposition rates in the humid tropics and that soil carbon levels, which can act as indicators of decomposition, are lower in tropical than in temperate areas (Zinke *et al.*, 1984). Two aspects are clear, firstly that the rates of biological activity and decomposition can themselves change rapidly (Swift *et al.*, 1979; Luizão and Schubart, 1987) and secondly, that the nutrient pathways are complex (Figure 13.3). Many parts of the system and its connecting pathways are poorly understood. For example, herbivore attack within the above ground biomass can often affect individual tree species (Janzen, 1983) although whole communities are rarely defoliated (Lugo *et al.*, 1974). The replacement of intact vegetation by monocultures has been known to result in insect population or microbial population explosions and severe damage to crops or plantation trees. High species diversity does, therefore, appear to give a measure of protection against grazing attack and microorganic infection (Orians *et al.*, 1974; UNESCO, 1978; Hubbell, 1979).
4. *Weathering and leaching*: The combination of high temperature, heavy rainfall and biotic interactions within the *tessera* often results in extremely high potential rates of weathering and nutrient removal. Although organic acids are usually less aggressive in the humid tropics than in areas outside, the acidity is nevertheless sufficient to result in the leaching or eluviation of most detachable nutrient ions in the upper soil horizons. It is

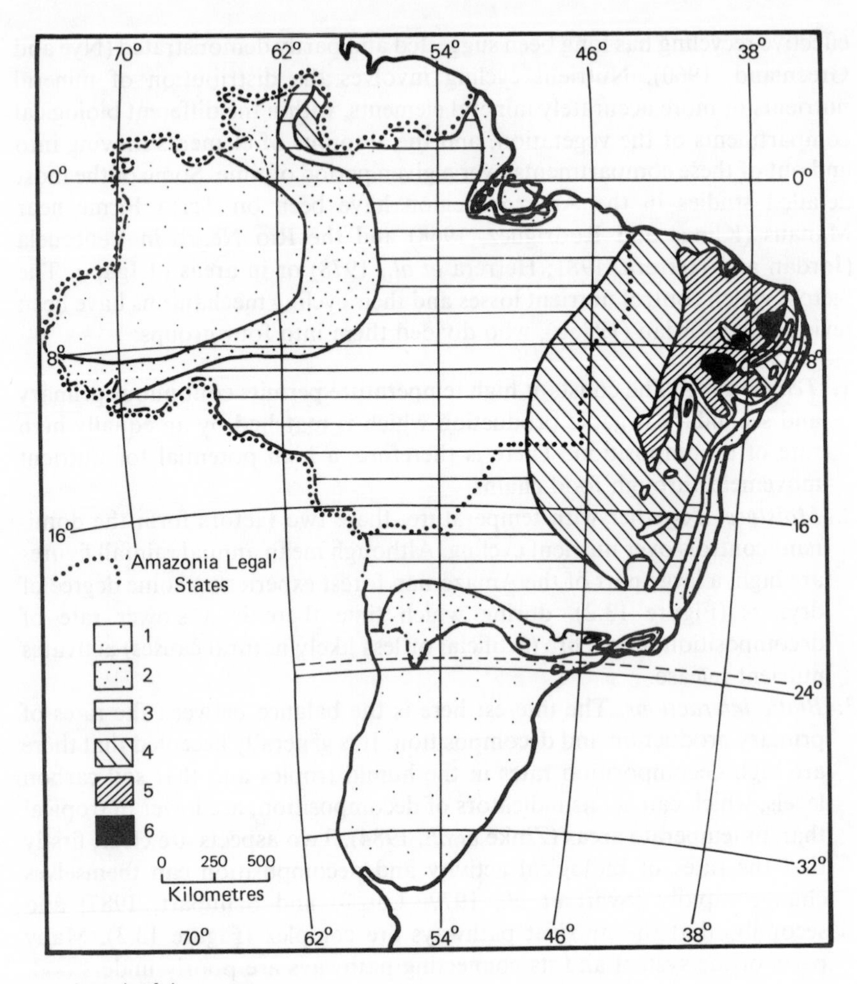

Length of dry season

1. Shorter than 1–2 months
2. 1–2 months
3. 3–4 months
4. 5–6 months
5. 7–8 months
6. 8 months or more

Figure 13.2 Rainfall seasonality in Brazil

worth remembering that, although there may not be mineral nutrient inputs from freshly weathered parent material below the soil surface, there is a constant atmospheric input in the form of dust (see Table 13.3). Parts of the Amazon rainforest system appear to have leaching rates less than or equal to the atmospheric input (Jordan, 1982a).

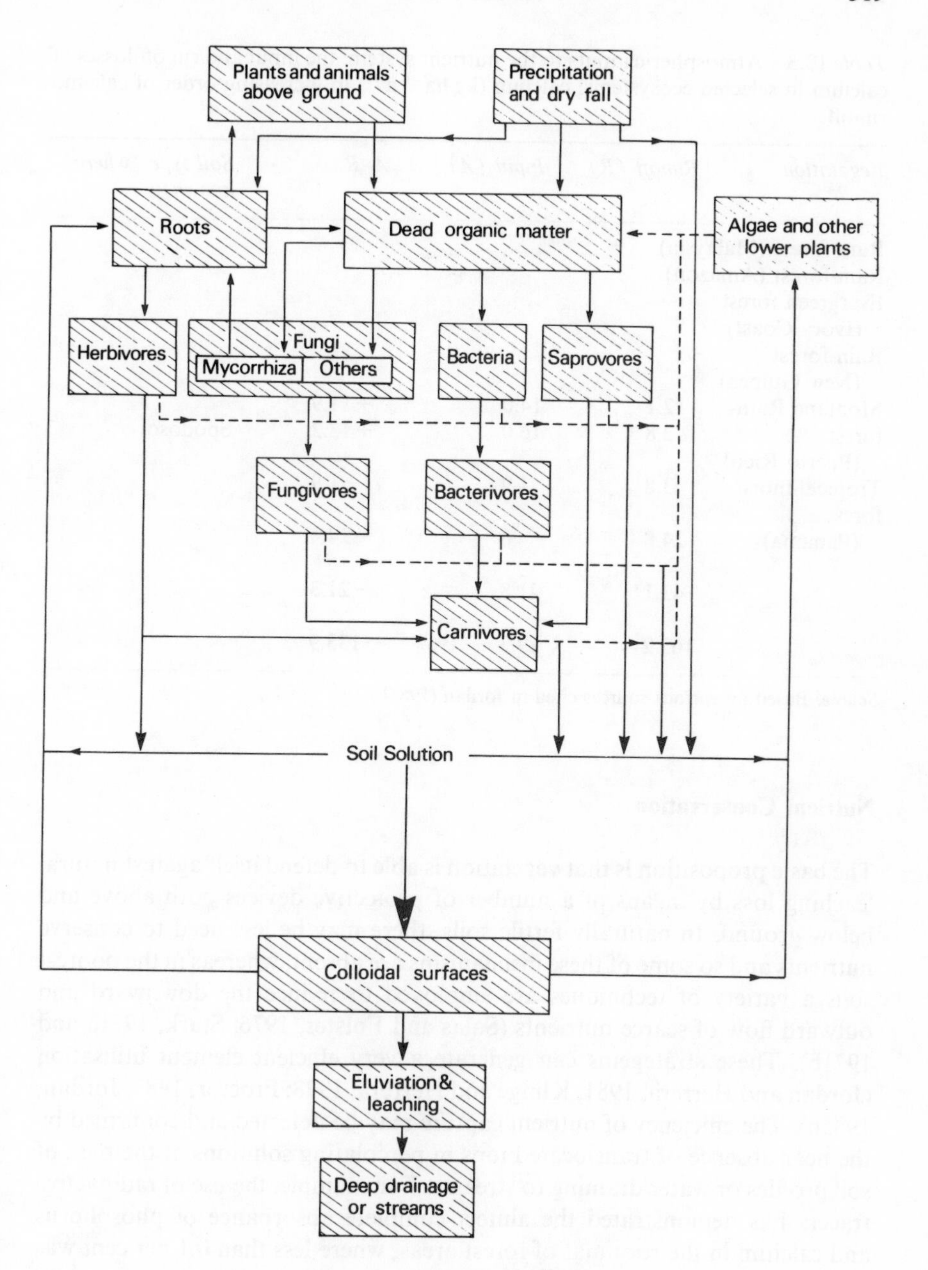

Figure 13.3: Nutrient pathways

Table 13.3 Atmospheric inputs to the nutrient system: the input and runoff losses of calcium in selected ecosystems, calcium (kg ha^{-1} a^{-1}) in increasing order of calcium runoff

Vegetation	*Runoff (R)*	*Input (A)*	*A–R*	*Soil type (where known)*
Rain-forest (Malaysia)				
Rain-forest (Amazon)				
Evergreen forest				
(Ivory Coast)				
Rain-forest				
(New Guinea)				
Montane Rain-	2.1	14.0	+11.9	
forest	2.8	16.0	+13.2	Spodosol
(Puerto Rico)				
Tropical moist	3.8	1.9	−1.9	
forest				
(Panama)	24.8	0	−24.4	
	43.1	21.8	−21.3	
	163.2	29.3	−133.9	

Source: Based on various sources cited in Jordan (1985b).

Nutrient Conservation

The basic proposition is that vegetation is able to defend itself against natural leaching loss by means of a number of protective devices both above and below ground. In naturally fertile soils, there may be less need to conserve nutrients and so some of these mechanisms are absent, whereas in the poorest soils a variety of techniques are employed to reduce the downward and outward flow of scarce nutrients (Salas and Folster, 1976; Stark, 1971a and 1971b). These strategems can generate a very efficient element utilisation (Jordan and Herrera, 1981; Klinge and Herrera, 1978; Proctor, 1983; Jordan, 1985b). The efficiency of nutrient capture may be detected and confirmed by the near absence of translocated ions in percolating solutions at the base of soil profiles or water draining to streams. For example, the use of radioactive tracers has demonstrated the almost complete absorbance of phosphorus and calcium in the root mat of forest areas, where less than 0.1 per cent was lost by leaching (Stark and Jordan, 1978).

The principal mechanisms for conserving nutrients may be grouped into four spatial components:

1. *The protective and absorptive nature of the above ground biomass*: A complete canopy, whether of forest or grassland, acts as a protective cover

for the soil reducing rainfall impact and erosion, retaining moisture, as well as absorbing and conserving nutrients (Salati, Vose and Lovejoy, 1986). A few examples will serve to illustrate some of the diverse mechanisms which can be found in the Amazon. In the humid forests, the epiphytic mosses, lichens and algae are particularly good examples of nutrient absorbers from rainwater and dust (Jordan *et al.*, 1979). Mats of living and dead briophytes, bromeliads, club mosses, lichens, ferns and epiphytes abound above the soil surface and roots can penetrate these mats even where they occur above ground. Many forest and savanna leaves also possess nutrient conservation devices. The scleromorphic evergreen character of leaves may help to resist insect attack (Klinge and Medina, 1979) and they may also be able to exist at lower nutrient contents than leaves in richer environments (Peace and Macdonald, 1981). They tend to be tough, long lived and protected against grazing by such means as hairs and thorns (Chapin, 1980; Hartshorn, 1978) particularly after the first vulnerable growth stages. Leaf drip tips may serve to remove excess leaf surface water and thereby prevent undue leaf epiphyte growth. As indicated earlier, the trees may be able to withdraw nutrients into the woody parts of the canopy before leaf abcission (Charley and Richards, 1983). It is possible that in some circumstances, ion substitution may occur in extremely oligotrophic conditions; for example Jordan (1985a) believes that silica may be able to replace the scarce phosphorus locked up in iron-aluminium complexes and thereby release phosphorus for plant growth. It is also likely that many of the indigenous plants possess chemical defences against herbivores and pathogens (Levin, 1976). Tree bark is nutrient poor and it has been suggested that this lowers the risk of microbial and insect attack (Jordan and Uhl, 1978). The range of protective devices in savanna areas of the Brazilian Amazon is less well known, and consequently the fate of added mineral elements in the form of fertilisers can only be estimated. Because of the differences in climate, the soil response is likely to demand solutions other than those devised from comparable research in the central cerrado.

2. *The protective and absorptive nature of the ground organic matter*: The litter layer also acts as a buffer similar to that of the above ground vegetation, against raindrop impact and against desiccation. The input of organic debris controls the food supply and activities of the decomposer organisms which in turn reflect the rate of organic decomposition.
3. *The absorptive nature of the root mat*: The dense root mat and its associated humus complex is situated close to the surface and filters nutrients from percolating solutions in well-drained soils (Jordan and Herrera, 1981; Stark and Jordan, 1978). Fine root extension has been shown to be rapid and very effective in scavenging available nutrients (Nye and Tinker, 1977; St John, 1983). Some species appear to have a facultative ability to allow a greater proportion of the photosynthate

production to go towards root growth in poor soils (Gerloff and Gabelman, 1983). The concentration of root decomposer organisms near to the soil surface and in the organic litter is obviously related to nutrient source.

4. *Nutrient conservation by mycorrhizae*: The role of mycorrhizae in humid tropical soils is well established (Ruehle and Marx, 1979; Janos, 1983; St John, 1985). At one time, the direct nutrient cycling hypothesis, requiring ectomycorrhizae to act as mutualistic organisms between the decaying organic matter and the plant roots, was believed to be a major aid in nutrient absorption over a wide range of conditions (Went and Stark, 1968). Such relationships have been confirmed as being very effective in the extremely impoverished soils of the tropical forest heaths (Singer and Araujo, 1979) or the Igapó forests (Singer and Araujo, 1986), but have been found to be less prevalent in richer soil environments. The precise mechanisms and nutrient pathways are still unclear (Herrera *et al.*, 1978; Cuevas and Medina, 1983). Many forests have been shown to possess endomycorrhizae penetrating the cortex of feeder roots (St John, 1980) and therefore unable to act as direct transfer organisms in the sense originally proposed. Herrera and Jordan (1982) envisaged a continuum of different mycorrhizal functions ranging from eutrophic conditions with relatively few symbiotic relationships to oligotrophic conditions with many. As St John and Coleman (1987) have suggested, there may be many complex pathways for nutrient transfer depending upon local conditions varying in time and space.

Since the conservation of nutrients is affected by their incorporation into living organisms, the implication of these mutualistic relationships is that management strategies should attempt to preserve as much as possible of the 'entire below ground ecosystem' (Jordan, 1985a) as well as the surface organic litter and a proportion of the above ground biomass.

The principle of nutrient cycling can be readily exemplified by considering one of the more oligotrophic situations in the Amazon where nutrient conservation is essential for the continued healthy growth of the native vegetation. In the *caatinga* (Herrera, 1985), there are a number of interesting nutrient protecting and recycling mechanisms. Some of these are above ground; for example the long-lived sclerophyllous leaves (which resist insect and herbivore attack and reduce leaf leaching) have been shown to translocate nitrogen reserves back to the main plant before senescence and leaf drop occurs. A further striking feature is that the low and critical levels of phosphorus in the organic litter are entirely taken up by the ectomycorrhizae which completely infect fine roots.

The efficiency of the nutrient cycling processes has also been assessed more quantitatively. Finn (1978) devised a cycling index from 0 (no cycling replacement) to 1 (complete return). In part of the southern Venezuelan Amazon, Jordan (1982b) was able to show that the index was between 0.6

and 0.8 depending upon the nutrient under consideration. This illustrates a remarkable overall cycling return of 60 to 80 per cent of the nutrient stock.

The impact of any vegetation change will obviously depend upon the intensity and extent of the disturbance and the time period over which it occurs. A natural gap in the forest caused by tree decay or storm damage, is similar to a small patch manually cleared and possibly lightly burnt during the course of shifting cultivation. Both represent minor disturbances which may not remove all of the trees within the affected area and may well leave a significant proportion of organic matter forming a ground protection. In any case, colonisation by pioneer plants will rapidly fill in the natural gaps or artificial plots abandoned after a few years of use. At the other extreme, large areas which have been mechanically cleared and burnt will lose virtually all of their natural regenerative mechanisms particularly if the surface soil has been scraped and compacted during the clearance. The impact of this upon soil properties is considered later and the point emphasised here concerns the biological nutrient conservation processes which are lost as the natural vegetation is disturbed. The short-term clearance of natural vegetation can therefore be viewed as the 'mining' of a concentrated layer of nutrients accumulated over a long period of time. The greater the scale or intensity of disturbance, the greater the replacement cost is likely to be in order to substitute for the lost conservation techniques.

PROPERTIES OF AMAZONIAN SOILS AND THEIR CONSTRAINING EFFECT ON DEVELOPMENT

There are a number of diagnostic soil properties which play a major part in determining the potential of areas for possible development. As indicated earlier, current knowledge of the soils over the whole region is limited by its reconnaissance nature. More detailed information is available for only a few restricted areas, usually close to where some form of development has occurred such as along roads, near to research stations and around land development schemes. Consequently, our understanding must necessarily be at a generalised and tentative level at present, although this does not detract in any way from the usefulness of interpretations at a broad scale as a first approximation (Dudal, 1980). Furthermore, many of the critical soil properties are confined to the upper soil horizons and surface layer and may, therefore, be markedly altered by different land use and management practices.

The principal soil constraints are indicated in Table 13.4. Most humid tropical soils are deficient in phosphorus. If a figure of 10 ppm is taken as the critical level, then around 90 per cent of the Amazon Basin is deficient. On the other hand, P-fixation is not high except where the topsoils have over 35 per cent clay and a high proportion of iron oxides (Dynia *et al.*, 1977;

Table 13.4 Gross estimates of major soil constraints to crop production in the Amazon Basin

Soil constraint[1]	*million ha*	*% of Amazon*
Nitrogen deficiency	437	90
Phosphorus deficiency	436	90
Aluminium toxicity	383	79
Potassium deficiency	378	78
Calcium deficiency	302	62
Sulphur deficiency	280	58
Magnesium deficiency	279	58
Zinc deficiency	234	48
Poor drainage and flooding hazard	116	24
Copper deficiency	113	23
High phosphorus fixation	77	16
Low cation exchange capacity	71	15
High erosion hazard	39	8
Steep slopes (> 30%)	30	6
Laterisation hazard if subsoil exposed	21	4
Shallow soils (< 50 cm deep)	3	<1

Note:
1. Nutritional deficiences of boron and molybdenum also have been noted in some Amazon Basin soils, but are not quantitatively estimable due to paucity of data.
Source: Nicholaides *et al.* (1984).

Cochrane and Sanchez, 1982). In any case, the severity of the deficiency depends on the crop. Less than 20 per cent of Amazon soils, mostly Oxisols with clay rich surface horizons (Sanchez, 1987) can fix large quantities of phosphorus.

Phosphorus fertilisation may not, therefore, be so much of a problem as it is in the cerrado soils further south (Serrão *et al.*, 1979; Benites, 1983) or in the stretches of savanna within the Basin, such as central Rondônia or North-East Roraima. Soil acidity, caused principally by Al-toxicity (defined as 60 per cent or more Al-saturation in the top 50 centimetres) affects over 73 per cent of the Basin and over 80 per cent of the area has pH levels less than 5.3. More usefully, from the point of view of agricultural development, many of the more promising well watered but well drained alluvial and high base status soils (including the *várzeas*) do not appear to have problems of aluminium toxicity. It is curious, in what appears at first sight to be a fairly uniformly humid area, that over 75 per cent of the region has a udic or periodic soil moisture regime and that there are drought stress conditions which affect over 50 per cent. Yet swampy areas abound in the Amazon, particularly around riverine areas, and flooding or poor drainage influences the character of a quarter of the Basin. Where flooding is not too frequent and severe, the sediment input may produce a high and relatively sustainable level of natural fertility. Over 50 per cent of the Basin suffers from low potassium reserves in the mineral soil. Burning does temporarily release nutrients from the biomass stock but extensive fertilisation would be required for continuous agricultural production. An effective CeC level of less than 4 cmol kg-1 (me/100g) is generally taken to indicate high potential leaching in well drained soils giving low nutrient retention and low available nutrient levels. CeC levels of this magnitude can induce deficiencies in other elements as a result of fertiliser applications. Such CeC levels are believed to occur in over 30 per cent of the topsoils of the Basin (Sanchez, 1987) and the figure may rise to over 40 per cent in subsoils according to Sanchez and Cochrane (1980). Deficiencies in all essential nutrient elements are reported, particularly nitrogen, magnesium silica, boron and copper in addition to those mentioned earlier.

Erosion is not considered to be a serious problem in the Amazon basin by Sanchez and his associates and it is suggested that susceptible soils cover only 10 per cent of the area. The most vulnerable soils tend to have subsurface changes in texture (with greater proportions of clay) and steep slopes (greater than 30 per cent). Taken together, these two factors affect soils which occupy about 7 per cent of the region. Erosion is a problem but to a lesser extent, in rolling and hilly land with 8 to 30 per cent slopes which occupies a further 20 per cent of the region. This still leaves nearly threequarters of the area with level or gentle slopes to 8 per cent) where potential erosion is believed to be minimal. However, bad management can still result in erosion even where potential erodibility is low. Severe gully erosion has been demonstrated along road cuttings, trails in settled areas and along cattle pathways in grazed areas (Ranzani, 1980). Relatively little precise measurement or monitoring over meaningful periods of time has been carried out so far in the Brazilian part of the Amazon.

To summarise this section on constraints to development resulting from characteristic soil properties, it is generally agreed that physical properties are largely favourable for agropastoral and forestry development provided that the drainage is adequate and soil conservation techniques are employed (Alvim, 1978, 1979, 1982; Sanchez, 1987). With appropriate management technologies equivalent to 'advanced' in the EMBRAPA system of land aptitude assessment (Ramalho Filho *et al.*, 1978), there is a low erosion risk resulting from the generally strong aggregate cohesion, and relatively few problems resulting from excessively sandy or excessively argillaceous conditions. Despite this, the definition of appropriate management is still unsatisfactory and fails to take into account polycultural and multiple land-use practices and a more ecologically sound approach to farming. There is also a need for many more monitored research plots measured over periods of perhaps 20 to 50 years. Slow secular or imperceptible erosion is not an aspect of soil management which can be discussed as a potential problem without long-term research, since it does not appear until well after the more spectacular forms of erosion have manifested themselves. It would appear that there are many more severe chemical constraints than physical, at least in the short term. They concern the widespread acidity, the low effective CeC, widespread deficiencies of most essential elements and deficiencies in a number of trace elements, low native or self-renewable fertility in cleared areas, frequent high phosphorus fixation and extensive aluminium toxicity. Only 7 per cent of the Amazon Basin has no sign of fertility limitation (Cochrane and Sanchez, 1982). It is almost certainly true that soil management needs to be site specific, even if broad generalisations derived from a limited number of research stations scattered throughout the whole of the Amazon region have been shown to be valid.

IMPACT OF LAND CLEARANCE METHODS ON THE SOIL ENVIRONMENT

The precise method of cutting and removing the natural vegetation has been shown to be of fundamental importance in the subsequent success of the replacement land use. Cochrane and Sanchez (1982) go so far as to say that land clearing is 'the most crucial step in affecting future productivity of farming systems'.

Land clearance may involve a range of activities from small-scale manual cutting of selected trees with no burning or physical removal of trees and shrubs from the site, to large-scale, mechanical destruction with total burning and/or mechanised clearing to wind rows, frequently involving scraping of the organic litter and topsoil (Bandy and Sanchez, 1986). On the whole, increasing the intensity of the disturbance and a greater efficiency at removal of the biomass creates increasing problems for future management.

(a) *Manual cutting and burning*: this is the most common land clearing practice for smallholders in the Amazon region and increases in scale from machetes or axes to chain saws (which can be taken as a criterion of increased prosperity, Leite and Furley, 1985). Various researchers have demonstrated the nutrient ash content following burning (e.g., Seubert *et al.*, 1977; Bandy and Sanchez, 1981; Smyth and Bastos, 1984), although the value can vary considerably with the effectiveness of burning, for instance the total biomass consumed and the temperature achieved. The implication is that the different species artificially planted as saplings or naturally regenerated from seeds, and utilised for natural forest succession fallows, may accumulate very different quantities of nutrients. Several studies have demonstrated the increasing availability of many elements in the period following burning (e.g., Brinkmann and Nascimento, 1973; Seubert *et al.*, 1977; Sanchez, 1976; Uhl *et al.*, 1982; Smyth and Bastos, 1984; Jordan, 1985b; and a number of papers in Lal, Sanchez and Cummings, 1986) (See Table 13.5). During the year after clearance and burning, there are large volatisation losses of nitrogen and sulphur, the soil organic matter levels decrease (until they find a new equilibrium after approximately 2 years), the pH of acid soils increases at the same time as the aluminium levels decrease (but these revert in time), and the surface soil temperature and moisture regimes fluctuate as radiation reaches the ground surface. The fertilizer value of the ash is less significant in more naturally fertile soils where, indeed, the ash may induce deficiencies in some elements. The incorporation of ash into loamy and sandy loam soils is also more effective than incorporation in

Table 13.5 Nutrient levels in ash from burning three types of vegetation at two locations in the Amazon

Location and soil	*Vegetation*	*Ash dry* (t ha^{-1})	*Nutrient level* (kg ha^{-1}) N	Ca	Mg	K	P	Zn	Cu	Fe	Mn	S	B
Manaus, Brasil (Typic Acrorthox)	Primary forest	9.2	80	82	22	19	6	0.2	0.20	58	2.3	—	—
	Secondary forest (12 years old)	4.8	41	76	26	83	8	0.3	0.10	22	1.3	—	—
	Kudzu fallow (4 years old)	1.5	24	16	6	15	3	0.1	0.03	7	0.4	—	—
Yurimaguas, Peru (Typic Paleudult)	Secondary forest (25 years old)	12.1	127	174	42	131	17	0.5	0.24	4	11.1	24.2	0.3
	Secondary forest (17 years old)	4.0	67	75	16	38	6	0.5	0.30	8	7.3	—	—
	Kudzu fallow (4 years old)	1.2	6	18	15	77	17	0.7	0.007	3	2.3	2.5	0.1

Source: (1987) based on Senbert *et al.* (1977); Smyth and Bastos (1984); Bandy and Sanchez (1981).

argillaceous soils. Less well documented, but of equal importance, is the effect of deforestation, clearance and burning on the biological cycling processes outlined earlier. The destruction of the microbial population and their nutrient scavenging role may have effects greater than the simple system of nutrient release indicated above.

(b) *Mechanised clearance*: manual clearing has the advantage of reducing soil compaction compared to mechanised land clearing. The use of bulldozers and heavy machinery, such as tree crushers (Toledo and Navas, 1986), has a particularly deleterious effect on compaction with marked increases in bulk density and reduction in aeration (Schubart, 1977). The effect of compaction can be readily seen in the failure of seedlings and transplants, where a puddled soil has prevented root respiration. To some extent, the effects of compaction can be offset by disking but this in turn disturbs the biological equilibrium. Topsoil scraping either by a sheer blade set too low or by dragging heavy trunks and roots is a more serious long-term problem. For example, Lal *et al.* (1975) demonstrated a 50 per cent reduction in crop yield with a 2.5 centimetre topsoil loss in Nigerian Alfisol.

There is scant evidence at present on the impact of land clearance methods within the Brazilian Amazon (Dias and Nortcliff, 1985; Nortcliff and Dias, 1988). For the nearest comparative information, is is necessary to look at the Yurimaguas Research Station results, in Table 13.6a and 13.6b (Alegre *et al.*, 1986; Alegre, Cassel and Bandy, 1986). A comparison between the effects of burning and bulldozing on an Ultisol has been examined by Seubert *et al.* (1977) and followed up at intervals in later years. The early sequence is illustrated in Figure 13.4. The base status of the soils can be seen to alter markedly. Exchangeable calcium and magnesium in the topsoil is increased sharply after burning and remains at about twice that measured in the bulldozed plots. The exchangeable potassium tripled in the topsoil after burning and is also higher in the bulldozed plots, probably resulting from the rapid mineralisation of leaves. In time, the potassium levels drop through leaching (as detected in the subsoil horizons). In the burnt plots, the exchangeable acidity is seen to drop sharply and then rise to less than the initial point after 10 months, whereas the bulldozed plots witnessed a steady increase with time. Aluminium saturation decreases in the burnt plots and remains fairly even in the bulldozed plots. The exchangeable aluminium appears to have been temporarily neutralised by the bases in the burnt ash. Burning and bulldozing also temporarily increases the available phosphorus levels. The organic phosphorus appears to be rapidly mineralised from unburnt leaves and residual organic matter within the subsoil but, over longer time periods, it would seem that the temporary injection of phosphorus lasts only as long as the organic supplies persist. On the whole, burning increases exchangeable bases and available soil

Table 13.6 Effect of land clearing methods
(a) On topsoil physical properties and organic carbon on a continuously cropped Ultisol in Yurimaguas, Peru

Clearing method	*Months after clearing*	*Infiltration rate* (mm h^{-1})	*Bulk density* (Mg m^{-3})	*Mean weight diameter* (mm)	*Organic* C (%)
Before clearing	0	324	1.16	0.48	1.04
Slash and burn	3	204	1.27	0.42	1.05
Straight blade	3	14	1.42	0.29	0.82
Shear blade	3	22	1.28	0.36	0.87
Slash and burn	23	107	1.32	0.38	1.03
Straight blade	23	15	1.42	0.36	1.02
Shear blade and disk	23	110	1.32	0.36	0.89

(b) Including post-clearing management, on the relative yield of five consecutive crops after clearing a 20-year-old secondary fallow on an Ultisol of Yurimaguas, Peru

	Percentage of cumulative maximum yields[1]	
Clearing methods	*No tillage, no fertiliser*	*Tilled, fertiliser*
Slash and burn	27	93
Straight blade	7	47
Shear blade	14	65
Shear blade + burning + heavy disk	28	89

Note:
1. Maximum yields of five consecutive crops in t ha^{-1}: upland rice, 4.0; soybean, 2.3; maize, 5.2; upland rice, 2.5; maize, 3.3

Source: Sanchez (1987) based on Alegre (1986).

phosphorus, decreases aluminium saturation and retards organic matter decomposition (because it left a significant residuum). On the other hand bulldozing, in addition to its compacting action and attendant problems, did not reduce the high aluminium levels and both phosphorus and potassium remained below critical values. A further point worth remembering is that bulldozing, in this example at least, cost about four times that of the traditional slash and burn method. The same sites were monitored over an 8-year period (Sanchez, Villachica and Bandy, 1983) but were cropped and fertilised during that period and cannot therefore be used as a measure of natural regeneration although the results are of significance in the later discussion on agricultural alternatives. Slash and burn techniques are the most conservative of nutrients and the least

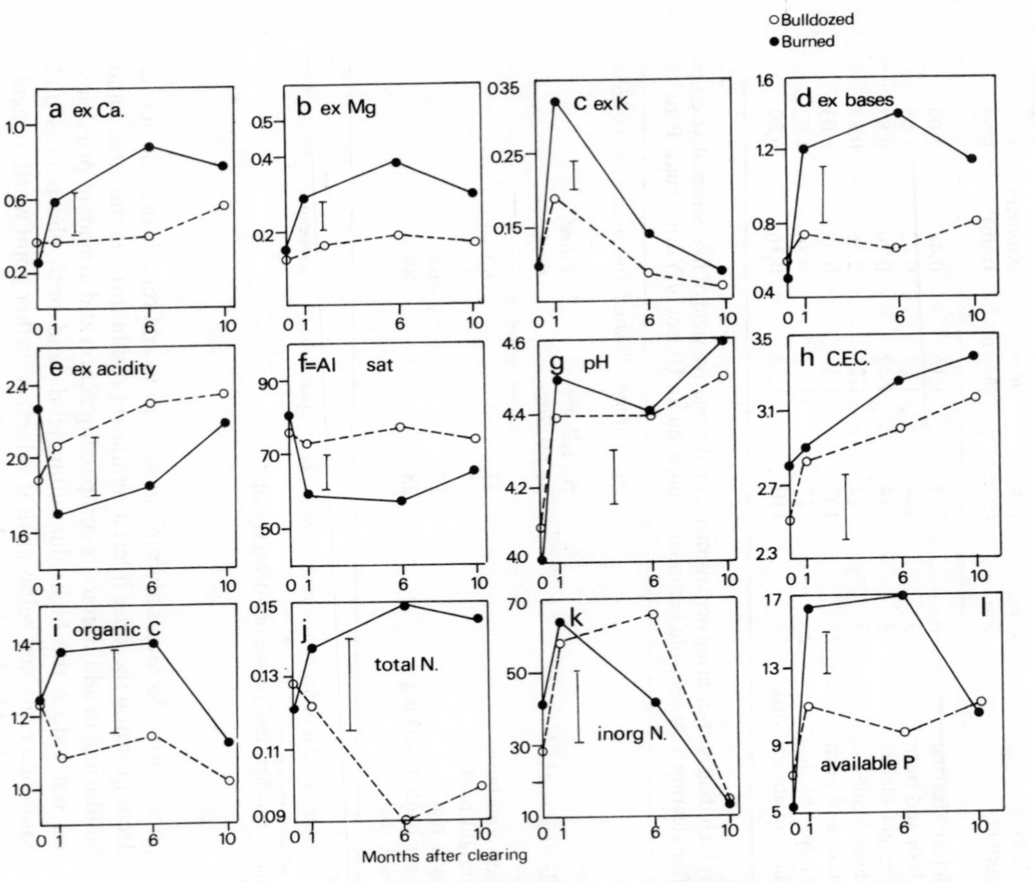

Figure 13.4 Impact of land clearance methods in the Brazilian Amazon
Source: Jordan (1987)

physically damaging to the soil, but such methods may not be a solution for larger scale commercial farming especially if there is a shortage of labour. On the basis of this evidence, it has been suggested (Sanchez, 1987) that where mechanised clearance is envisaged, by using a shear blade followed by burning and heavy disking, the impact on infiltration, bulk density, and organic matters is kept to a minimum.

ALTERNATIVE LAND USE SYSTEMS AND THEIR ADAPTABILITY TO SOIL CONSTRAINTS

The substitution of different forms of land use for natural forest or savanna has created a range of problems, both environmental and socio-economic, and the role played by the soil resources varies in each locality (see Fearnside, 1985; Prance and Lovejoy, 1985; Fearnside, in press). The most widespread traditional land use has been migratory agriculture (shifting cultivation), using slash and burn techniques to clear the native vegetation. Most sedentary farming families or groups have utilised low input technologies at small scales of operation (smallholdings of less than around 100 hectares) or high input technologies at medium and larger scales. Wetland areas, especially the riverine *várzeas*, have been utilised for many years for rice, subsistence and small-scale cashcrop production. Clearance for pasture is generally large scale, but varies considerably in the level of management sophistication and capital input. The principal aim of the more advanced agricultural techniques is to provide sustainable management options for the different soil resources available although, with a few promising exceptions, this cannot be said to have been achieved at present within the Brazilian Amazon. In addition, there have been areas replanted with exotic or selected indigenous tree and shrub species for silviculture and plantations with crops such as rubber and cocoa. The land use systems that have evolved in the Amazon up to the present do not have the diversity, particularly within the smallholder units, that is to be found in the older settled areas of the humid tropics (Norman, 1979, 1984).

(a) *Shifting cultivation*: Migratory clearance of forest and short-lived cultivation of the cleared patches represents a traditional form of subsistence, particularly on Oxisols and Ultisols where it is by far the most dominant form of food production (Sanchez *et al.*, 1982a). Although it is ecologically sound, assuming a small scale of operation and an adequate regeneration fallow (Szott and Palm, 1986), it supports relatively few people and requires a large area in relation to yield. Where demographic pressures are increased, as in the area around Bragantina near Belém (Alvim, 1979; Sanchez *et al.*, 1982a), there is a temptation to extend the period of cultivation whilst reducing the fallow. This results in the well known drop in crop yield and soil fertility (Scott, 1987). It is arguable whether a case could be made to allow shifting cultivation to occupy more than a small proportion of the Amazon if more productive and sustainable forms of agriculture could be developed, despite the fact that it is the traditional form of livelihood for many people. As Alvim (1978) has commented, shifting cultivation has no capacity to contribute to any improvement in the way of life of the farmer or, as Sanchez *et al.* (1982a) put it 'shifting cultivation guarantees perpetual poverty for those who practice it'.

Shifting cultivation and the recovery of the native vegetation has been studied in detail by Jordan and his colleagues in an area near San Carlos de Rio Negro, near to the Venezuelan border with Colombia and Brazil (Jordan, 1982b, 1987; Saldarriaga, 1987). This serves as a good illustration of a form of soil usage which, with many local variations, is widespread in South America (Denevan, 1981; Watters, 1971; Moran, 1977, 1981; Denevan *et al.*, 1983; Savage *et al.*, 1982). Shifting cultivation has evolved over thousands of years, during which time the indians and latterly the *caboclos* have been able to identify plant species for use in a multitude of medical, food and household functions, and to protect or plant the most useful species (Meggers, 1985). Whereas initially, farmers and small communities were virtually self-supporting and had relatively little need to trade with or contact other groups, a growing exchange with a market economy tends to grow with time (Gross *et al.*, 1979). Near San Carlos, experimental plots were established on Oxisols covering low hills. Dry weight biomass and nutrient status were measured prior to clearing and burning following local practice (Jordon, 1987). Figure 13.5 illustrates the nutrient depletion and there was a major loss of nutrients over the total ecosystem. However, losses were partly compensated by the decomposition of organic litter left on the surface and the soil nutrient levels remained fairly high – despite the fact that crop production declined sharply. It is likely that nutrient availability decreased rather than the total nutrient stocks and that there was a short but crucial period of soil erosion before a protective vegetation cover was re-established. The effect of erosion after only one cultivation cycle can be dramatic, as exemplified in *milpa* clearance on sloping karst in Central America (Furley, 1987). Pioneer weeds and other plants characteristic of the early stages of succession did not appear to suffer from nutrient deficiencies in the same way as the crop plants. In the short term therefore, the impact of 1 hectare forest clearance has not seriously impaired forest re-growth although the extent to which the original forest can be precisely reconstructed is difficult to assess.

Saldarriaga (1987) looked at the longer-term recovery following shifting cultivation in the Upper Negro, taking plots at different stages of abandonment to establish a chronosequence over about 100 years.

Making the assumption that the sites represent the model of a single succession, the data suggest that it takes around 80 years to build up the biomass to a value of 50 per cent of that in the original forest. Atmospheric input apparently increased total ecosystem stocks of calcium and potassium. Phosphorus also increased, possibly by mobilisation of soil reserves and nitrogen also accumulated largely on account of biological fixation. The implication of this research is that a very long time, well in excess of 100 years, which represents a period far longer than most normal forest fallow, is required to return the forest to its original nutrient equilibrium. Even then, the diversity and species composition of the vegetation is likely to be very different from the initial state.

At present there is little population pressure on shifting cultivation in the Brazilian Amazon, except in localised areas of older land colonisation. Consequently there is little pressure on the authorities to pay attention to what is currently or potentially, a serious form of land degradation, or to concentrate on agronomic research of a long-term nature. Unfortunately the degradation becomes obvious only when it is too late (or too costly) to effect an appropriate repair whilst at the same time maintaining the structure of the local settlements and their socio-economic stability.

(b) *Crop production at low input technologies*: Whereas the traditional form of shifting cultivation is essentially subsistence farming with little or no soil management, it is argued that low cost technologies can stabilise smallholder farming practice by increasing productivity and improving standards of living.

Low input technology minimises chemical inputs and avoids the use of machinery wherever possible but tries to maximise the efficiency. There are several strategies which can be tried to achieve this. Sanchez and Salinas (1981) outlined three main lines of approach: first, by using crop varieties which are tolerant of the acid and other soil constraining conditions; secondly, by returning crop residues to the soil and managing the soil organic matter more successfully; and thirdly, by developing a fallowing system which is intermediate in nature between shifting cultivation and continuous agricultural production:

1. Knowledge of the soil limitations discussed earlier in this chapter has been steadily increasing over the past decade, although a number of aspects (for example manganese toxicity and the part played by some trace deficiencies) are still far from clearly understood. Many economically important plant species are tolerant of acid conditions in the tropics and many have originated there. Tolerance covers many aspects of individual plant reactions to limiting soil factors, and in a general way describes how a species contends with high levels of aluminium and

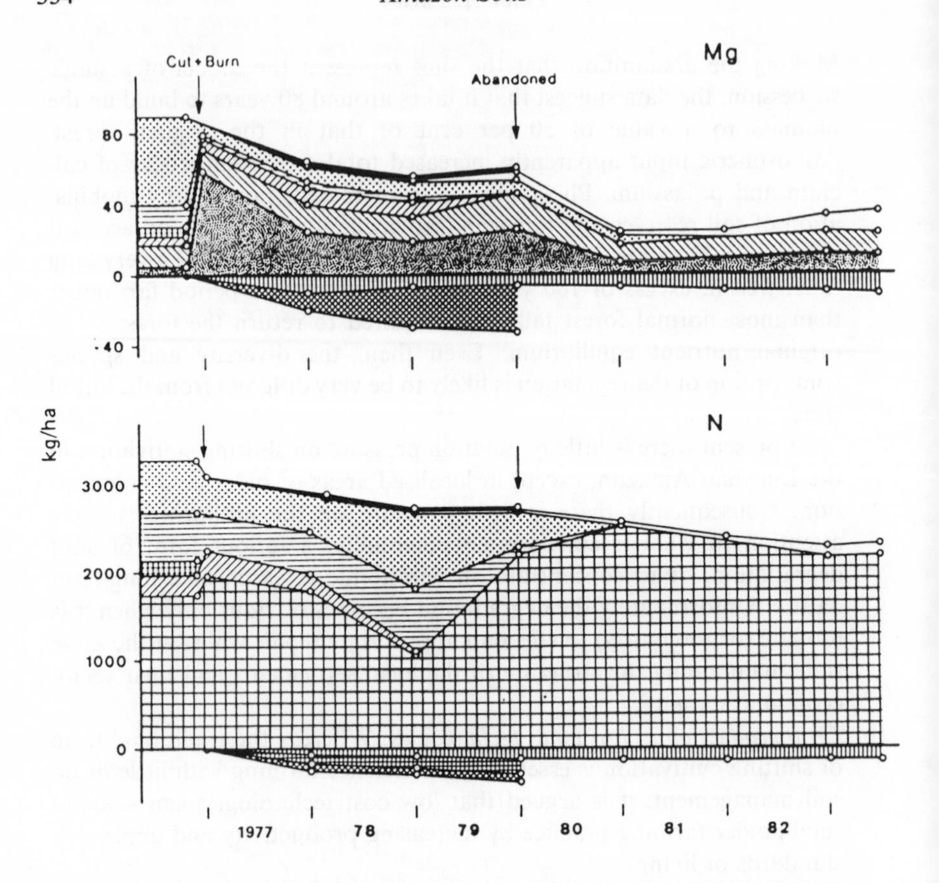

manganese and deficiencies in major nutrients, such as calcium, magnesium and phosphorus and micronutrients, such as copper and zinc. Nearly 400 species are tolerant of either acid soils, aluminium toxicity or lateritic (plinthitic) soils. Sanchez and Salinas (1981) have modified the list and suggested a number of new and potentially important food crops which could be used.

Detailed tests of varieties have been undertaken at Yurimaguas (Nicholaides and Piha, 1985; Benites and Nurena, 1985). These have shown that, for example, upland rice and cow peas have a high degree of acid tolerance; sweet potatoes and peanuts have a moderate tolerance but several other likely varieties failed to show tolerance. This reinforces the earlier point about the necessity for site specific trials of promising crops. Yields declined after six successive harvest periods and it is suggested that the low input system, in this example at least, may represent a transition technology between shifting cultivation and continuous intensive agriculture.

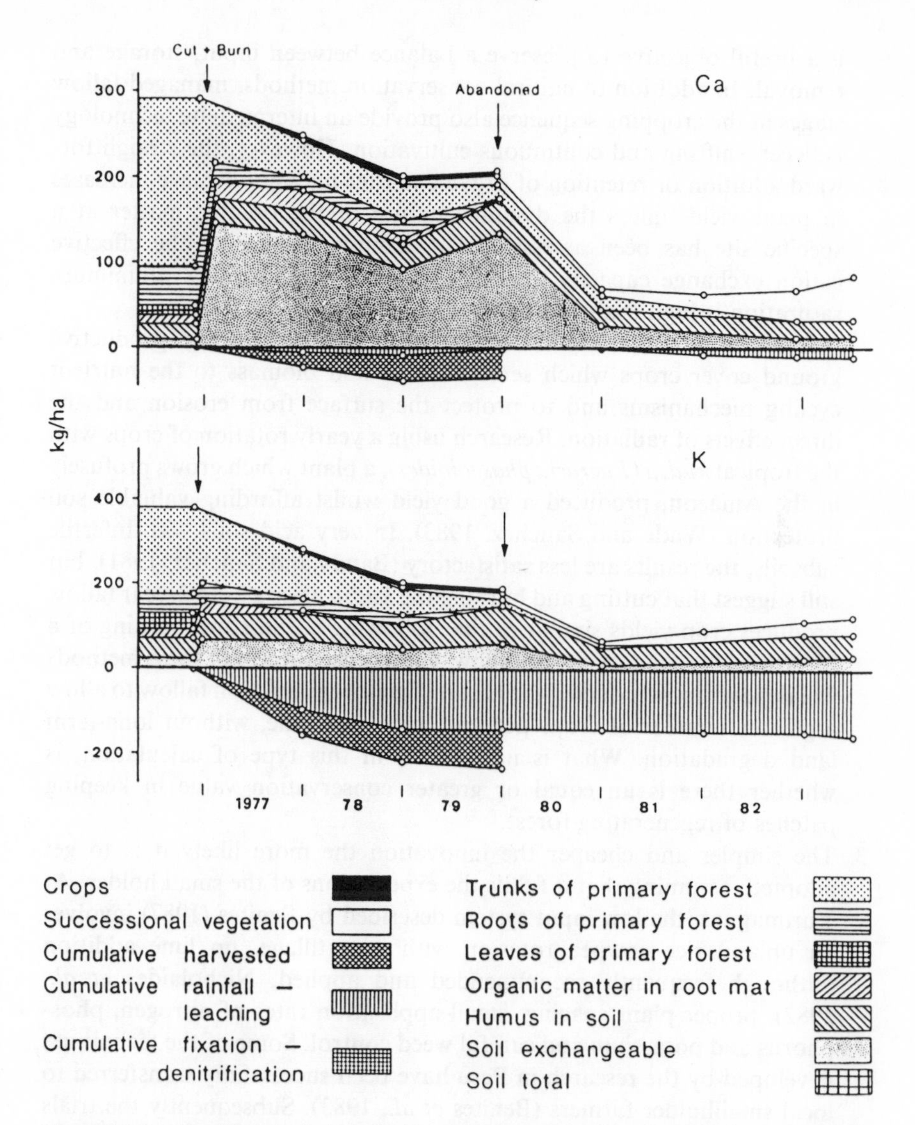

Figure 13.5 Nutrient depletion during shifting cultivation, San Carlos, over a six-year period of observation

2. The maintenance of organic litter, the retention of tree stumps with the rooting systems intact, together with the retention of unused crop residues and other devices to increase the continuous slow decomposition and release of nutrients is obviously a helpful nutrient conservation technique. Humid tropical soils are not as low in organic matter as often believed (Sanchez *et al.*, 1982c) but, with high rates of decomposition, it

is a useful objective to preserve a balance between input, storage and removal. In addition to natural conservation methods, managed fallow stages in the cropping sequence also provide an intermediate technology between shifting and continuous cultivation. However, the straightforward addition or retention of organic matter may not achieve increases in plant yield unless the dynamic nature of the organic matter at a specific site has been assessed. For example, increasing the effective cation exchange capacity may be of questionable value if aluminium saturation is high (Sanchez and Miller, 1986).

A useful development has been the use of fast growing, productive ground cover crops which serve both to add biomass to the nutrient cycling mechanisms and to protect the surface from erosion and the direct effects of radiation. Research using a yearly rotation of crops with the tropical *kudzu* (*Pueraria phaseoloides*), a plant which grows profusely in the Amazon, produced a good yield whilst affording valuable soil protection (Wade and Sanchez, 1983). In very acid soils with infertile subsoils, the results are less satisfactory (Bandy and Sanchez, 1981), but still suggest that cutting and burning the *kudzu* after a 2- to 3-year fallow produces crop yields similar to that following cutting and burning of a 25-year-old secondary forest fallow. Clearly the success of these methods will depend upon reducing the natural forest regeneration fallow to allow a greater number of crops per unit area over time, without long-term land degradation. What is not certain in this type of calculation, is whether there is an equal or greater conservation value in keeping patches of regenerating forest.

3. The simpler and cheaper the innovation the more likely it is to get adopted, assuming that it fulfils the expectations of the small holder. At Yurimaguas, the low input system described by Benites (1987) involves an upland rice–cowpea rotation with zero tillage, no lime addition (although frequently recommended and applied, Nicholaides *et al.*, 1982), proper plant spacing, small application rates of nitrogen, phosphorus and potassium and careful weed control. Some of the techniques developed by the research in Peru have been successfully transferred to local smallholder farmers (Benites *et al.*, 1983). Subsequently the trials have been extended to central Amazonia at the Manaus EMBRAPA station following an agreement in 1980 with North Carolina State University forming the scientific link. Additional worldwide links now exist through the acid tropical soils network and IBSRAM (International Board for Soil Research and Management).

This more concerted research is likely to make considerable strides in using many of the Oxisols and Ultisols over the next one or two decades. The emphasis of such research is likely to be upon the dual purpose of providing improved yields on staple subsistence crops whilst developing cash crops capable of improving the longer-term prosperity of the whole farming community. At present however, the full cost of protecting and conserving the environment from land degradation over the long term is not usually given sufficient attention.

Continuous Cropping Including Agroforestry and Silviculture

When the fallow period, whether of natural regeneration or of introduced plants, is reduced to a minimum, the system is in continuous production. However, this may involve different types of rotation and agroforestry or other multiple land use techniques.

To achieve one or more crops per year, given the environmental and soil constraints considered earlier, fertiliser inputs, the use of machinery and more advanced management techniques become essential. The scale of operation is usually increased, although one of the ideas behind the increased research into the use of acid soils is to try to develop methods that can also be used at an intermediate farming scale (approximately 50 to around 400 hectares). Paddy rice has been grown traditionally on the fertile alluvial *várzea* soils. With recent advances in improved varieties and spacing, farmers may be able to double the yields of a few years ago. Some of the soil constraints to developing such wetland soils are reviewed by Sanchez and Buol (1985).

Research on the relationships between soil properties and perennial crops is in its infancy within the Brazilian Amazon. A useful body of information exists on cocoa especially on Alfisols around the CEPLAC station in Bahia (Alvim, 1979, 1982; Alvim and Cabala-Rosand, 1984) and extended to their research operations in Rondônia and elsewhere. More information on agroforestry techniques exists in other areas of the humid tropics and has been collated by ICRAF (International Council for Research in Agro-Forestry) in Nairobi with some assessment in Sanchez, 1979; Bishop, 1982; Peck, 1982; Valencia, 1982; and Denevan *et al.*, 1983. A recent review of the uncertain future directions of forestry in the Brazilian Amazon is given in Rankin (1985). Multiple cropping techniques have not yet been researched in any detail in the Amazon region with the exception of some preliminary work in the Yurimaguas area (Valverde and Bandy, 1982; Wade and Sanchez, 1984). These techniques include alley and intercropping and a host of other alternatives including the incorporation of trees into the rotation (for instance *Gmelina arborea*, the peach palm, *Guilielma gasipaes* or the oil palm, *Elaeis guineensis* in the wetter areas). The incorporation of ranching as a part of a multiple land-use system as opposed to the current monoculture methods is also being examined, although not specifically in the Amazon (see Reynolds, 1988). Additionally, it is helpful to think of silviculture as a perennial cropping system to examine whether a blend of forestry, grazing and cropping may prove to be a better utilisation of the soil resources than single land use systems. One of the largest tree plantations ever installed in the tropics at Jarí indicated the problems of a monospecific system where, in addition to the well known social and economic difficulties, there were problems of soil fertility and nutrient deficiency as well as fungus outbreaks (*Ceratocystis fimbriata* on *Gmelina*) and leaf cutter ant problems (Russell, 1987; Fearnside and Rankin, 1982).

A technology for providing continuous annual crops has been developed. By choosing appropriate cultivars for the soil conditions, the NCSU work at Yurimaguas has shown that basic food crops can be successfully grown with reasonable yields and without expensive capital inputs. The research programme is continuing to investigate intensive crop production on the fertile alluvial soils, low input crop production on well drained Oxisols and Ultisols, with legume based pastures and agroforestry in the areas of more accidented topography containing steeper slopes. Three crops a year have been possible in level, well drained Ultisols with a sandy loam topsoil over a clay loam subsoil and with pH values of 4, high initial levels of aluminium toxicity, deficiencies in phosphorus and potassium, and most other nutrients. As indicated earlier, these are conditions typical of much of the Basin. The most promising crop combinations (Sanchez *et al.*, 1982, 1982a; Buol and Sanchez, 1986) consist of rotations of upland rice, corn and soybean, or peanuts as a substitute for corn. Such rotations fit in with the rainfall pattern in the area and avoid the pathogen build up reported in areas of monoculture. The fertiliser requirements are considerable (Figure 13.6) and involve considerable management expertise although the transfer of technology could be achieved at simpler levels (Table 13.7a). The results can be interpreted to demonstrate that continuous production is feasible in the Amazon and would require levels of fertilisation comparable to those of Ultisols in the South-East USA. What the results cannot show, however, is the long-term effects of changing the land use so drastically, either on the climate if at a large enough scale, or on long-term imperceptible land degradation. Neither can they take into account the impact of implanting such techniques in areas of completely different socio-economic background and tradition. They also make a number of assumptions, such as the presence of a ready market and transport network and other aspects of the economic infrastructure which certainly do not yet exist over much of the Brazilian Amazon. They are, nevertheless, extremely interesting and relevant to the discussion of sustainable development because they indicate the probability that a large proportion of the Amazon Basin could technically be used for continuous agriculture without invoking large-scale engineering or other large capital costs. After 8 years of continuous cultivation and 20 fertilised crops, the soil properties in the example discussed were in many ways better than the initial state (Table 13.7b).

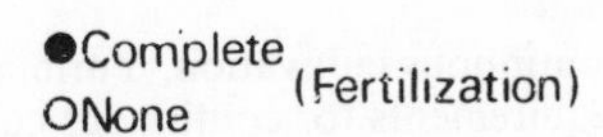

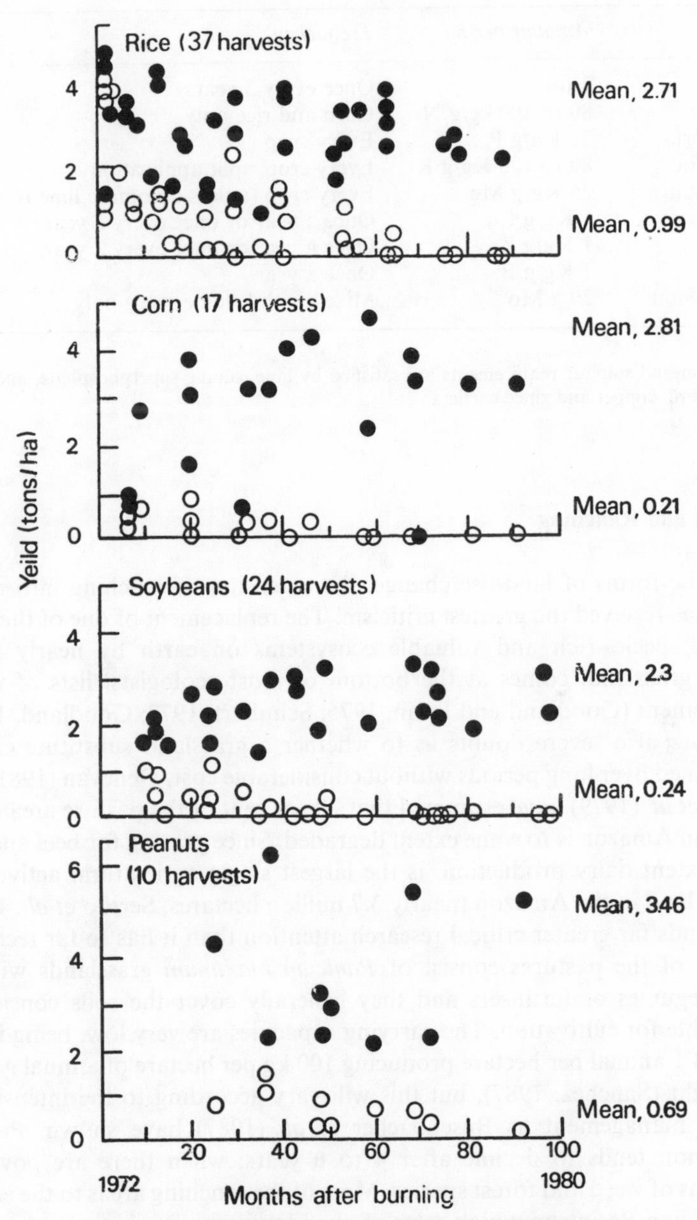

Figure 13.6 Fertiliser requirements in Amazonia
Source: Sanchez (1987), based on Serrão *et al.* (1979)

Table 13.7 Continuous cultivation, Yurimaguas
(a) Fertiliser requirements for continuous cultivation with annual rotations

Input[1]	*Amount per ha*	*Frequency*
Lime	3 tons	Once every 3 years
Nitrogen	80 to 100 kg/g N	Corn and rice only
Phosphorus	25 Kg/g P	Every crop
Potassium	80 to 100 Kg/g K	Every crop, split application
Magnesium	25 Kg/g Mg	Every crop (unless dolomitic lime is used)
Copper	1 Kg/g Cu	Once a year or once every 2 years
Zinc	1 Kg/g Zn	Once a year or once every 2 years
Boron	1 Kg/g B	Once a year
Molybdenum	20 g Mo	Mixed with legume seeds only

Note:
1. Calcium and sulphur requirements are satisfied by lime, simple superphosphate, and magnesium, copper and zince carriers.

Pasture and Ranching

Of all the forms of land-use change, the extension of ranching into native forest has received the greatest criticism. The replacement of one of the most diverse, species-rich and valuable ecosystems on earth by nearly single species grassland, comes at the bottom of most ecologists' lists of useful development (Goodland and Irwin, 1975; Schubart, 1977; Goodland, 1980). There are also severe doubts as to whether a grassland substitute can be maintained over long periods without considerable cost. Denevan (1981) and Serrão *et al.* (1979) have estimated that 20 per cent of the pasture area in the Brazilian Amazon is to some extent degraded. Since grazing for beef and to a lesser extent dairy production, is the largest single agricultural activity on cleared land in the Amazon (nearly 3.7 million hectares, Serrão *et al.*, 1979), it demands far greater critical research attention than it has so far received.

Most of the pastures consist of *Panicum maximum* grasslands without either legumes or fertilisers and they generally cover the soils considered unsuitable for cultivation. The carrying capacities are very low, being in the order of 1 animal per hectare producing 100 kg per hectare of annual gain in liveweight (Sanchez, 1987), but this will vary according to the intensity of pasture management as Buschbacher *et al.* (1987) have shown. Pasture production tends to decline after 4 to 6 years, when there are powerful invasions of weed and forest species. Most of the ranching areas to the east of the Amazon Basin show high rates of abandonment after between 4 and 10 years. In these conditions, Amazonian pastures can virtually be considered as a variant on shifting cultivation (Alvim, 1979). On the other hand, some of the pasture invader species could be better and more usefully managed. For

Table 13.7 *cont.*

(b) Changes in topsoil (0 to 15 cm) properties after 8 years of continuous cultivation: 20 crops and complete fertilisation

Time	pH	*Organic matter* (%)	*Exchangable, millequivalents per 100 cm²*				*CEC*	Al *satura-tion* (%)	*Available* (*parts per* million) *cm²*				
			Al	Ca	Mg	K			P	Zn	Cu	Mn	Fe
Before clearing	4.0	2.13	2.13	0.26	0.15	0.10	2.78	82	5	1.5[1]	0.9[1]	5.3[1]	650[1]
94 months after clearing	5.7	1.55	0.06	4.98	0.35	0.11	5.51	1	39	3.5	5.2	1.5	389

Note:
1. 30 months after clearing.
Source: Based on Sanchez *et al.* (1982a).

example, Hecht (1979) has shown that some of the incoming shrub legumes are useful macro and micro nutrient accumulators.

Initial work on ustic soils of the pastures in the eastern part of the Basin (e.g., Falesi, 1976; Falesi *et al.*, 1982; Serrão *et al.*, 1979), suggested that many properties were maintained at reasonable levels and were readily improved without much difficulty or cost. With appropriate fertiliser application and better pasture management, it was claimed that the soil physical properties and to some extent the soil chemical character was improved by conversion of forest to grassland (Sanchez, 1977b; Serrão and Humma, 1982; Toledo and Serrão, 1982). It was further claimed that the improvements were sufficiently encouraging that such areas could be considered as transitional to perennial crop production. Since then, further work has to some degree supported and in other aspects rejected this idea of the soil dynamics (Baena, 1977; Fearnside, 1979, 1985; Hecht, 1983). Confusion partly arises from the great variability in results, reflecting the difficulties of comparing the evidence over enormous areas with limited samples and different time periods. However, with more careful management than has normally been the case, it does appear the pasture implantation can maintain the benefits of burning cleared vegetation for a few years, especially the levels of calcium and magnesium. Burning also helps to diminish the level of aluminium toxicity and keeps sufficient available phosphorus for about 4–5 years and thereby retards the decline in soil fertility on cleared land.

The decline in pasture productivity has been correlated with nitrogen and phosphorus deficiencies and the poor adaptability of Panicum maximum (which is very responsive to changes in the phosphorus availability). A comparable picture has emerged from research on Oxisol pastures at Paragominas in northern Mato Grosso. The productivity decline here appears to be related to available phosphorus levels falling below 4 ppm and, as may be expected in view of the P-fixation described earlier, is faster in more argillaceous surface horizons than in loamy surface soils (Sanchez and Uehara, 1980) (see Figure 13.7). Work by Serrão and his associates demonstrated that manual forest clearance, burning (both initially or to reclaim abandoned pastures), application of P-fertiliser at around 25 kg per hectare and the planting of aluminium-resistant species (e.g. *Brachiaria humidicola* and *Pueraria phaseoloides*) helped to sustain new pastures and rejuvenate old grasslands. It also markedly increased carrying capacities (Serrão, 1981; Sanchez and Homma, 1982). Several other promising grass and legume species can be identified in CIAT's Tropical Pastures Programme. It would appear from research in Brazil and elsewhere that in the Amazon Basin, the maintenance of a grass–legume pasture would require annual fertiliser inputs of around 25 kg per hectare P and K and probably the addition of Mg and S. Better managements than prevalent today would be a necessity since the strategies developed on the well drained *cerrado* or in the drier Eastern and Southern savannas are not directly applicable to the more constantly humid and poorly drained parts of the Amazon Basin (Cochrane, 1979).

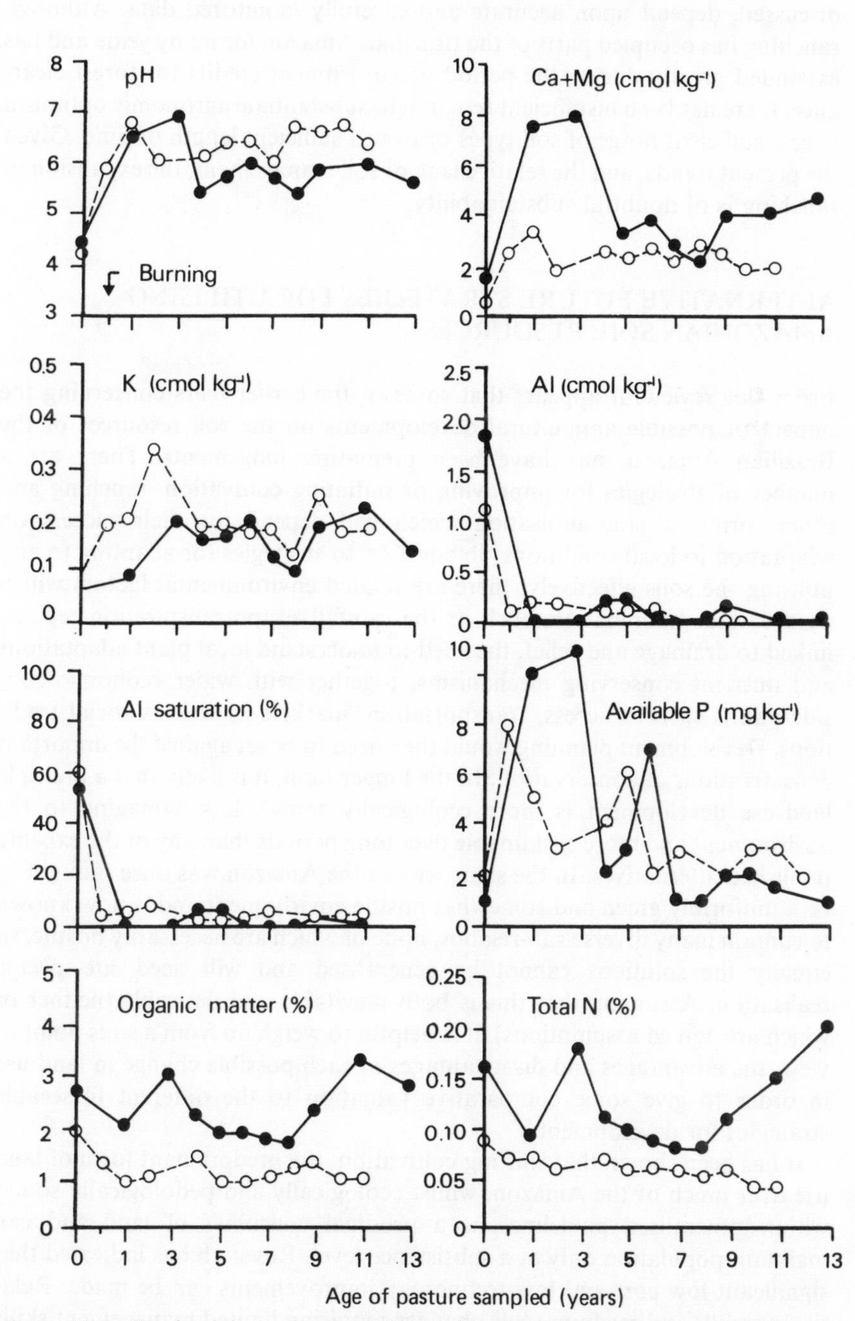

Figure 13.7 Productivity and soil nutrients of *panicum maximum* pastures of known age

It is once again apparent that solutions to the controversies concerning the possible use of the region for pasture, as for the other changes in land use discussed, depend upon accurate and carefully monitored data. Although ranching has occupied parts of the Brazilian Amazon for many years and has expanded greatly during the period of government credits for forest clearance, there has been insufficient research to substantiate agronomic optimism over a sufficient range of soil types or over a sufficient length of time. Given the present trends, and the relative lack of soil management, the expansion of ranching is of doubtful substainability.

ALTERNATIVE FUTURE STRATEGIES FOR UTILISING AMAZONIAN SOIL RESOURCES

From this review, it appears that some of the earlier fears concerning the impact of possible agricultural developments on the soil resources of the Brazilian Amazon may have been premature judgements. There are a number of strategies for improving or initiating cultivation, ranching and other forms of land utilisation which will depend for their success on adaptation to local conditions. In addition to strategies for adapting to and utilising the soils effectively, there are related environmental factors which need careful examination, such as the rainfall–evapotranspiration regime, linked to drainage and relief, the need to understand local plant adaptations and nutrient conserving mechanisms, together with wider economic considerations such as access, transportation, markets as well as social traditions. Development planning would then need to be set against the important considerations of conservation. In the longer term, it is likely that a multiple land-use development is more ecologically sound, less damaging to the environment and more sustainable over long periods than any of the existing published alternatives. In the same way as the Amazon was once thought to be a uniformly green and somewhat hostile environment and is now known to contain many diverse sub-regions, none of which are necessarily hostile, so equally the solutions cannot be generalised and will need site-specific realisation. Assuming that this is both inevitable and desirable (neither of which are agreed assumptions), it is helpful to weigh up from a soils point of view, the advantages and disadvantages of each possible change in land use in order to give some comparative valuation to the different forseeable strategies for development.

It has been shown that shifting cultivation, the predominant form of land use over much of the Amazon, whilst ecologically and pedologically sound within generalised guidelines, is a prodigal consumer of land and can maintain population only at a subsistence level. Research has indicated that significant low cost and low technology improvements can be made. Relatively small- and medium-scale changes requiring limited management skills and a minimum input of fertilisers and, where possible, utilising local knowledge of the environment, do appear to offer a potential for increasing and stabilising crop production. However, this smallholder approach has not been able to attract aid funding to a significant extent in the past and improvements cannot be achieved without disturbing the way of life of the existing farming community. Development appears to necessitate a change from a migratory to a sedentary way of life and may in the end simply represent a transition to intensive crop cultivation. It would, for example, completely destroy the traditional way of life of many indians and possibly also that of adapted long-term colonists and *caboclos*. Such groups of people cannot easily be integrated in formal improvement schemes. The smallholders involved in more recent colonisation projects are intended to be at a more advanced or intermediate technical level, but frequently the land-use within the colonist plots has reverted to a type of shifting cultivation either within the plots or unofficially utilising temporary clearings outside. This group could be helped by the type of research under way at Yurimaguas, and currently being developed near Manaus and elsewhere in the Brazilian Amazon. It is not intended to act as a stimulus for a new wave of colonisation schemes, although a demonstrable success would undoubtedly trigger fresh influxes of land hungry migrants.

It has been believed for the past few decades that the typical acid soils (Oxisols and Ultisols) of the Amazon could not be developed for continuous, sustainable agriculture. Limited patches of naturally fertile soil gave rise to initial optimism which then turned to extreme pessimism as colonisation plots failed, pastures became degraded and the problematic nature of the soil constraints became more evident. As illustrated earlier (Table 13.4), the major part of the Brazilian Amazon consists of nutrient-poor, acidic soils, often with a problem of aluminium toxicity and with a quarter of the soils either poorly drained or flooded. Nevertheless a package of strategies has been devised which offers the promise of more soundly based production in well drained areas with slopes below 5 to 8 per cent (Sanchez, 1977a). These include choosing the most appropriate methods of land clearance, the use of adapted cultivars together with well-managed fertilisation programmes that can be fashioned to fit different soil–landscape combinations. It has been calculated that such techniques are of an equivalent cost to those which were required to bring similar soils into profitable cultivation in the South-East of the USA. However, such a comment camouflages the enormous difference between the two environments in terms of social background and the level of support which can be given to farmers. Relatively few farming enterprises in the Brazilian Amazon can afford the cost of development at the level required, nor is there the infrastructure, technical and financial back-up to maintain a comparable level. This advanced stage of soil and land management may be possible for a limited number of developments in more accessible areas with favourable soils, but it is not an overall solution either at present or in the perceptible future. A more realistic medium-term possibility is that improvements may emerge from a number of nucleii where expertise and facilities have been developed. Outside technical help and finance might be more usefully spread over a number of smaller-scale operations rather than at the massive regional aid scale, which has not so far proved successful.

There is little doubt that continuous agriculture could be made to work within both forest and savanna areas of the Brazilian Amazon. Whether it is cost effective, and economically viable at different scales of farm enterprise, socially acceptable, environmentally safe and, from a soil point of view, sustainable over a long term is less certain. As Fearnside (1986) has commented, development planning could usefully examine the concepts of human carrying capacity. Using a strategy whereby the most naturally fertile and accessible soils are used first and then progressing to soils needing increased levels of management and capital input, and leaving aside entirely for the present the 'inapt' soils as identified by the Projeto RADAM surveys, there is considerable scope for improved crop, animal and timber yields. This could still leave adequate land for conservation without the pressures of excess demands on the environment as envisaged in the more pessimistic of the ecological assessments. For example in Rondônia, Furley (1980) calculated that 50 per cent of the soils were too poor to warrant development at the present time, but this still leaves both a small percentage of good soils as well as a considerable group of soils which could be put to continuous use by the methods discussed above. This presupposes a resolution of the problems of transgressing traditional indian territories.

The idea of multiple land-use development along more acceptable ecological lines has gained momentum. It already seems to work in small examples, for instance at Tomé-Assú in Pará (Jordan, 1987). Detailed research on agroforestry techniques has not been monitored long enough to demonstrate conclusively that yields of crops, shrubs and trees are increased consistently over long periods whilst conserving the soil environment. A good deal is known however, of the value of cover crops to protect the soil surface against drying, oxidation and erosion and also of the complexity of natural nutrient cycling systems. The importance of conserving organic matter and its effective decomposition involving numerous living organisms is also known if not, as yet, well understood. (Swift, 1984; Swift and Sanchez, 1984). This knowledge indicates that multiple above ground layering of plants, with ground protection to keep biological processes uninterrupted in the litter and surface soil horizons, is likely to maintain an ecological balance better than clear felling or monocultural land uses. If such a strategy could include the conservation of large blocks of forest, perhaps following the ideas of the size of minimum conservation areas (Lovejoy, 1984), and also incorporate a sustained yield forestry element, then it is likely that the system would be acceptably close to being self-sustaining.

Two models of this approach may be given as examples. Jordan (1985b) has looked at the problem of nutrient loss during land clearance, particularly on slopes. He envisaged a method of progressive strip cutting (Figure 13.8a) to minimise soil erosion and loss of cations and to maximise the natural spread of seeds and mycorrhizal spores into the disturbed area. Jordan's suggestions imply that small-scale disturbances in terms of intensity, size and duration can withstand the impact and maintain productivity whilst retaining an ability to recover. He also demonstrated that native adapted species are more likely to be effective in maintaining the productive capability of a site and in restoring soil fertility. Nutrient supply is critical and conservation is emphasised in this type of approach. Such a model is devised for sloping forested areas and extrapolation from the sites studied would require knowledge of the relationships between the soil conditions and the natural vegetation (Cochrane, 1986).

At a larger soil landscape scale, Sanchez *et al.* (1982a) explore the different strategies available over a group of differing landscape components (Figure 13.8b). This is a similar approach to the terrain analysis methodology familiar in other parts of the tropics. The undesirable effects of large-scale insensitive clearing for limited land uses could in this way be avoided. Paddy rice and permanent crop cultivation is recommended on alluvial terraces with a low flood hazard. Shifting cultivation would not be encouraged to expand but could be continued if and where any change in land-use would cause unreasonable disturbance to the existing local population. Continuous crop cultivation with pasture, possibly incorporated into an agroforestry scheme would be best adapted to the level and gently sloping Oxisols and Ultisols. The remaining areas would be appropriate for agroforestry with or without ranching, for forestry, or designated as Reserves. Such areas would therefore be confined to the rolling and dissected hill-lands having slopes up to 30 per cent. Before any development, implying forest destruction, is permitted to go ahead, it would be helpful for planners to consider the alternative strategies available, possibly utilising a check system, as illustrated in Figure 13.8c.

On account of its size and because of the limited number of scientists and planners working in the field, the Brazilian Amazon has been developed in large tracts and at a sub-regional scale. The role of the smallholder or peasant farmer or even a group of farmers or indian tribe, is rarely considered at this scale. Most agronomic research is concerned with commercial agriculture. Research to improve the standards of living of the smallholder tends to assume that shifting cultivation and more sedentary subsistence agriculture are systems which are both inefficient and imply perpetual poverty. Provided that the wishes of the individual or local community, in terms of customs and preferred way of life, are respected, there is no technical reason why greater productivity and some form of cash crop farming should not be encouraged. For this to happen on a regional scale, however, considerable advisory and financial support, along with the creation of an

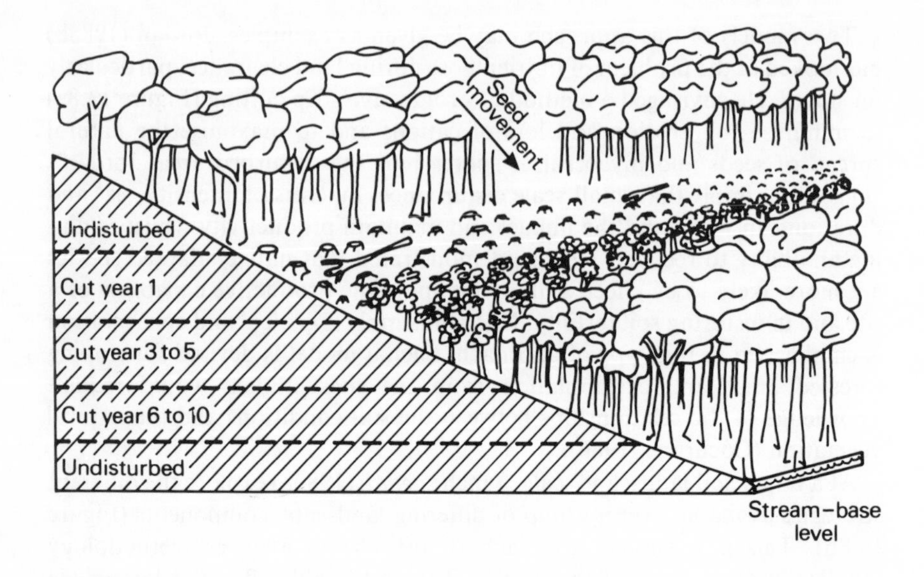

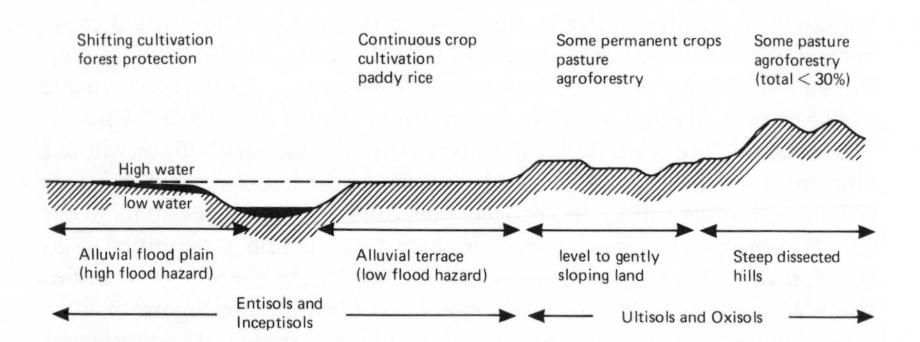

Source: Jordan (1985b)
a Strip-cutting and soil erosion
Source: Sanchez *et al.* (1982a)
b Settlement strategies for alternative landscapes
c Planners' guide for alternative land-use in Amazonia

Figure 13.8 Multiple land-use development

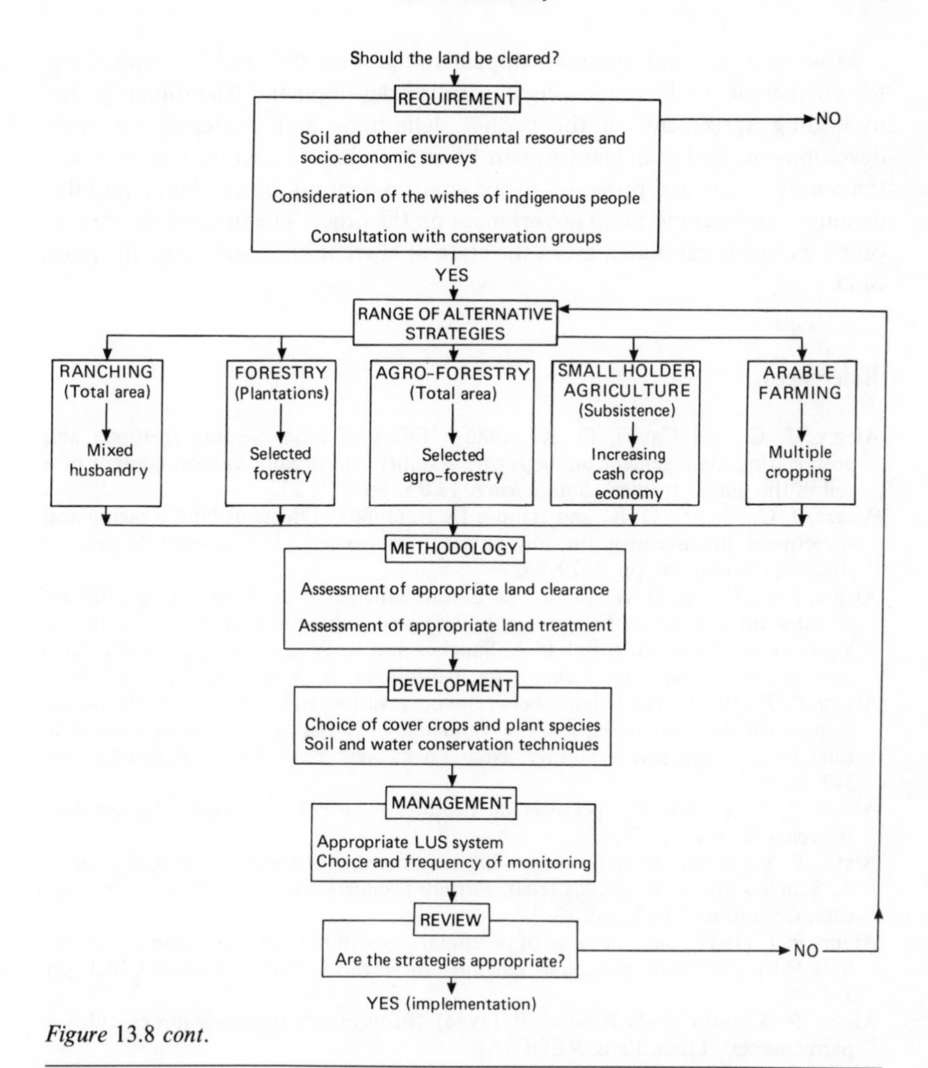

Figure 13.8 *cont.*

infrastructure and markets, would be required to cater for growth in production and the rising expectations of farmers. At present, there is little sign that the smallholder is likely to gain this support. So in summary, there appears to be growing optimism concerning the possibilities for utilising the limited soil resources of the Brazilian Amazon, but to do so would require a substantive change in the level of government support, particularly for smallholder farmers. Such support would seem to need both technical understanding of the environment as well as economic incentives and social safeguards. Over the long term, a policy combining conservation with development seems to be best achieved by adopting multiple land use strategies.

Most scientists and observers appear to agree on the need for conserving the environment whilst allowing controlled development. The difficulty lies in finding agreement on the precise definitions and strategies for such development. Soil management can be seen to have a pivotal role between the requirements and pressures of the environment on the one hand, and the demands of developers and government on the other. The future balance can only be crudely estimated given the state of current knowledge and the pace of change.

References

Alegre, J. C. and Cassel, D. K. (1986), 'Effect of land clearing methods and postclearing management on aggregate stability and organic carbon content of a soil in the humid tropics', *Soil Science*, 142(5) pp. 289–95.

Alegre, J. C., Cassel, D. K. and Bandy, D. E. (1986), 'Effects of land clearing and subsequent management on soil physical properties', *Soil Science Society of America Journal*, 50, pp. 1379–84.

Alegre, J. C., Cassel, D. K., Bandy, D. E. and Sanchez, P. A. (1986), 'Effect of land clearing on soil properties on an Ultisol and subsequent crop production in Yurimaguas, Peru', in R. Lal, P. A. Sanchez and R. W. Cummings, Jr (eds), *Land clearing and development in the tropics*, Rotterdam: A. A. Balkema, pp. 163–80.

Alvim, P. T. (1977), 'The balance between conservation and utilization in the humid tropics, with special reference to the Amazonia of Brazil', in G. T. Prance and T. S. Elias (eds), *Extinction is Forever*, New York: New York Botanical Garden, pp. 347–52.

Alvim, P. T. (1978), 'Perspectivas de produção agrícola na região Amazônica', *Interciencia*, 3(4) pp. 243–51.

Alvim, P. T. (1979), 'Agricultural production potential of the Amazon region', in P. A. Sanchez and L. E. Tergas (eds), *Pasture production in acid soils of the tropics*, Cali, Colombia: CIAT, pp. 13–23.

Alvim, P. T. (1982), 'An appraisal of perennial crops in the Amazon Basin', in Hecht, S. B. (ed.), *Amazonia: agriculture and land use research*, Cali, Colombia: CIAT, pp. 311–28.

Alvim, P. T. and Cabala-Rosand, P. (1984), 'Proyecto de investigación en cultivos permanentes', Lima, Peru: REDINAA.

Baena, A. R. C. (1977), 'The effect of pasture (*Panicum maximum*) on the chemical composition of the soil after clearing and burning on typical highland forest', M.S. thesis, Iowa State University, Ames, Iowa.

Bandy, D. E. and Sanchez, P. A. (1981), 'Managed *kudzu* fallow as an alternative to shifting cultivation in Yurimaguas,' *Agronomy Abstracts*, p. 40.

Bandy, D. E. and Sanchez, P. A. (1986), 'Post clearing management alternatives for sustained production in the Amazon,' in R. Lal, P. A. Sanchez and R. W. Cummings (eds), *Land clearing and development in the tropics*, Rotterdam: A. A. Balkema, pp. 347–61.

Beek, J. and Bramão, L. (1969), 'Nature and geography of South American soils' in *Biogeography and ecology in South America*, vol. I, The Hague: N. V. Junk, pp. 87–112.

Benites, J. R. (1983), 'Soils of the Peruvian Amazon: their potential for use and development', in J. F. Wienk, and H. A. de Wit (eds), *Management of low fertility acid soils of the American humid tropics*, San Jose, Costa Rica: IICA, pp. 85–93.

Benites, J. R. (1987), 'Transfer of acid tropical soil management technology', in IBSRAM (International Board for Soil Research and Management), *Management of acid tropical soils for sustainable agriculture*, Bangkok, Thailand, pp. 245–60.
Benites, J. R., Bandy, D. E., Nicholaides, J. J., Piha, M. I. and Sanchez, P. A. (1983), 'Successful soil management technologies and their transfer to small farms in the Peruvian Amazon', in R. D. Williams (ed.), *Communication of weed science technologies in developing countries*, Corvallis, Oregon: International Plant protection Center, pp. 117–40.
Benites, J. R. and Nurena, M. A. (1985), 'Integrated low input cropping systems', TROPSOILS Triennial Technical Report, 1981–84, Raleigh, North Carolina: North Carolina State University, pp. 149–51.
Bishop, J. P. (1982), 'Agroforestry systems for the humid tropics east of the Andes', in S. B. Hecht (ed.), *Amazonia: agriculture and land use research*, Cali, Colombia: CIAT, pp. 403–16.
Bornemisza, E. and Alvarado, A. (1975), *Soil management in tropical America*, Raleigh, North Carolina: North Carolina State University.
Brinkmann, W. L. F. and de Nascimento, J. C. (1973), 'The effects of slash and burn agriculture on plant nutrients in the Tertiary region of Central Amazonia', *Turrialba*, 28, pp. 284–90.
Buol, S. W. and Sanchez, P. A. (1978), 'Rainy tropical climates: physical potential, present and improved farming systems', *Proceedings, 11th Soil Science Congress* (Edmonton) 2, pp. 292–312.
Buol, S. W. and Sanchez, P. A. (1986), 'Red soils in the Americas: morphology, classification and management', in *Proceedings, International Symposium on Red Soils*, Institute of Soil Science, Academia Sinica, Beijing, China: Science Press, pp. 14–43.
Buschbacher, C., Uhl, C. and Serrão, E. A. S. (1987), 'Large scale development in Eastern Amazonia', in C. F. Jordan (ed.), *Amazonian Rainforests*, New York: Springer Verlag, pp. 90–9.
CEPLAC (Centro de Pesquisas do Cacau, Itabuna, Bahia) (1973), De Silva, L. F., Carvalho Filho, R. and Santana, M. B. M. 'Solos do Projeto Ouro Prêto', *Boletim técnico*, 23.
CEPLAC (1976), da Costa Pinto Dias, A. C. and de Melo, A. A. O, 'Solos do Projeto Ouro Prêto', *Boletim técnico*, 45.
CEPLAC (1977), Leão, A. C. and Carvalho Filho, R., 'Solos do Projeto Burareiro', *Boletim técnico*, 52.
Charley, J. L. and Richards, B. N. (1983), 'Nutrient allocation in plant communities: mineral cycling in terrestrial ecosystems', in O. L. Lange *et al.* (eds), *Physiological Plant Ecology*, vol. IV, 'Ecosystem Processes: mineral cycling, production and man's influence', Berlin: Springer Verlag.
Chapin, F. S. (1980), 'The mineral nutrition of wild plants', *Annual Review of Ecology and Systematics*, 11, pp. 233–60.
Cochrane, T. T. (1979), 'An ongoing appraisal of the savanna ecosystem of tropical America for beef cattle production', in P. A. Sanchez and L. E. Tergas (eds), *Pasture production in the acid soils of the tropics*, Cali, Colombia: CIAT, pp. 1–12.
Cochrane, T. T. (1984), 'Amazonia: a computerized overview of its climate, landscape and soil resources', *Interciencia*, 9, pp. 298–306.
Cochrane, T. T. (1986), 'The distribution, properties and management of acid mineral soils in tropical South America', in *Proceedings, International Symposium on Red Soils*. Institute of Soil Science, Academia Sinica, Beijing, China: Science Press, pp. 77–89.
Cochrane, T. T. and Sanchez, P. A. (1982), 'Land resources, soils and their management in the Amazon region: a state of knowledge report', in S. B. Hecht

(ed.), *Amazonia: agriculture and land use research*, Cali, Colombia: CIAT, pp. 137–209.

Cochrane, T. T., Sanchez, P. A., Azevedo, L. G., Porras, J. A. and Garver, C. L. (1985), *Land in tropical America*, vols. 1–3, Cali, Columbia: CIAT.

Correa, J. C. (1984), 'Recursos edáficos do Amazônas, EMBRAPA–UEPAE, Brazil: Manaus.

Cuevas, E. and Medina, E. (1983), 'Root production and organic matter decomposition in a terra firme forest of the Upper Rio Negro Basin', in *Wurzelokologie und ihre Nutzanwendung*, International Symposium, Gumpenstein Irdning, Austria, pp. 653–66.

Denevan, W. M. (1981), 'Swiddens and cattle versus forest: the imminent demise of the Amazon rain forest reexamined', in V. H. Sutlive, N. Altshuler, and M. D. Zamora (eds), *Where have all the flowers gone? Deforestation in the Third world*, Studies in Third World Societies, 13, Williamsburg, Virginia: College of William and Mary, pp. 25–44.

Denevan, W. M., Treacy, J. M., Alcorn, J. B., Padoch, C. *et al.* (1983), 'Indigenous agroforestry in the Peruvian Amazon: Bora Indian management of swidden fallows', in J. Hemming (ed.), *Change in the Amazon Basin. Man's impact on forests and rivers*, Manchester: University of Manchester Press, pp. 137–55.

Dias, A. C. P. and Nortcliff, S. (1985), 'Effects of two land clearing methods on the physical properties of an Oxisol in the Brazilian Amazon', *Tropical Agriculture (Trinidad)*, 62, pp. 207–12.

Dudal, R. (1980), 'Soil-related constraints to agricultural developments in the tropics', in *IRRI (International Rice Research Institute), Soil related constraints to food production in the tropics*, Los Banos, Laguna, Philippines, pp. 23–27.

Dynia, J. F., Moreira, G. N. C. and Bloise, R. M. (1977), 'Fertilidade de solos da região da Rodovia Transamazônica, 2 Fixação de fósforo em Podzólico Vermelho-Amarelo e Terra Roxa Estruturada Latossólica', *Pesquisa agropecuária Brasileira*, 12, pp. 75–9.

EMBRAPA (Empresa Brasileira de Pesquisa Agropecuária) (1981), 'Mapa de solos do Brasil 1:5,000,000', SNLCS (Serviço Nacional de Levantamento e Conservação de Solos), Rio de Janeiro.

Falesi, I. C. (1972), 'O estudo atual dos conhecimentos sobre os solo da Amazônia brasileira', *Boletim Técnico Instituto Pesquisa Agropecuária Norte*, 54, pp. 17–67.

Falesi, I. C. (1976), 'Ecossistema de pastagem cultivada na Amazônia brasileira', *Boletim técnico*, 1, EMBRAPA–CPATU, Belém.

Falesi, I. C., Baena, A. R. C. and Dutra, S. (1982), 'Consequências da explotação agropecuária sobre as condições fiscas e químicas dos solos das microrregiões do nordeste paraense'. CPATU, *Boletim Pesquisa*, 14, Belém, EMBRAPA.

Fearnside, P. M. (1979), 'Cattle yield prediction for the Transamazon Highway of Brazil', *Interciencia*, 4, pp. 220–5.

Fearnside, P. M. (1985), 'Agriculture in Amazonia', in G. T. Prance and T. E. Lovejoy (eds) *Key Environments: Amazonia*, Oxford: Pergamon Press, pp. 393–418.

Fearnside, P. M. (1986), *Human Carrying Capacity of the Brazilian Rainforest*, New York: Columbia University Press.

Fearnside, P. M. (in press), 'An ecological analysis of predominant land uses in the Brazilian Amazon', in A. Anderson (ed.), *Alternatives to deforestation*, Belém.

Fearnside, P. M. and Rankin, J. M. (1982), 'The new Jarí: risks and prospects of a major Amazonian development', *Interciencia*, 7(6), pp. 329–39.

Finn, J. T. (1978), 'Cycling index: a general definition for cycling in compartment models', in D. C. Adriana and I. L. Brisbin (eds), *Environmental chemistry and cycling processes*, Technical Information Center, Washington, D.C.: US Department of Energy, pp. 138–64.

Furley, P. A. (1980), 'Development planning in Rondônia based on natural renewable resource surveys', in F. Barbira-Scazzocchio (ed.), *Land, people and planning in contemporary Amazônia*, University of Cambridge: Centre of Latin American Studies, Occasional Publications, 3, pp. 37–45.

Furley, P. A. (1986), 'Radar surveys for resource evaluation in Brazil: an illustration from Rondônia, in M. J. Eden and J. T. Parry (eds), *Remote sensing and tropical land management*, Chichester: Wiley, pp. 79–99.

Furley, P. A. (1987), 'Impact of forest clearance on the soils of tropical cone karst', *Earth Surface Processes*, 12, pp. 523–9.

Furley, P. A. (ed.) (1988), 'Biogeography and Development in the Humid tropics', Special Issue, *Journal of Biogeography*, 15.

Gerloff, G. C. and Gabelman, W. H. (1983), 'Genetic basis of inorganic plant nutrition', in A. Lauchli and R. L. Bielski (eds), *Encyclopedia of plant physiology*, New Series, vol. 15B, Inorganic plant nutrition, Berlin: Springer Verlag, pp. 453–80.

Goodland, R. J. A. (1980), 'Environmental ranking of Amazonian development projects in Brazil', *Environmental Conservation*, 7, pp. 9–26.

Goodland, R. J. A. and Irwin, H. S. (1975), *Amazon jungle: green hell to red desert?*, Amsterdam: Elsevier.

Gross, D. R., Eiten, G., Flowers, N. M. *et al.* (1979), 'Ecology and acculturation among native peoples of Central Brazil', *Science*, 206, pp. 1043–50.

Hartshorn, G. S. (1978), 'Tree falls and tropical forest dynamics', in P. B. Tomlinson and M. H. Zimmerman (eds), *Tropical trees as living systems*, Cambridge: Cambridge University Press, pp. 617–38.

Hecht, S. B. (1979), 'Spontaneous legumes on developed pastures in the Amazon and their forage potential', in P. A. Sanchez and L. E. Tergas, (eds), *Pasture production in acid soils of the tropics*, Cali, Colombia: CIAT, pp. 65–79.

Hecht, S. B. (1981) 'Deforestation in the Amazon Basin: magnitude, dynamics and soil resource effects', in V. H. Sutlive, N. Altshuler and M. D. Zamora (eds), *Where have all the flowers gone? Deforestation in the Third World*, Studies in Third World Societies, 13, Williamsburg, Virginia: College of William and Mary, pp. 61–108.

Hecht, S. B. (1982), 'Agroforestry in the Amazon Basin: practice, theory and limits of a promising land use', in S. B. Hecht (ed.), *Amazonia: agriculture and land use research*, Cali, Colombia: CIAT, pp. 31–72.

Hecht, S. B. (1983), 'Cattle ranching in eastern Amazonia', Ph.D. thesis, Department of Geography, Berkeley, California: University of California.

Herrera, R. (1985), 'Nutrient cycling in Amazonian forests', in G. T. Prance and T. E. Lovejoy (eds), *Key environments: Amazonia*, Oxford: Pergamon Press, 95–108.

Herrera, R., Jordan, C. F., Klinge, H. F. and Medina, E. (1978), 'Amazon ecosystems: their structure and functioning with particular emphasis on nutrients', *Interciencia*, 3, pp. 223–32.

Hubbell, S. P. (1979), 'Tree dispersion, abundance and diversity in a tropical dry forest', *Science*, 203, pp. 1299–1309.

Irion, G. (1978), 'Soil infertility in the Amazonia rainforest', *Naturwissenschaften*, 65, pp. 515–19.

Janos, D. P. (1983), 'Tropical mycorrhizas, nutrient cycles and plant growth', in S. L. Sutton, T. C. Whitmore and A. C. Chadwick (eds), *Tropical rain forest: ecology and management*, Oxford: Blackwell Scientific, pp. 327–45.

Janzen, D. H. (1983), 'Food webs: who eats what, why, how and with what effects in a tropical rainforest?', in F. B. Golley (ed.), *Ecosystems of the world*, vol. 14A, *Tropical rain forest ecosystems: structure and function*, Amsterdam: Elsevier, pp. 167–82.

Jordan, C. F. (1982a), 'The nutrient balance of an Amazonian rainforest', *Ecology*, 63, pp. 647–54.
Jordan, C. F. (1982b), Amazonian rainforests', *American Scientist*, 70, pp. 394–401.
Jordan, C. F. (1985a), 'Soils of the Amazon Forest', in G. T. Prance and T. E. Lovejoy (eds), *Key environments: Amazonia*, Oxford: Pergamon Press, pp. 83–94.
Jordan, C. F. (1985b), *Nutrient cycling in tropical forest ecosystems*, New York: Wiley.
Jordan, C. F. (1987), *Amazonian rainforests*, Ecological Studies 60, New York, Springer Verlag.
Jordan, C. F., Golley, F., Hall, J. D. and Hall, J. (1979), 'Nutrient scavenging of rainfall by the canopy of an Amazonian rain forest', *Biotropica*, 12, pp. 61–6.
Jordan, C. F. and Herrera, R. (1981), 'Tropical rainforests: are nutrients really critical?', *American Naturalist*, 117, pp. 167–80.
Jordan, C. F. and Uhl, C. (1978), 'Biomass of a tierra firme forest of the Amazon Basin', *Oecologia Plantarum*, 13, pp. 387–400.
Klinge, H. C. (1965), 'Podzol soil in the Amazon Basin', *Journal of Soil Science*, 16, pp. 95–103.
Klinge, H. C. (1971), 'Matéria orgânica e nutrientes na mata de terra firme perto de Manaus', *Acta Amazônica*, 1(1), pp. 69–72.
Klinge, H. C. (1975), 'Root mass estimation in lowland tropical rainforests of central Amazonia, Brazil, III, Nutrients in fine roots from giant humus podsols', *Tropical Ecology*, 16, pp. 28–39.
Klinge, H. C. and Herrera, R. (1978), 'Biomass studies in an Amazon caatinga forest in southern Venezuela, I, Standing crop of composite root mat in selected stands', *Tropical Ecology*, 19, pp. 93–110.
Klinge, H. C. and Medina, E. (1979), 'Rio Negro caatingas and campinas, Amazonas states of Venezuela and Brazil', in R. L. Specht (ed.), *Ecosystems of the world*, vol. 9A, *Heathlands and related shrublands*. Amsterdam: Elsevier, pp. 483–8.
Klinge, H. C. and Rodriguez, W. A. (1968), 'Litter production in an area of terra firme forest, 1, Litterfall, organic carbon and total nitrogen contents', *Amazonia*, 1, pp. 287–302.
Lal, R. (1987), *Tropical ecology and physical edaphology*, Chichester: Wiley.
Lal, R., Kang, B. T., Moorman, F. R. *et al.* (1975), 'Problemas de manejo de suelos y posibles soluciones en Nigéria Occidental', in E. Bornemisza and A. Alvarado (eds), *Soil management in tropical America*, Raleigh, North Carolina: North Carolina State University, pp. 380–417.
Lal, R., Sanchez, P. A. and Cummings, R. W. (eds) (1986), *Land clearing and development in the tropics*, Rotterdam: A. A. Balkema.
Leite, L. L. and Furley, P. A. (1985), 'Land development in the Brazilian Amazon with particular reference to Rondônia and the Ouro Prêto Colonization project', in J. Hemming (ed.), *Change in the Amazon Basin*, vol. 2, Manchester: Manchester University Press, pp. 119–39.
Levin, D. (1976), 'The chemical defences of plants to pathogens and herbivores', *Annual Review of Ecology and Systematics*, 7, pp. 121–59.
Lovejoy, T. E. (1984), 'Application of ecological theory to conservation planning', in F. di Castri, F. W. G. Baker and M. Hadley (eds), *Ecology in Practice*, Part 1, Ecosystem Management, Dublin: Tycooly International Publishers, pp. 402–13.
Lugo, A. E., Brinson, M., Cerame Vivas, M. *et al.* (1974), 'Tropical ecosystem structure and function', in E. G. Farnworth and F. B. Golley (eds), *Fragile Ecosystems*, New York: Springer Verlag, pp. 67–111.
Luizão, F. J. and Schubart, H. O. R. (1987), 'Litter production and decomposition in a terra firme forest of Central Amazonia', *Experientia*, 43, pp. 259–65.

McNeil, M. (1964), 'Lateritic soils', *Scientific American*, 211(5), pp. 96–102.
Meggers, B. J. (1985), 'Aboriginal adaptation to Amazonia', in G. T. Prance and T. E. Lovejoy (eds), *Key environments: Amazonia*, Oxford: Pergamon Press, pp. 307–27.
Moorman, F. R. and van Wambeke, A. (1978), 'The soils of the lowland rainy tropical climates, their inherent limitations for food production and related climatic restraints', *Transactions, 11th International Soil Science Congress* (Edmonton), 2, pp. 292–312.
Moran, E. F. (1977), 'Estratégias de sobrevivência: o uso de recursos ao longo da rodovia Transamazônica', *Acta Amazônica*, 7, pp. 363–79.
Moran, E. F. (1981), *Developing the Amazon*, Bloomington, Indiana: Indiana University Press.
Moran, E. F. (ed) (1983), *The dilemma of Amazonian development*, Boulder, Col.: Westview Press.
National Research Council (1982), *Ecological aspects of development in the humid tropics*, Washington, D.C.: National Academy Press.
NCSU (North Carolina State University) (1972, 1973, 1974, 1975, 1976/77, 1978/79, 1980/81), *Research on tropical soils*, Annual Reports, Soil Science Department, Raleigh, North Carolina: North Carolina State University.
Nicholaides, J. J. (1983), 'Crop production systems in the Amazon Basin', in E. F. Moran (ed.), *The dilemma of Amazonian development*, Boulder. Col.: Westview Press.
Nicholaides, J. J., Sanchez, P. A., Bandy, D. E., Villachica, J. H., Coutu, A. J. and Valverde, C. S. (1982), 'Fertilizer management for continuous crop production on Ultisols of the Amazon Jungle Basin of Peru', *Plant Nutrition*, 1982, vol. 2, Farnham Royal, England: Commonwealth Agricultural Bureau, pp. 425–30.
Nicholaides, J. J., Bandy, D. E., Sanchez, P. A. and Villachica, J. H. (1984), 'From migratory to continuous agriculture in the Amazon Basin', *Improved production systems as an alternative to shifting agriculture*, FAO Soils Bulletin, 53, Rome: FAO/UN, pp. 141–68.
Nicholaides, J. J., Bandy, D. E., Sanchez, P. A., Benites, J. R. *et al.* (1985), 'Agricultural alternatives for the Amazon Basin', *BioScience*, 35(5), pp. 279–85.
Nicholaides, J. J. and Piha, M. I. (1985), 'A field method for selecting cultivars with tolerance to aluminium and high yield potential', *Conference on aluminium tolerance in sorghum*, Cali, Columbia: CIAT.
Norman, N. J. T. (1979), *Annual cropping systems in the tropics*, Gainesville, Florida: University of Florida Press.
Norman, N. J. T. (1984), *The ecology of tropical food crops*, Cambridge: Cambridge University Press.
Nortcliff, S. and Dias, A. C. P. (1988), 'Soil conditions following forest clearance in the Amazon Basin', in P. A. Furley (ed.), 'Biogeography and development in the humid tropics', special issue, *Journal of Biogeography*, 15.
Nye, P. H. and Greenland, D. J. (1960), *The soil under shifting cultivation*, Technical Communication, 51, Commonwealth Bureau of Soils, Farnham Royal, England: Commonwealth Agricultural Bureau.
Nye, P. H. and Greenland, D. J. (1964), 'Changes in the soil after clearing tropical forest', *Plant and soil*, 21, pp. 101–12.
Nye, P. H. and Tinker, P. B. (1977), *Solute movement in the soil-root system*, Berkeley, California: University of California Press.
Orians, G., Apple, J., Billings, R. *et al.* (1974), 'Tropical population ecology', in E. Farnworth and F. B. Golley (eds), *Fragile Ecosystems*, New York: Springer Verlag.
Peace, W. J. H. and MacDonald, F. D. (1981), 'An investigation of the leaf anatomy,

foliar mineral levels and water relations of trees of a Sarawak forest', *Biotropica*, 13, pp. 100–9.

Peck, R. B. (1982), 'Forest research activities and the importance of multistrata production systems in the Amazon Basin', in S. B. Hecht (ed.), *Amazonia: agriculture and land use research*, Cali, Colombia: CIAT, pp. 373–86.

Prance, G. T. (1978), 'The origin and evolution of the Amazon flora', *Interciencia*, 3(4), pp. 207–22.

Prance, G. T. and Lovejoy, T. E. (eds) (1985), *Key environments: Amazonia*, Oxford: Pergamon Press.

Proctor, J. (1983), 'Mineral nutrients in tropical forests', *Progress in Physical Geography*, 7, pp. 422–31.

Prance, G. T. and Lovejoy, T. E. (eds) (1985), *Key environments: Amazonia*, Oxford: Pergamon Press.

Proctor, J. (1983), 'Mineral nutrients in tropical forests', *Progress in Physical Geography*, 7, pp. 422–31.

Projeto RADAMBRASIL (1972–78), *Levantamento da Regiaõ Amazônica*, vols 1–12, Ministério das Minas e Energia, Departamento Nacional de Produção Mineral, Rio de Janeiro.

Ramalho Filho, A., Pereira, E. G. and Beek, K. J. (1978), *Sistema de avaliação da aptidão das terras*, EMBRAPA–SNLCS, Ministério da Agricultura–SUPLAN, Brasília. D. F.

Rankin, J. M. (1985), 'Forestry in the Brazilian Amazon', in Prance, G. T. and T. E. Lovejoy (eds), *Key environments: Amazonia*, Oxford: Pergamon Press, pp. 369–92.

Ranzani, G. (1980), 'Erodibilidade de alguns solos do Estado do Amazonas, *Acta Amazônica*, 10, pp. 263–9.

Reynolds, S. (1988), 'Some factors of importance in the integration of pastures and cattle with coconuts (Cocos nulifera)', in P. A. Furley (ed.), *Biogeography and development in the humid tropics*, Special issue, *Journal of Biogeography*, 15, pp. 31–40.

Ruehle, J. L. and Marx, D. H. (1979), 'Fiber, food, fuel and fungal symbionts', *Science*, 206, pp. 419–22.

Russell, C. E. (1987), 'Plantation Forestry', in C. E. Jordan (ed.), *Amazonian Rain Forests*, New York: Springer Verlag, pp. 76–89.

Salas, G. de las, and Folster, M. (1976), 'Bioelement loss on clearing a tropical rain forest', *Turrialba*, 26(2), pp. 179–86.

Salati, E., Vose, P. B. and Lovejoy, T. E. (1986), 'Amazon rainfall, potential effects of deforestation and plans for future research', in G. T. Prance (ed.), *Tropical rain forests and the world atmosphere*, AAAS (American Association for the Advancement of Science), Selected Symposium, 101, Boulder, Col.: Westview Press, pp. 61–74.

Saldarriaga, J. G. (1987), 'Recovery following shifting cultivation', in C. F. Jordan (ed.), *Amazonian Rain Forests*, New York: Springer Verlag, pp. 24–33.

Sanchez, P. A. (1976), *Properties and management of soils in the tropics*, New York: Wiley.

Sanchez, P. A. (1977a), 'Advances in the management of Oxisols and Ultisols in tropical South America', in *Proceedings of the International Seminar on Soil Environment and Fertility Management in Intensive Agriculture*, Society of Soil Science and Manure, Tokyo, pp. 535–66.

Sanchez, P. A. (1977b), 'Manejo de solos da Amazônica para produção agropecuária intensiva', *Boletim International Sociedade Brasileira Ciência do Solo*, 2(3), pp. 60–3.

Sanchez, P. A. (1979), 'Soil fertility and conservation considerations for agroforestry systems in the humid tropics of Latin America', in H. O. Mongi and P. A. Huxley (eds), *Soils research in agroforestry*, ICRAF (International Centre for Research in Agroforestry) Nairobi, pp. 79–124.

Sanchez, P. A. (1981), 'Soils of the humid tropics', in *Blowing in the wind: deforestation and long range implications*, Studies in Third World Societies, 14, Williamsburg, Virginia: College of William and Mary, 347–410.

Sanchez, P. A. (1982), 'A legume based pasture production strategy for acid infertile soils of tropical America', in American Society of Agronomy, *Soil erosion and conservation in the tropics*, Special Publication, Madison, Wisconsin, pp. 97–120.

Sanchez, P. A. (1987), 'Management of acid soils in the humid tropics of Latin America', in IBSRAM (International Board for Soil Research and Management), *Management of acid tropical soils for sustainable agriculture*, Bangkok, Thailand, pp. 63–96.

Sanchez, P. A. and Buol, S. W. (1985), 'Agronomic taxonomy for wetland soils', in IRRI (International Rice Research Institute), *Wetland soils: characteristics, classification and utilisation*, Los Banos, Laguna, Philippines, pp. 207–27.

Sanchez, P. A. and Cochrane, T. T. (1980), 'Soil constraints in relation to major farming systems in tropical America', in IRRI (International Rice Research Institute), *Priorities for alleviating soil related constraints to food production in the tropics*, Los Banos, Laguna, Philippines, pp. 197–239.

Sanchez, P. A. and Miller, R. H. (1986), 'Organic matter and soil fertility management in acid soils of the tropics', *Proceedings, 13th International Congress of Soil Science*.

Sanchez, P. A. and Salinas, J. G. (1981), 'Low input technology for managing Oxisols and Ultisols in tropical America', *Advances in Agronomy*, 34, pp. 279–406.

Sanchez, P. A. and Tergas, L. E. (eds) (1979), *Pasture production in acid soils of the tropics*, Cali, Colombia: CIAT.

Sanchez, P. A. and Uehara, G. (1980), 'Management considerations for acid soils with high P-fixation capacity', in F. E. Kashwney (ed.), *Phosphorus in agriculture*, Madison, Wisconsin: Soil Science Society of America.

Sanchez, P. A., Bandy, D. E., Villachica, J. H. and Nicholaides, J. J. (1982a), 'Soils of the Amazon Basin and their management for continuous crop production', *Science*, 216, pp. 821–7.

Sanchez, P. A., Couto, W. and Buol, S. W. (1982b), 'The fertility capability soil classification system; interpretation, applicability and modification', *Geoderma*, 27, pp. 283–309.

Sanchez, P. A., Gichuru, M. P. and Katz, L. B. (1982c), 'Organic matter in major soils of the tropical and temperate regions', *Proceedings, 12th International Congress of Soil Science*, vol. 1, New Delhi, India, pp. 99–114.

Sanchez, P. A., Villachica, J. H. and Bandy, D. E. (1983), 'Soil fertility dynamics after clearing a tropical rainforest in Peru', *Soil Science Society of America Journal*, 47, pp. 1171–8.

Savage, J. M., Goldman, C. R., Janos, D. P. *et al.* (1982), *Ecological aspects of development in the humid tropics*, Washington, D.C.: National Academy Press.

Schubart, H. O. R. (1977), 'Critérios ecológicos para o desenvolvimento agrícola das terras firmes da Amazônia', *Acta Amazônica*, 7, pp. 559–67.

Scott, G. A. J. (1987), 'Shifting cultivation where land is limited', in C. F. Jordan (ed.), *Amazonian Rain Forests*, New York: Springer Verlag, pp. 34–45.

Serrão, E. A. S. (1981), 'Pasture research results in the Brazilian Amazon', *Proceedings of the 14th International Grasslands Congress*, Boulder, Col.: Westview Press, pp. 746–50.

Serrão, E. A. S. and Humma, A. K. O. (1982), 'Recuperação e melhoramento de pastagens cultivadas em área de floresta Amazônica', Documento 17, EMBRAPA–CPATU.

Serrão, E. A. S, Falesi, I. C., Veiga, J. B. and Texeira, J. F. (1979), 'Productivity of cultivated pastures in low fertility soils of the Amazon in Brazil', in P. A. Sanchez and L. E. Tergas (eds), *Pasture production in acid soils of the tropics*, Cali, Colombia: CIAT, pp. 195–225.

Seubert, C. E., Sanchez, P. A. and Valverde, C. (1977), 'Effects of land clearing methods on soil properties and crop performance in an Ultisol of the Amazon jungle in Peru', *Tropical Agriculture* (Trinidad), 54, pp. 307–21.

Singer, R. and Araujo, I. J. S. (1979), 'Litter decomposition and ectomycorrhiza in Amazonian forests', *Acta Amazônica*, 9, pp. 25–41.

Singer, R. and Araujo, I. J. S. (1986), 'Litter decomposition and ectomycorrhizal basidiomycetes in an igapo forest', *Plant systematics and evolution*, 153, pp. 107–17.

Sioli, H. (1980), 'Forseeable consequences of actual development schemes and alternative ideas', in F. Barbira-Scazzocchio (ed.), *Land, People and Planning in Contemporary Amazônia*, Occasional Publications, 3, University of Cambridge: Centre of Latin American Studies, pp. 252–68.

Smyth, T. J. and Bastos, J. B. (1984), 'Alterações na fertilidade em um latossolo amarelo álico pela queima da vegetação', *Revista Brasileira Ciência do Solo*, 8, pp. 127–32.

Soil Survey Staff (1975), *Soil Taxonomy*: a basic system of soil classification for making and interpreting soil surveys, Handbook no. 436, Washington, D.C.: US Department of Agriculture.

Sombroek, W. G. (1966), *Amazon Soils: a reconnaissance of the Brazilian Amazon region*, Wageningen, Netherlands: Centre for Agricultural Publications and Documentation.

Sombroek, W. G. (1979), *Soils of the Amazon Region*, International Soils Museum, Annual Report, Wageningen, Netherlands.

St John, T. V. (1980), 'Root size, root hairs and mycorrhizal infection: a re-examination of Baylis' hypothesis with tropical trees', *New Phytologist*, 84, pp. 483–7.

St John, T. V. (1983), 'Response of tree roots to decomposing organic matter in two lowland Amazonian rainforests', *Canadian Journal of Forest Research*, 13(2), pp. 346–9.

St John, T. V. (1985), 'Mycorrhizae', in G. T. Prance and T. E. Lovejoy (eds), *Key environments: Amazonia*, Oxford: Pergamon Press, pp. 277–83.

St John, T. V. and Coleman, D. C. (1987), 'The role of mycorrhizae in plant ecology', *Canadian Journal of Botany*, 61, pp. 1005–14.

Stark, N. (1971a), 'Nutrient cycling, I, Nutrient distribution in some Amazonian soils', *Tropical Ecology*, 12, pp. 24–50.

Stark, N. (1971b), 'Nutrient cycling, II, Nutrient distribution in Amazonian vegetation', *Tropical Ecology*, 12, pp. 177–201.

Stark, N. (1978), 'Man, tropical forests and the biological life of a soil', *Biotropica*, 10, pp. 1–10.

Stark, N. M. and Jordan, C. F. (1978), 'Nutrient retention by the root mat of an Amazonian rainforest', *Ecology*, 59, pp. 434–7.

Swift, M. J. (ed.) (1984), *Soil biological processes and tropical soil fertility*, Biology International, Special Issue, 5.

Swift, M. J. and P. A. Sanchez (1984), 'Biological management of tropical soil fertility for sustained productivity', *Nature and Resources*, 20(4), pp. 2–10.

Swift, M. J., Heal, O. W. and Anderson, J. M. (1979), *Decomposition in terrestrial ecosystems*, Studies in Ecology 5, Berkeley, California: University of California.

Szott, L. T. and Palm, C. A. (1986), 'Soil and vegetation dynamics in shifting cultivation fallows', *1st Symposium on the humid tropics*, 1, pp. 360–79, Belém: EMBRAPA–CPATU.

Toledo, J. M. and Navas, J. (1986), 'Land clearing for pastures in the Amazon', in R. Lal, P. A. Sanchez and R. W. Cummings (eds), *Land clearing and development in the tropics*, Rotterdam: A. A. Balkema, pp. 97–116.

Toledo, J. M. and Serrão, E. A. S. (1982), 'Pastures and animal production in Amazonia', in S. B. Hecht (ed.), *Amazonia: agriculture and land use research*, Cali, Colombia: CIAT, pp. 281–309.

Uhl, C., Jordan, C. F., Clark, H. *et al.* (1982), 'Ecosystem recovery in Amazon caatinga forest after cutting, burning and bulldozer clearance treatments', *Oikos*, 38, pp. 313–20.

UNESCO (1978), *Tropical forest ecosystems: a state of knowledge report*, Paris: UNESCO/UNEP/FAO.

Valencia, J. E. (1982), 'Investigaciones silviculturales y agroforestales adelentadas por CONIF', in S. B. Hecht (ed.), *Amazonia: agriculture and land use research*, Cali, Colombia: CIAT, pp. 407–12.

Valverde, C. S. and Bandy, D. E. (1982), 'Production of annual food crops in the Amazon', in S. B. Hecht (ed.), *Amazonia: agriculture and land use research*, Cali, Colombia: CIAT, pp. 243–80.

Van Wambeke, A. (1978), 'Properties and potentials of soils of the Amazon Basin', *Interciencia*, 3, pp. 233–42.

Vieira, L. S. and Tadeu, P. C. dos Santos (1987), *Amazônia: seus solos e outros recursos naturais*, São Paulo: Editora Agronómica Ceres Ltd.

Wade, M. K. and Sanchez, P. A. (1983), 'Mulching and green manure applications for continuous crop production in the Amazon Basin', *Agronomy Journal*, 75, pp. 39–45.

Wade, M. K. and Sanchez, P. A. (1984), 'Productive potential of an annual cropping scheme in the Amazon', *Field Crops Research*, 9, pp. 253–63.

Watters, R. F. (1971), *Shifting cultivation in Latin America*, FAO Forestry Development, paper 17, Rome: FAO.

Went, F. W. and Stark, N. M. (1968), 'Mycorrhiza', *BioScience*, 18, pp. 1035–9.

Whitmore, T. C. and Prance, G. T. (eds) (1987), *Biogeography and Quaternary history in tropical America*, Oxford Monographs in Biogeography, 3, Oxford: Clarendon Press. Oxford.

Zinke, P. J., Stangenberger, A. G., Post, W. M. *et al.* (1984), *Worldwide organic soil carbon and nitrogen data*, Oak Ridge National Laboratory, Environmental Sciences Division Publication, 2212, Washington, D.C.: US Department of Energy.

14 Environmentally Appropriate, Sustainable Small-farm Strategies for Amazonia

Chris Barrow

> In our impatience with 'backward' small farmers and in our haste ... rapidly [to] 'commercialise' them, we have overlooked key aspects of their farming systems that could enhance our efforts to increase food production and improve rural well-being (Harwood, 1979, p. xiii).

INTRODUCTION

Many of the efforts to establish food or commodity crop production in Amazonia have, at best, been short-lived successes. For example, the infamous 'Zona Bragantina', a roughly 30,000-square kilometre area of eastern Pará settled in the early part of this century by small farmers. The Belém–Bragança Railway (opened 1908 and closed in 1936) permitted the settlement of the Zona Bragantina, where settlers used to conditions in the more arid North-East or the South of Brazil cleared the forest and tried to farm; most ended up practising shifting cultivation and then had to abandon the degraded land. Today much of the Zone is still covered in secondary regrowth scrub (*capoeira*) and yields less than it did before the settlers cleared it. Efforts by big business to establish large-scale commodity crop production have also fared badly. The Fordlândia and Belterra rubber plantations, established along the Rio Tapajós in the late 1920s, did not produce for long. Labour difficulties, pests and crop disease meant that they could not compete economically with South-East Asian plantations. Few of the large cattle ranches which have spread in the last few decades can be said to be sustaining high levels of grazing on their largely unimproved pastures. Ambitious attempts by an American multi-millionaire, Daniel Ludwig, and his successors, to establish paper pulp production using plantations of fast growing exotic trees, and large-scale, intensive rice production on a 1.4 million hectare landholding along the Rio Jarí are surrounded in intrigue and debate about profitability. A few years ago, Ludwig sold the enterprise to a consortium of companies, having reputedly spent over US$ 1,000 million trying to establish production.

Sustainable development strategies may be defined as those which are environmentally sustainable over the long term, are consistent with social values and institutions, and encourage 'grassroots' participation in the development process (Barbier, 1987, p. 102). Agricultural sustainability requires that a reasonable level of nutrients be maintained in the soil; to do this losses due to leaching, erosion and removal of harvested crops must be balanced. Soil compaction, excessive salt build-up, weed growth, multiplication of insects and other pests and crop, livestock and human diseases must be controlled. The farmers are likely to require cooking fuel, basic amenities, access to markets and a supply of meat or fish. A strategy which can meet these demands also has to continue to be acceptable to the people who practise it. It is also necessary that the society in the long run control its numbers. If at all possible a sustainable agriculture strategy should avoid damage to surrounding lands or drainage systems; for example, pesticides which might damage a region's rivers should be avoided.

There can be no single sustainable agriculture strategy for Amazonia's small farmers; an appropriate one will have to be found to suit each local situation. As far as possible, outside inputs should be minimised, unless the authorities can reliably supply them at a price farmers can afford. Pesticides and herbicides and other agrochemicals should be used with caution. The call for the authorities to support and encourage the development of environmentally appropriate, sustainable small-farm strategies is not new. It was being made in the 1930s, partly as a consequence of the 'Zona Bragantina' fiasco, but has grown since the 1970s as road building and settlement (officially-sponsored and spontaneous) led to increased forest clearance and, in many cases, short-lived agricultural production and land degradation (Skillings and Tcheyan, 1979).

Any examination of agricultural strategies practised in a region as large as Amazonia must involve generalisation, and will omit much. With this caveat in mind it is possible to recognise broad categories of constraints, which might conspire to hinder the spread of sustainable agriculture, especially sustainable small-farm strategies, in Amazonia. These include the following:

1. Lack of suitable crops/techniques/strategies
2. Lack of access to farm extension
3. National development priorities which do not help small farmers
4. Reluctance or resistance to change shown by small farmers
5. Economic disincentives – lack of credit, inputs, difficult communications, poor market prices
6. Lack of security for farmers
7. Environmental factors

Availability of Suitable Crops, Techniques and Strategies

Strategies for obtaining a sustained supply of crops, forest products and game have long been known to Amerindians and *caboclos* (poor rural peoples, generally of mixed Amerindian–Portuguese stock). The *caboclo* practices are similar to those of Amerindians except that they gather fewer products and have shorter fallows in their shifting cultivation. The shifting cultivation of these people can give a sustained subsistence crop of cassava, rice, maize, beans and fruits if there is enough forest land to allow long forest fallows (say of 20 or more years' duration). Their protein is provided by river fish and game; sustaining this supply also requires a low population density to be maintained. Some shifting cultivators may manage to gather forest products for sale (rubber, waxes, Brazil-nuts, rosewood oil, açaí fruit) in addition to growing subsistence crops. Unfortunately, where there is road, and to a lesser extent river access, settlement has led to population densities which prohibit adequate forest fallows and which exceed the capacity of the land to provide game and fish. So Amerindian and *caboclo* strategies must increasingly be modified.

There are non-traditional practices which could be used to achieve sustainable small-farm cultivation in Amazonia. A recent discussion on the identification and diffusion of sustainable agriculture in the humid tropics is given by Charlton (1987). Most of the modifications or alternative methods will almost certainly have to be started by, and will probably need on-going support from, some form of farm extension agency. These might be State, Church or private bodies, and their input is going to be crucial.

Access to Farm Extension

There is a well-developed network of agricultural research, technical assistance and farm extension bodies with offices in larger towns in Brazilian Amazonia (EMBRAPA/EMATER/EMBRATER/ASTER). In principle at least, improved small-farm strategies can be spread *if* technical assistance and farm extension organisations and development authorities are adequately funded and encouraged actively to promote and support them. It is crucially important that any body, whether researching possible small-farmer agriculture strategies or involved in farm extension, provision of credit, and so on, *understands the farmer's perspective* (Charlton, 1987). Unfortunately, many extension personnel, due to attitude, transport and funding problems, contact only the more accessible settlers. Amazonia's agricultural extension services must put less reliance on demonstration plots in experimental stations and the publication of pamphlets, and take new techniques and inputs to the small farmers.

National Development Priorities and Small Farmers

Despite many government-supported settlement schemes for small farmers, and research into techniques and strategies suitable for them, small farmers have been badly neglected. At least from the 1960s, large-scale ranching, mineral resources and hydroelectricity development, and more recently large-scale commodity crop production, have had more support.

Some settlement of Amazonia has been encouraged for strategic reasons; in the words of Janzen (1973, p. 1216) 'Although seldom openly acknowledged much of the motive for governmental manipulation of tropical agriculture is political. Occupancy implies ownership . . . and this is not an uncommon sentiment with respect to the Amazon basin'. In the past, other Latin American countries, and even powers outside the South American continent, have shown an interest in Amazonia's resources. Brazil's concern has understandably been to establish a claim to territory by encouraging rapid settlement. How well the settlers utilise resources and practise agriculture has not been a first priority so far.

Reluctance and Resistance to Change

It is possible to identify examples of reluctance or resistance to change which may hinder the achievement of sustainable agriculture. For example, it is puzzling that many of those who have settled in Amazonia in the 1960s and 1970s have migrated from areas that 50 or even 20 years ago were forested so the environment should hardly be alien to them, yet their farming frequently fails. Moran (1980) suggests that 'chronic mobility prevents the development of site-specific agronomic knowledge'. For some reason(s) settlers are either unable or are unwilling to stay in one place long enough to adapt to local conditions and sustain production. Somehow, economic and social incentives must be provided to reduce this onward migration. There is also a tendency for settlers (owners of large ranches and small farmers) to scorn the *caboclo*, and so fail to learn from their successes.

In some parts of Amazonia small farmers appear to believe that cattle ranching has more 'status' than cultivation. In Ecuadorian and Colombian Amazonia, at least, the goal of most settlers is to convert to ranching at some point in the future (Hiraoka and Yamamoto, 1980, p. 442). In Brazilian Amazonia, it is not unusual for settlers themselves to plant pasture, once they have cropped the land and exhausted the soil, because the pasture land sells for a better price. This practice is common along the Transamazon Highway in Pará and along the Cuiabá–Porto Velho Highway in Roraima (Fearnside, 1983), p. 141).

Economic Disincentives

A major problem is the distance and therefore cost of transportation of produce. Commodity prices have been far from attractive and stable in recent years, especially the prices of rice, cocoa and coffee. Acquisition of inputs like fertiliser, seeds and pesticide can be difficult. Financial support for small farmers is a problem almost everywhere: with little crop surplus for sale and the high probability of failure, extending credit can be a risky business. Small farmers with few funds to weather out short-term setbacks, who seek and fail to obtain credit, or who obtain it and get into debt, often accept whatever they are offered by land speculators in order to be able to afford to move on. Many settlers arrive in Amazonia penniless, so they have to work as labourers, they neglect their own plots and they cannot afford to wait for (environmentally more appropriate) perennial crops to mature.

Lack of Security

Lack of security of tenure is one of the major disincentives for Amazonian small farmers, and one which is unlikely to decline without major land reform and improvements in legislation – and, more important, its enforcement. (It is possible to divide small-farmer settlers into two categories: *posseiros*, squatters without documentary title to their land, and *colonos*, those relocated on small plots by the government who should have legal rights; the former are more vulnerable). Small farmers suffer because they lack satisfactory title to their land and the legal and political support needed to ensure that such rights are respected by those seeking to evict them.

The Environment

It is possible to divide Amazonia into two broad ecological categories: *terra firme* (drylands which are not periodically or permanently flooded), and *várzeas*/wetlands (the former are areas that are periodically flooded). These two types of environment offer different opportunities and present different challenges and hindrances to agriculture. In Amazonia's warm, humid tropical environment, if the *terra firme* forest cover is removed soil fertility usually declines rapidly and erosion and/or the formation of impermeable crusts follows. The humid conditions mean that nutrients leach from the soil rapidly, especially if it is cleared of vegetation cover. Fertiliser also has to be applied with care or its benefits will be short-lived. Pests and weeds proliferate and make cultivation difficult, soils dry out during seasonal droughts, yet crops are difficult to dry and store due to high humidity. Farmworkers and livestock are prone to disease, and communications along mainly unsurfaced roads are poor. There may be periods when pasture is scarce or human food is in short supply, as a consequence of seasonal drought, heavy rain or flooding. These 'hungry seasons' may debilitate small-farmers or their livestock enough to cause them to mismanage, sell or abandon their land (Townsend, 1985). Environmental difficulties are considerable, but not insurmountable, as some of the case studies later in this chapter show.

TERRA FIRME AGRICULTURE

For millenia, Amerindians, and for the last few centuries *caboclos*, have lived by gathering forest products (fruit, vegetables, game) and by practising shifting cultivation. These peoples have tended to settle along the rivers and streams, where they can fish and hunt aquatic game, and where periodic flooding often replenishes soil fertility. In the past turtles and their eggs were an important part of the diet of these original settlers (Bates, 1863).

Since the 1930s, 'modern' settlers have generally cleared *terra firme* forest and grown mainly annual crops (rain-fed rice, maize, beans, cassava and yams). After a few harvests, as fertility declines and/or weeds and pests become an increasing problem and as the game in the area is hunted out, there is little protein for the small farmer and his family, who are forced to abandon their landholding and move on to clear a new plot. If few areas of the original vegetation remain, proper tree regeneration is unlikely and the land acquires a cover of scrubland which offers fewer opportunities for wildlife to survive and does not moderate runoff as well as the natural forest cover did. In the mosaic of scrubland, cleared farmland and forest patches left in the wake of clearance, recurrent fires tend to further open up the vegetation so that the end result is likely to be poor quality pasture with a few fire-resistant, hardy shrubs and trees.

In contrast to the limited range of annual crops grown by 'modern' settlers, Amerindian or *caboclo* farmers grow a greater diversity of crops including many fruit and other perennial shrub or tree crops. The thicker plant cover helps prevent soil damage, makes forest regeneration and ultimately the recovery of soil fertility more likely, and better controls runoff. 'Modern' small farmers have generally been less adept at identifying areas of soil suitable for cultivation than Amerindians and *caboclos*. The latter often seek out specific 'indicator' species in the natural forest cover which suggest the potential for cultivation is good (modern soil surveys may help counter this problem).

Some settlers, particularly those on government-sponsored colonisation schemes, have planted perennial tree or shrub crops: cocoa, rubber, citrus or coffee; jute has also been quite an important crop in the past. These may be more ecologically appropriate than annual crops, but the growers have still had to face the fact that in Amazonia the risk of crop disease, insect damage and the cost of transport mean that to be a success an agricultural enterprise must make a relatively high return on investment – perhaps in excess of 15 per cent. CEPLAC, the Brazilian cocoa promotion and research body, does fund small-farmer producers, and has been particularly active in Rondônia. However, the price of cocoa on the world market has fluctuated considerably in recent years, as has coffee. In the Tomé-Açú region of Pará and in the Ecuadorian and Colombian Amazon, these two perennials have long been important small-farmer cash crops. The problem remains that the growers face risks unless world commodity prices stabilise at a satisfactory level.

Attempts to Sustain Agricultural Production on *Terra Firme* Lands

The Tomé–Açú Mixed Agricultural Cooperative

Few attempts by small farmers to crop the *terra firme* have been sustained: one exception has been that of the Japanese colonists at Tomé–Açú (and its more recent 'satellite' settlements) about 150 kilometres from Belém (Pará). Some of the colonists' land had been farmed and abandoned by Brazilian settlers yet has since been kept in production by the Tomé–Açú farmers for decades. The Colony was linked by road to the rest of Brazil in the mid-1970s and this opened up new opportunities which have helped to offset some environmental setbacks suffered in recent years.

The first Japanese agricultural cooperative was established in the Tomé–Açú region in 1929 (Tanaka, 1957), initially growing rain-fed rice and cocoa. In 1931, a vegetable producing cooperative was established and within five years exported produce to the city of Belém. During the post-1945 period, black pepper (*Pimenta nigrum*) production was begun. Pepper sales gave sufficient profit for the colonists to found the Cooperative Agrícola Mista de Tomé–Açú (CAMTA) in 1949. By 1955, CAMTA was exporting pepper to the USA and Europe and the Colony had achieved municipality status. A 'satellite' settlement was founded in 1961 and within 12 years supported 116 families (each farming between 20 and 80 hectares); a third settlement (Aiaçú) was added in 1977.

Pepper, although labour-intensive will flourish on poor soil and gives a good return. However, in the late 1970s, price fluctuations and fungus disease (*Fusarium solani* and *F. piperi*) began to trouble production. The response of CAMTA farmers was to diversify and to reorientate their marketing. Initially, farmers in the Cooperative responded to the disease problem by growing pepper on widely-scattered 25-hectare plots to reduce transmission of the fungal spores. By the early 1980s, things had deteriorated to the point that growers began to abandon pepper. Pepper, once infected, takes about four years to die, during that period yields gradually decline. CAMTA farmers responded by planting passion fruit on the dying pepper vines and posts – this gives a crop within about eight months, continuing for about five years. At the same time they planted cocoa, a disease resistant hybrid variety because cocoa is also threatened by a fungal disease – *vassoura de bruxa* (*Crinipellis perniciosa*) (Cardoso, 1974). The declining pepper crop plus the passion fruit provided finance until the cocoa came into fruit (at about six years of age). This was especially welcome because in the late 1970s and early 1980s interest rates on rural credit were very high. By 1981, CAMTA was producing a cocoa crop worth at that time US$ 1,000,000, and rising.

While visiting CAMTA in 1981, it was apparent that the Cooperative was responding to the opening of new roads and was experimenting with a range of new crops and crop rotations. There was a growing market for beans in North-East Brazil, CAMTA was therefore experimenting with a maize–bean or cowpea rotation to try and sustain bean production without needing to clear new land each year. Intensive poultry–egg production and pig rearing were also being promoted. The eggs and meat were trucked to Belém and the Centre–South of Brazil via the newly-opened road system (5,000 chicken carcasses per month in late 1981). The waste from the chicken slaughter and preparation is fed to the pigs and the pig and chicken manure is applied to arable crops so helping to sustain production. It seems likely that the idea of chicken and pig production was stimulated by experiments made in eastern Pará in the early 1970s (Sioli, 1980, p. 227). When frosts damaged papaya production further south in Brazil in the late 1970s, farmers at Tomé–Açú responded by growing the popular Hawaiian variety rather than the less palatable traditional varieties for sale to the then 'vacant' city markets like São Paulo. There was also an expansion into citrus cultivation and experiments in cultivating guaraná (*Paullinia sorbilis*) for which there was, and still is, an expanding soft-drinks market.

The impression gained from visiting Tomé–Açú was that farmers worked land of only moderate or even poor fertility, that they had responded to setbacks with determination and considerable care, though there may have been some element of good luck. The key to success was unclear but seems to have been the support given by a close-knit cooperative group, a group that paid attention to, and had invested in, agricultural experimentation, marketing and market research, and which kept itself informed of promising new developments. How much this cooperation and enterprise results from Japanese cultural attitudes and/or support for one another, is unclear. However, there seemed to be nothing unusual about Tomé–Açú agricultural practices that would prevent their being replicated elsewhere in Amazonia by other migrant groups with different cultural backgrounds.

'Zona Bragantina' Experiments

In the early 1970s in the 'Zona Bragantina' (the region east of Belém centred on the municipality of Bragança), Japanese agronomists had some success with a mixed cropping strategy. This, it was hoped, would replace the degenerate shifting cultivation common in the region and would sustain production. The strategy was to clear a plot by felling and fire, plant rain-fed rice (seeding the ash or soil relatively densely) between cassava planted in rows 1.5 metre × 1.5 metre apart, with guaraná or coffee planted at 6 metre intervals and Brazil-nut (*Bertholetia excelsa*) at 18 metre intervals. The coffee and guaraná fruit in about six years, the rice utilises the nutrients in the ash and provides profits and food to help the settlers so that they can weed the other crops until they begin to yield. The Brazil-nut provides a valuable and steady annual harvest, but not for 15 years. This fairly straightforward agroforestry strategy is an improvement on shifting cultivation because the tree crops are able to extract nutrients from deep below ground, but for how long cropping can be sustained without some input of fertiliser is uncertain (probably several decades at least), (Sioli, 1980, p. 226). While not a permanent solution, such a strategy is a considerable improvement on past settlers practices in the 'Bragantina' region.

The Yurimaguas Fertiliser and Rotation Experiments

In the early 1970s on-farm trials of fertiliser use and crop rotation took place at Yurimaguas (Peruvian Amazonia), a locality with typically low-fertility soils. The fertiliser and crop rotation system, which included rice, maize, groundnuts and soya continued for 21 successive harvests, giving satisfactory yields and actually improved soil. On nearby control plots, cropped without application of fertiliser and rotation, only three crops were possible (Charlton, 1987, p. 154). The experiment indicates that, given appropriate inputs and agricultural expertise, sustainable cultivation is possible on some relatively poor Amazonian soils. The difficulties remain of getting regular supplies of fertiliser to small farmers at a suitable price and of ensuring that they have assistance to apply it in the correct manner. These difficulties are unlikely to be resolved without considerable support from agricultural extension authorities plus, in many regions, improved communications and transport systems.

Pasture Improvement in Roraima

There are extensive regions of grassland in Amazonia, for example, on Marajó Island (Pará), and in Roraima. The spread of ranching in Pará, Rondônia and other regions, means that there is an increasing area of man-made grassland which seldom provides good quality grazing for long. Experiments at the EMBRAPA Agua Boa Experimental Station near Boa Vista (Roraima – visited by the author in 1987) suggest that there may be promising ways of improving pasture in Amazonia. Scrub or grassland is cleared, burnt and ploughed, but because the soils are generally nutrient-deficient and have a high aluminium content, crops tend to be shallow-rooted and vulnerable to drought. The cost of fertiliser and gypsum treatment is high. However, by seeding with a mix of rain-fed rice, a grass intended to give improved grazing (*Brachiaria humidicola*), and an exotic forage legume *Stylosanthes capitata*, EMBRAPA feel the costs can be overcome. The rice gives a crop within about 120 days which is sufficient to pay for soil preparation and fertiliser. The grass and *Stylosanthes* provides good quality forage, which in trials and local farmers' experience resists overgrazing and allows greater carrying capacity. Trials have so far run for about five years with little sign of pasture degradation (personal observation, August 1987). The disadvantage is that this pasture improvement will mainly benefit the larger landowner, not the small farmer, unless the latter can form cooperatives or perhaps produce forage for larger ranchers. Another possibility for small farmers might be a crop rotation using 'elephant grass' (*Pennesetum purpureum*), or the shrub legume *Tephrosi* sp. which could be cut and sold as forage to large ranches.

The Use of Irrigation to Improve Terra Firme *Agriculture*

A promising strategy for small farmers might be the use of water from seasonally-flooded depressions or streams dammed with a simple earth-bank to support dry season irrigation of rice or other crops. In some localities simple gravity irrigation may be possible; elsewhere pump-irrigation would be needed. The costs of the latter would be a hindrance, and may necessitate small farmers being given access to credit or forming cooperative groups. Given some application of fertiliser, rice, vegetation and other crop production should be sustainable and should lead to much less soil erosion than with present practices.

The Monteverde Pasturage System

In the Monteverde region of Costa Rica there have been some interesting farming developments, like at Tomé-Açú, on poor soils under tropical forest. The system at Monteverde has been to only partially clear the forest, using no fire. The trees that are left shade the ground and the cut vegetation is strewn on the ground to give it further protection. The area part-cleared is seeded with pasture-improving grasses which benefit from the decomposition of the cut debris. Farmers divide their cleared land into thirty portions and graze cattle on each for only a very short time before moving them to the next. In this way overgrazing is avoided and the land is fertilised by the cattle dung. In effect each plot gets grazed for only twelve days a year with a month to recuperate between grazings (Caufield, 1986, p. 117). There is no reason why a version of this system could not be used by large ranchers to improve their carrying capacity; if they were to do so it might reduce their constant demand for more pasture.

Possible Routes to Sustainable Small-farm Agriculture

Intercropping

There has been considerable interest in intercropping world-wide. By growing a suitable mix of crops which benefit each other by deterring pests, giving shade, providing support for climbing plants and, above all, fixing and making available nitrogen to nearby crops intercropping gives better and/or more secure yields and helps to sustain production.

Alley Cropping
Alley cropping is the planting of rows of crops alongside a row of plants (usually perennial shrubs or trees) which fix nitrogen – and therefore can provide green manure, and woodfuel.

VÁRZEA/WETLAND AGRICULTURE

Before the 1950s much of the settlement and agricultural development in Amazonia had been alongside the rivers (Monteiro, n.d.; Moran, 1981). Signs of sustained cultivation along the rivers of Amazonia include dark 'anthropogenic' soil deposits which have been dated as early as 5,500 years old (Sternberg, 1956). Although fewer peoples practise drained field cultivation today, there is a long history of such wetland cultivation in South and Central America. The classic is the *chinampas* system still used in Mexico, in which the cultivator excavates channels, throwing up the soil between to form a raised-bed. The channels help to drain the land and can provide water for irrigation in dry periods; more important, mud and weed from the channels can be added to the raised-bed at regular intervals to sustain fertility (Barrow, 1987c; Charlton, 1987, p. 163; Darch, 1985). This may have potential for Amazonian small farmers as a sustainable farming strategy that could be used in wetland areas unlikely to attract ranchers or large-scale cultivators.

There is some confusion over terminology applied to Amazonian wetlands. Limnologists and ecologists generally recognise three categories of Amazonian river, according to water quality. *Whitewater* rivers are turbid and rich in suspended sediments, mainly clays (for example, the Solimões, Purús, Juruá, Branco). *Blackwater* rivers drain regions where nutrient-poor sandy soils have developed above erosion-resistant igneous or metamorphic rocks. It is suggested that intensive leaching of humic compounds from these poor soils stains the rivers flowing over them a dark colour. Because they carry little suspended load these rivers are poor in fish and deposit little silt to form floodlands (for example, the Negro, Tapajós, Juruana, Cururú). *Clearwater* rivers carry little or no material suspended or in solution and originate from regions of Palaeozoic and Plio-Pleistocene sedimentary rocks (for example, the Xingú), (Stark and Holley, 1975; Barrow, 1985). While some ecologists loosely apply the term *várzea* to almost any seasonally inundated area, limnologists reserve the term for floodlands formed alongside *whitewater* rivers, although they can also form along the lower courses of the other two types of river, but in such situations are less well-developed. In some parts of Amazonia (for example, the Tocantins) the term *vazantes* is applied to areas subject to flooding by rainfall which has failed to infiltrate, as opposed to river waters.

Both the extent and value of Amazonia's *várzeas* are still debated: most estimates suggest that there are between 64,000 and 128,000 square kilometres, or 1 to 2 per cent of Amazonia, (Denevan, 1984, p. 322; Barrow, 1985, p. 114). Information on the flooding regimes, soil characteristics, etc. can be found in Barrow (1985, 1987a, 1987b). The timing and duration of flooding vary enormously from *várzea* to *várzea*, and the quality and quantity of alluvium deposited each time the wetlands are flooded is also variable. A mid-Amazonian *várzea* may receive 15 centimetres or more of relatively fertile sediment each year; the floodlands of the Rio Branco (Roraima) get a significant cover of silt each flood, but it is less fertile. And some rivers flood but deposit virtually no silt. So, like the *terra firme*, soils in Amazonian floodlands vary from site to site and a problem is the identification of areas where soil dries out for sufficient time and is fertile, ideally gets a periodic fresh supply of fertile silt, is an area which is not likely seriously to erode during future floods and which has a soil texture amenable to cultivation. Where *várzea* soils are good and get periodic deposits of fertile silt, there is great potential for sustainable agriculture. Even where the soils are less fertile there are opportunities for simple irrigated cultivation. Cultivators have access to fish and game, whereas on the *terra firme* fish is less easy to come by and game animals tend to get hunted out very soon after settlement. Access by river can also be easier in many parts of Amazonia.

Várzea Cultivation in the Mid-Tocantins

During field visits in 1985 the author saw a number of *várzea* small farms along the mid- and lower-Tocantins from about 20 kilometres below Tucuruí to the Pará 'estuary'. There were a number of cultivation strategies, each suited to a particular combination of soil type and flood regime (and subject to considerable local variation). In the mid-Tocantins, higher *várzeas altas* with sandy soils were first to be exposed and were planted in April/May with alternating rows of beans (four varieties of *Phaseolus* species were noted including the varieties 'sempre verde' and 'feijão branco') and melon (*Cucumis sp.*), separated by windrows of cleared 'capim' grass (*Paspalum fasciculuum*), which were either burnt or used as mulch. A typical holding of this type would be between 0.5 and 3.5 hectares in size, farmed by one man with perhaps two assistants (usually family). There was no mechanisation, only hoe and adze, although some farmers were using knapsack sprayers and pesticides against insects. As the water falls the 'backslope' of the *várzea* levee is progressively planted, the lowest, wettest land fringing the permanently flooded *igapó* forest is planted in June and July. The soils of these lower-lying slopes towards the *igapó* have more silt and less sand and are planted with maize and cassava.

A 'Successful' Mid-Tocantins Várzea *Small Farmer*
On one 3.5 hectare holding, the farmer (a Brazilian *posseiro*) claimed a yield of 700 kg per hectare a year of beans and, that in spite of no use of fertilisers, his yields from the same land had remained steady for 12 years. Neither had he any weed problems. Insect pests were a nuisance rather than a problem and rodents and birds seemed to do little damage to the crops. The small farmer also had two other plots of land, each of roughly 3.0 hectares, at different localities. Presumably this spread his risks of crop failure due to erosion or other causes (the last severe flood affecting crops had been in 1980). Some of the bean and most of the melon crop was sold. This farmer was a particularly successful individual and he had achieved his success without aid from agricultural extension services which, he said, had never contacted him. He obtained good quality seeds from a friend in southern Brazil and he showed a willingness to experiment. Doubtless, given suitable agricultural extension aid, he could further improve his yields. As a *posseiro*, this farmer feared that larger landowners might pressure him off of his land.

Other farmers were even less secure than the 'successful farmer'; a number of small farmers in the vicinity were sharecroppers or had no legal proof of land ownership, worked smaller plots and had lower yields. The organisation GETAT had recently evicted some small farmers in the area for underutilizing their *várzea* plots.

Some *várzeas* are being cultivated (mainly for rice) using machinery – for example, at Jarí (Amapá), at Itacoatiara (Amazonas), in the region southeast of Belém, in parts of Marajó Island (Pará) and along the Rio Branco (Roraima). In parts of Amazonia (for example, Roraima), where labour is expensive and hard to find, large growers will use mechanised cultivation and small-farmers must compete using family labour if they are to sell rice at a competitive price.

Although bold, practical proposals for sustainable small-farm development in *várzeas* were made by Camargo (1948, 1949, 1958) they have somehow failed to attract the support they deserve. *Várzea* development received little more attention until the late 1970s when two authors wrote on the potential: Katzman (1976) and Petrick (1978). In 1981 the PROVARZEAS Programme (*O Programa Nacional de Aproveitamento Racional de Várzeas Irrigáveis*) was launched in the Centre–South and North-East of Brazil. In the early 1980s, PROVARZEAS programmes were initiated in Amazonia, overseen by the Brazilian Ministry of Agriculture, and there are now over twenty State irrigation programmes which concentrate on easily-developed natural wetlands. Some of the PROVARZEAS support is directed at small farmers in Amazonia, but much goes to farmers working wetlands south of Amazonia, who have easier access to markets. According to an

EMATER staff member interviewed in 1985, that organisation has been encouraging and supporting settlers to take up rubber planting in the *várzeas* of Lower Amazonia. There were also quite a few small farmers between Cametá and Belém who were market gardening on *várzea* land growing beans, cabbage, maize, chillies, banana and sweet potatoes. Many of these cultivators did not own their land and, although no one else was known to have title to it and they did not pay rent, there was the risk that, if they were really to improve the holding, it might be taken from them.

Cultivation of Várzeas *of the Lower Pará, Guamá, Caeté and Mojú That are Subject to Regular 'Tidal' Flooding*

There has been quite a lot of development of small-farm 'polder' rice production in Lower Amazonia where freshwater 'tides' (river water backed up by ocean tides at the mouth of the Amazon) flood the low *várzea* beaches and mudbanks (*praias*), in some cases as frequently as twice a day. Simple earthen bunds and flap gates are all that are needed for 'polder' rice cultivation, which allows one or more crops of irrigated rice a year as long as salt accumulation is not a problem. It is near the Atlantic that salinity may build up in 'polders', but this can usually be countered by leaving the land fallow to be washed free of salts for one growing season in every five, and in such areas one rice crop a year is usual. Most of the harvest is sold in nearby Belém, and in spite of fluctuating market prices for rice the small farmers appeared in 1981, 1983 and 1985 to make enough, for there was a continued expansion in the number practising this type of production.

An extensive research literature and a number of educational pamphlets have been produced mainly by EMBRAPA and FCAP in Belém. Much of this effort is directed at the main areas where this type of rice production is carried on in Pará, such as the municipality of Igaripé-Mirim, Marajó Island, near Bragança and in the Tailândia area. The main rice crop here is planted in December and is harvested in June–July. Typically, yields are around 5,000 kg per hectare a harvest (7,200 has been obtained in experimental stations) and up to three crops a year are possible (but in practice one or two are more likely) (EMATER, 1981; 1983). These are excellent yields; for a typical *terra firme* settler growing rain-fed rice a yield of 1,500 kg per hectare a year would be good, and 750 kg per hectare a year not unusual.

In Lower Amazonia one problem often faced by the small farmer wishing to start *várzea* rice farming is the difficulty of removing tree stumps from the ground. This means that mechanisation is hindered, but it also makes it less attractive for large-scale producers, so there is less likely to be conflict between small farmers and larger landowners than with many other agricultural strategies. One risk to *várzea* rice-growing small farmers is that large-scale, mechanised rice producers elsewhere, for example, in the *cerrados* south of Amazonia, might produce a lot of rice in the future and drive down market prices. Things might be further improved if better rice varieties become available. There has been a lot of effort to develop better-yielding, disease-resistant and fertiliser-responsive hybrids in east and north-west Amazonia, but what would be useful is a tall-growing flood-resisting rice. Such a rice could be grown in areas where floodwater was less predictable and/or slow to subsidise and would open-up more areas of *várzea* for rice production.

Várzea *Rice Farming in Roraima*

The author visited several large-scale rice schemes in *várzeas* of the Rio Branco in August 1987. Demand for rice is growing in the region and this has encouraged a number of landowners to start large-scale mechanised rice cultivation, some aided by PROVÁRZEAS funding. The Rio Branco *várzeas* are not as fertile as some of those elsewhere in Amazonia but the soil withstands the weight of machinery and repeated ploughing, is easy to clear and grade, and is productive provided an annual dressing of chemical fertiliser is applied. The usual technique is to pump-irrigate the *várzeas* during the October to March dry season, supplying the water simply by flooding levelled 'polders' using a diesel pumpset. Only one harvest a year is possible but yields are quite good (up to 5,000 kg per hectare a year).

Given access to credit and suitable land, there is no reason why small farmers could not also produce rice or other crops using tractors provided by a cooperative or hand-guided ricefield cultivators in the Rio Branco *várzeas*. However, it seems unlikely that larger landowners would welcome such competition for the more extensive stretches of *várzeas*. Assuming that to be the case, small-farmers will have to use the smaller patches of *várzeas*, which are less suited to large-scale, mechanised cropping, or seek aid to establish pump-irrigation on *terra firme* land using water from wet hollows or dammed streams.

EMBRAPA (Belém) has made some progress with introducing the aquatic plant *Azolla* as a green manure for fertilising rice 'polders' (Barrow, 1985). Were this to prove successful and become a widespread practice, it might help small farmers avoid use of expensive chemical fertilisers with their associated pollution risk. More importantly, it would give small farmers, especially those forced to cultivate less-fertile *várzeas* (like those of the Rio Branco), a free source of fertiliser. One problem which hinders both small farmers and large-scale rice production in lower Amazonia is the need to remove tree stumps from the *várzeas*; this seems to be less of a problem in the *várzeas* elsewhere in Amazonia, and certainly in Roraima stump removal is a minor problem.

CROPS WITH POSSIBLE POTENTIAL FOR AMAZONIAN SMALL FARMERS

A *várzea* plant which may have some potential for small farmers is the semi-aquatic *aninga* (*Calthea lutea*), a large-leaved perennial herb about 2 m tall which grows on the lower *várzeas* and fringes of wetter *igapó* forest throughout Amazonia. Its leaves yield a wax for which there is a good market (*cauassú* wax); there is also thought to be some potential for producing paper from the plants (National Academy of Sciences, 1975, p. 138; Anon., 1982). The collection from wild *aninga* plants is easy and cultivation should present few problems. Two harvests a year are possible after the first two years during which the plants mature; this gives an annual yield of 80 kg per hectare of crude wax. Growing on muddy, semi-submerged land which is of little use for anything else, land which receives a periodic supply of fresh silt during floods, it should be possible to sustain production indefinitely. The wax travels well and has a high value-to-weight ratio, making it an ideal crop for small farmers in remote areas, the only disadvantage being that *aninga* leaves are attractive and toxic to livestock.

Two cash crops, widely grown between the 1930s and 1960s, were jute (*Corchorus capsularis*) and the less important *malva* (*Pavonia malocaphylla*), and on dry ground *ucama* (*urena lobata*). Though still grown, they have become difficult to market and have stagnated since 1965, largely because the market is depressed by competition from synthetic textiles (Barrow, 1985, p. 121). Cane-sugar is still grown in *várzea* areas, near Cametá and Abaetetuba, where it supports local liquor distilleries, but it seems unlikely that demand for the crop will increase much in the future. The market for groundnut, soya, ramie (*Boehmeria nivea*), chick peas (*Cicer anetinum*) and, above all, beans should remain promising and small farmer strategies would be wise to include some production of one or more of these crops (Nogueira, 1981).

Intercropping of beans and maize in the Solimões–Amazonas region *várzeas* has been found to yield 1,042 and 5,700 kg per hectare a year respectively, discourage insect pests and help maintain fertility (EMBRAPA, 1979). There has been some interest in the production and export of tropical fruit, either as fresh fruit or as pasteurised or frozen puree, and in the production of milk and cheese using buffalo. The latter has been expanding in the Belém and Marajó region and may well become more widespread. Buffalo are easier to keep, more robust and can be used for haulage and pulling ploughs as well as producing milk, which should make them attractive to small farmers (Barrow, 1985, p. 122). The potential of livestock like the *capybara* (*Hydrochaerus hydrochaeris*) and river turtles has been noted (Barrow, 1985, p. 123), but there appears to be little chance of improved market demand for the former in Brazil and there is little recent experience of rearing techniques for the latter.

Table 14.1 lists Amazonian palm species which may have potential. So far,

Table 14.1 Amazonian palm species with possible potential as small-farm crops

Palm species (Brazalian name)		*Value*
Açai	*Euterpe edulis*	Fruit/palm heart
	E. oleraceae	
Burití	*Mauritia flexuosa*	Edible oil
	M. vinifera	
Curitirana	*Mauritia martiana*	Edible oil
Pupunha	*Bactris gosipaes*	Fruit
Peach palm	*Guilielma gasipaes*	Fruit
	G. speciosa	
Tucum	*Astrocaryum tucuma*	Oil
	A. vulgare[1]	
Piaçaba	*Orbignia piassaba*	Oil
Bacaba	*Oenocarpus bacabu*	Edible oil
	O. distichus	
	O. multicaulis	
	O. minor	
	O. mapora	
Pissava	*Atalia speciosa*	
	Atalia junifera	
Cohune	*Leopoldina piasaba*	Fibre
Seje	*Iriartea polycarpa*	Edible oil
	Jessenia polycarpa	
	J. bataua	
	J. mataua	
Babaçú	*Orgygnia martiana*	Edible oil/fuel oil
	O. speciosa	
——	*Elaeis oleifera*	Oil
——	*Manicaria saccifera*	Fibre/fruit

Note:
1. Over 50 species.
Source: Goodland, Irwin and Tillman (1978); Balik (1982); National academy of Sciences (1975).

there seems to have been little effort to develop cropping strategies that incorporate these plants, yet there is probably considerable promise. The *buriti* palm in particular is a source of fibre, starch, oil and 'palm hearts'.

The tree *Copaifera multijuga* which yields *copaiba* oil, a vegetable oil which needs little processing before it can be used in diesel engines, is native to Amazonia. Tapped at least twice a year, one tree can yield at least 25 litres at each tapping (*The Times*, 8 June 1987, p. 14).

CONCLUSIONS

There appear to be possibilities for developing sustainable agricultural strategies for small farmers in Amazonia, both in *terra firme* and wetland environments. But their adoption and successful practice will depend very much on the availability of agricultural extension services, on access to credit and on improved security of tenure for cultivators. There has so far been little progress, particularly in solving the latter problem. The human misery, the checks to development and the ecological damage caused by land disputes in and around the Tocantins basin have been vividly described by Branford and Glock (1985). The challenge is not so much what strategy small farmers should practice, rather it is: How can they be placed in a position where they can follow satisfactory ways? Already too many opportunities for pursuing that goal have been lost.

Much agricultural research has concentrated on maximizing production of commercial crops. For small farmers in areas relatively remote from markets whose first priority is to secure survival, it would be better if research were focussed on means of obtaining secure, sustained production, rather than maximisation of yields. Once such production were achieved then farmers might turn their attention to producing a saleable crop in addition to their subsistence crops. Ideally, such a saleable crop would store and transport easily, have a good value-to-weight ratio and be in relatively constant demand. Ideally, production would demand the minimum of inputs, in particular fertilisers, pesticides and herbicides (Groh, 1986). The problem is that much of the present agricultural research effort is not directed towards such goals.

In addition to lack of security of landholdings, Skillings (1984) noted other constraints on development: the weak data base (which makes planning and administration difficult), the scarcity of public services and infrastructure, the high cost of transport, narrow regional market and scarcity of entrepreneurial, managerial and technical talent. To these must be added the failure of agricultural extension bodies to make adequate contact with small farmers.

Redclift (1987, p. 19) presents a table of agroecosystems as a function of their *productivity* (yield per unit of land/labour/inputs); *stability* (degree to which productivity remains constant in the face of environmental or socio-economic disturbances); *sustainability* (the agroecosystem's ability to maintain productivity – if it cannot, decline may be slow and possibly difficult to recognise, or it may be sudden and perhaps catastrophic); and *equitability* (the degree to which agricultural products are shared amongst members of a household, village or region).

In Amazonia, traditional Amerindian/*caboclo* shifting cultivation/gathering has low productivity, low stability, high sustainability and high equitability. However, it breaks down if population increases too much or forest land is in limited supply. The 'degenerate' rice–maize–cassava shifting cultivation practices of many settlers and those of many large-scale cattle ranchers are low in all four respects. Crop rotation and a crop mixture which includes tree crops could be the route to improved productivity, sustainability and stability (it seems to be working at Tomé–Açú).

Large-scale ranching might be improved through re-seeding pasture or by

Monteverde-type rotational grazing. If so, there might be less need constantly to expand pasturage and so a reduction of conflict with small farmers may take place. The 'successful' mid-Tocantins small farmer who has reasonable productivity and sustainability has lower stability, indeed he may be threatened by hydroelectric development impacts.

Small farmers adopting a similar strategy would be well-advised to hold more than one plot of land or have some form of insurance against periodic crop loss or flood erosion. The large-scale rice farmers of the Rio Branco *várzeas* face less risk and their stability is better because the Rio Branco tends to flood in a predictable manner. But as the silt deposited is not very fertile, sustainability can be maintained only through periodic input of fertilisers. Nor is their production system equitable.

If the effort were made to support Amazonia's small farmers their lot could doubtless be much improved Redclift (1987, p. 76) notes that in 1979 incentives to large-scale cattle ranching in Brazil cost the Brazilian Government US$ 63,000 (at 1979 rates of exchange) for *each* ranching job created – what could such a level of funding do for small farmers? Probably a very great deal. Barbier (1987, p. 105) describes a project in Honduras which successfully transformed an unsustainable smallholder agroecosystem into a sustainable one through 'appropriate agricultural technology, training, and erosion control – including intercropping of "green manure" crops with the tradition corn.' *The cost was US$13 per cultivator.*

Yet, as Redclift (1987, pp. 119–20) notes, relatively little attention or money has been spent on developing indigenous systems of agriculture. He argues that, 'The existence of new scientific inputs cannot correct this situation, since the decision to give priority to non-renewable development is a political decision' (Redclift, 1987, p. 142). By investing enough in sustainable small-farm agriculture, the Brazilian government could reduce forest clearance, soil erosion, land conflict, growing rural unrest and poverty. The support for large-scale ranching has led to speculative development which, while it profits the speculator, is inefficient in producing cattle or other crops and hurts the small farmers. The situation is going to be difficult to change for, in addition to enjoying tax incentives and low interest loans in times of high inflation, large-scale cattle ranching probably has the lowest set-up costs of any of the presently feasible Amazonian land-uses. Added to which land values rise fast – between 1970 and 1975 alone, pastureland in Amazonia rose in value by an average of 38 per cent a year. Even if burdened by interest on loans and producing only 1.5 or less head of cattle per hectare (typical figures for Amazonian ranches), there are still profits to be made in ranching (Fearnside, 1983, p. 143).

To summarise, the 'development scenario' in Amazonia militates against the promotion of sustainable small-farmer agriculture. There seem to be several possible opportunities for easing the problem: (1) Divert small farmers to carry on, hopefully sustainable, agriculture in areas or environments that are less attractive to those who compete against them. (2) Protect the small farmer through better legislation and better 'policing' so that the unwanted attention of competitors is discouraged. (3) Find and promote environmentally appropriate, sustainable agricultural strategies which are also 'socially appropriate and socially sustainable' (Charlton, 1987, p. 154). (4) Unless farm extension services are strengthened and improved, little progress is likely. Extension staff have to understand small farmers' needs and attitudes and have to get out into the rural areas. The latter problem could be solved by increasing funding; the former requires attitudinal changes which may be slow to come. (5) Provide large-scale ranchers and growers with better strategies which allow them to carry more cattle or grow more crops on the land they have, thereby reducing the need for them to expand. (6) Reduce opportunities for land speculation and cut tax benefits to large landowners. (7) Give small farmers a stronger voice in the policy-making arena.

(1) to (4) above are palliative measures which, even if they work, will have little more than local impact unless adopted on a very wide-scale. There are two crucial questions: (1) do any of the suggested case study strategies have real potential and, (2) how can their adoption be promoted and supported on a sufficiently broad front to provide secure, sustainable, satisfactory livelihoods for enough of Amazonia's small farmers?

In answer to the first question, some of the strategies (including the Tomé–Açú and Bragantina approaches) have worked in regions with at best mediocre quality soils, in spite of setbacks and lack of adequate ongoing government support. Thus there must be potential for them to succeed in large areas of Amazonia. The strategies outlined are by no means a complete selection of possibilities.

To find suitable strategies for widespread adoption will need support and time. To spread and establish selected strategies will require much more support and time. How much time remains depends upon the rates of land degradation, population increase, in-migration and rising discontent among the peoples of Amazonia. It would probably be sensible for the authorities to spend the next ten years on pilot studies to identify strategies, and then to spend, say, four five-year plan periods promoting and supporting the selected strategies.

Those small farmers who have settled Amazonia, apart from those managed by INCRA, have been left too much to their own devices. In the past, large ranchers could quickly open up a region and establish a presence; they were also able to lobby the central government. The Brazilian government now has less to gain from supporting further expansion of cattle ranching, and has much to gain from supporting small farmers. Unfortunately, the small farmers are less easily heard in Brasília, São Paulo, and Rio de Janeiro than the influential businessmen and ranchers. But it must be increasingly clear that, if the small farmers are not supported, land will certainly continue to be degraded and poverty will certainly continue to increase. Sustained small-farm agriculture is also more productive than large-scale ranching could ever be. But what might prompt the authorities to act is the mobilisation and political emancipation of the small farmers. The civilian 'New Republic' ('Nova Republica') would be rash to ignore the plight of so many increasingly restless peasants.

References

Anon. (1982), *New Scientist*, 21 October p. 156.

Balik, M. J. (1982), 'Palmes neotropicales: nuevos fuentes de aceites comestibles', *Interciencia*, 7(1), pp. 25–9.

Barbier, E. B. (1987), 'The concept of sustainable economic development', *Environmental Conservation*, 14(2), pp. 101–10.

Barrow, C. J. (1981), 'Development of the Brazilian Amazon', *Mazingira*, 14(1), pp. 36–47.

Barrow, C. J. (1985), Development of the varzeas of Brazilian Amazonia, in J. Hemming (ed.), *Change in the Amazon Basin, Vol. 1: Man's Impact on Forests and Rivers*, Manchester: University of Manchester Press, pp. 118–28.

Barrow, C. J. (1987a), 'The environmental impacts of the Tucuruí Dam on the Middle and Lower Tocantins River Basin, Brazil', *Regulated Rivers*, 1(1), pp. 49–60.

Barrow, C. J. (1987b), 'The impact of hydroelectric development on the Amazonian environment: with particular reference to the Tucuruí Project', *Journal of Biogeography*, 14( ), pp. 67–78.

Barrow, C. J. (1987c), *Water Resources and Agricultural Development in the Tropics*, London: Longman.

Bates, H. W. (1986), *A Naturalist on the River Amazon*, London: John Murray.

Branford, S., and Glock, O. (1985), *The Last Frontier: Fighting Over Land in the Amazon*, London: Zed Books.

Camargo, F. C. de (1948), 'Terra e Colonização no Antigo e Novo Quaternário na Zona da Estrada de Ferro de Bragança, Estado do Pará, Brasil', *Bol. Mus. Paraense E. Goeldi*, 10, pp. 123–47, Belém.

Camargo, F. C. de (1949), 'Reclamation of the Amazonian floodlands near Belém', in *Proc. UN Scientific Conference on Conservation and Utilization of Resources*, Lake Success, New York, New York: United Nations, pp. 598–602.

Camargo, F. C. de (1958), 'Report on the Amazon region, problems of humid tropical regions', in UNESCO, *Humid Tropics Research: Problems of Humid Tropical Regions*, Paris: UNESCO, pp. 11–24.

Cardoso, M. (1974), 'A pimenta do reino', *Agronómico*, 26(1), pp. 19–24.

Caufield, C. (1986), *In the Rainforest*, London: Heineman; Picador (Pan Books).

Charlton, C. A. (1987), 'Problems and prospects for sustainable agriculture in the humid tropics', *Applied Geography*, 7(2), pp. 153–74.

Darch, J. P. (1985), *Drained Field Agriculture in Central and South America*, British Archaeological Reports, International series, 189, Oxford.

Denevan, W. M. (1984), 'Ecological heterogeneity and horizontal zonation of agriculture in the Amazon floodplain', in M. Schmink and C. H. Wood (eds), *Frontier Expansion in Amazonia*, Gainesville, Florida: University of Florida Press.

EMATER (1981), 'Sistema de Producão para Arroz em Várzeas: Municípios de Bragança, Augusto Correa e Viseu (revisadas), Sistemas de Producao Bólétim, 332, Belém.

EMATER (1983), 'Sistema de Producão para Arroz em Várzeas (microregio 16)', *Série Sistema de Producao Boletim, 3*, Belém.
EMBRAPA (1979), *Informativo* (March) Manaus, p. 6.
Fearnside, P. M. (1983), 'Land-use trends in the Brazilian Amazon as factors in accelerating deforestation', *Optima*, 31(2), pp. 141–8.
Fearnside, P. M. (1986), 'Agricultural plans for Brazil's Grande Carajás Program, lost opportunity for sustainable development?', *World Development*, 14(3), pp. 385–409.
Goodland, R. J. A., Irwin, H. S. and Tillman, G. (1978), 'Ecological development for Amazonia', *Ciência e Cultura*, 30(3), pp. 275–89.
Groh, D. (1986), 'How subsistence economies work', *Development: Seeds of Change*, 3, pp. 23–30.
Harwood, R. R. (1979), *Smallfarm Development: Understanding and Improving Farming Systems in the Humid Tropics*, Boulder, Col.: Westview Press.
Hecht, S. B. (1985), 'Environment, development and politics: capital accumulation and the livestock sector in Eastern Amazonia', *World Development*, 13(6), pp. 663–84.
Hiraoka, M. and Yamomoto, S. (1980), 'Agricultural development in the upper Amazon of Ecuador', *Geographical Review*, 70(4), pp. 423–45.
Janzen, D. H. (1973), 'Tropical agroecosystems', *Science*, 182, pp. 1212–19.
Katzman, M. T. (1976), 'Paradoxes of Amazonian development in a resources starved world', *The Journal of Developing Areas*, 10, pp. 445–60.
Monteiro, S. T. (n.d.), 'Anotaçoes para uma história rural do médio Amazonas', EMATER–Amazonas, Manaus (mimeo).
Moran, E. F. (1980), 'Mobility and resource use in Amazonia', in Barbira-Scazzocchio, F. (ed.), *Land, People and Planning in Contemporary Amazonia*, Centre for Latin American Studies, Cambridge: University of Cambridge, pp. 46–57.
Moran, E. F. (1981), *Developing the Amazon*, Bloomington, Indiana: Indiana University Press.
National Academy of sciences (1975), *Underexploited Tropical Plants with Promising Economic Value*, Washington, D.C.: National Academy of Sciences.
Nogueira, O. L. (1981), 'Cultiva de feijão do Caupi no Estado do Amazonas', *EMBRAPA/UEPAE Circular Técnico*, 4, Manaus.
Petrick, C. (1978), 'The complimentary function of floodlands for agricultural utilization: the Brazilian Amazon region', *Applied Sciences and Development*, 12, pp. 24–46.
Redclift, M. (1987), *Sustainable Development: exploring the contradictions*, London: Methuen.
Sioli, H. (1980), 'The development of Amazonia: a re-evaluation', in Barbira-Scazzocchio, F. (ed.), *Land, People and Planning in Contemporary Amazonia*, University of Cambridge: Centre for Latin American Studies, Occasional Publications, 3, pp. 257–68.
Skillings, R. F. and Tcheyan, N. O. (1979), *Economic Development Prospects of the Amazon Region of Brazil*, Johns Hopkins University, School of Advanced International Studies, Washington, D.C.
Skillings, R. F. (1984), 'Economic Development of the Brazilian Amazon: opportunities and constraints', *The Geographical Journal*, 150(1), pp. 48–54.
Stark, N. M. and Holley, C. (1975), 'Final report on studies of nutrient cycling on white and black water areas in Amazonia', *Acta Amazônica*, 5(1), pp. 57–76.
Sternberg, H. O. (1956), '*A Agua e o Homem na Várzea do Careiro*', Tese de Concurso a Cátedra de Geografia do Brasil da Faculdade Nacional de Filosofia do Brasil, Rio de Janeiro.

Tanaka, K. (1957), 'Japanese immigrants in Amazonia and their future', *Kobe University Economic Review*, 3, pp. 1–23.

Townsend, J. (1985), 'Seasonality and capitalist penetration in the Amazon Basin', in J. Hemming (ed.), *Change in the Amazon Basin, Vol. II: the frontier a decade after colonization*, Manchester: Manchester University Press, pp. 140–57.

15 Economic Values and the Environment of Amazonia

David Pearce and Norman Myers[1]

INTRODUCTION

The continuing deforestation of Brazilian Amazonia has generated increasing national and international debate. Underlying that debate is both a *conflict of values* and a dispute over the *necessary basis for economic development*.

Value conflicts arise because the irreversible consequences of the destruction of primary forest causes loss of preservation value. Many people are upset by the destruction of ecosystems as a whole, by individual species losses, and by the displacement of aboriginal and other populations. That upset, a loss of welfare, is formally equivalent to an economic loss since economic benefits and costs are measured in terms of welfare gains and losses which may or may not have a corresponding cash flow (Pearce, 1986).

Disputes over the necessary basis for economic development are epitomised by the recent discussion of the principles of *sustainable development* (World Commission on Environment and Development, 1987; Pearce, 1988). One of the features of the sustainable development paradigm is a requirement that, in broad terms, the stock of natural capital – forests, soil, water – be kept constant. The idea behind the constant natural capital concept is that future generations should have roughly equal access to that stock, serving the basic objectives of intertemporal equity – i.e., fairness between generations. Such a requirement poses all kinds of problems about the relationship between natural and man-made capital, the extent of substitutability between them, the factual nature of their productivities, and so on. These are discussed at length in Pearce (1988) and do not concern us unduly here. An alternative critique of value theory which evaluates natural capital in a different way to what we suggest here is provided by Bunker (1988). But the natural capital characteristic of sustainable development thinking partly explains concern over Amazonian deforestation which is equivalent to systematically reducing one part of the capital stock of the nation and, arguably, not compensating for it in full by accumulation of man-made capital. Moreover, sustainable development suggests that certain traditional objectives of economic growth can be served *without* the capital depletion process. The issue, then, is whether Brazilian Amazonia can be developed without deforestation.

In this chapter we review the state of natural capital depletion in Amazonia and look at some underlying causes. These we find in policy measures employed by successive Brazilian governments. Without implying that economic reform is the *only* policy measure required, we show that reformulation of basic macroeconomic incentives is needed. But the international debate is only partly about the best design of sustainable development policy, the benefits of which accrue to Brazil and the neighbouring nations. It is also about the loss of values in the developed world. This raises the issues of responsibility for conservation and slowing the process of deforestation. Simply put, if the rich nations derive value from a stock of assets owned by relatively poor nations, they should pay for the benefits obtained. In the economics jargon, Amazonian forests are conferring all kinds of external benefit on others and the optimal use of the forests is better served if those benefits are appropriated by the nation with the property rights. We investigate the nature of the economic values in question, and raise the issue of how far international compensation is feasible.

NATURAL CAPITAL IN AMAZONIA

Since the great bulk of Amazonia occurs in Brazil, we shall deal with this portion first, before going on to consider the sizeable sector that lies in Venezuela, Colombia, Ecuador, Peru and Bolivia.

The Legal Amazon region of Brazil encompasses an area of about 5 million square kilometres, making up 57 per cent of national territory. A large proportion of this region, some 2.86 million square kilometres, was still covered with moist forest in 1980. In addition, there was roughly 0.5 million square kilometres of transitional forest, while some 0.3 million square kilometres had little or no forest. Thus the tropical forest zone embraced almost 3.7 million square kilometres, or about 67 per cent of the Legal Amazon; and moist forest proper covered almost 2.9 million square kilometres, or about 57 per cent.

Despite some deforestation since 1980, Brazil still contains a far greater proportion of the tropical forest biome than any other nation – roughly three times more than each of the next two nations, Indonesia and Zaire. In addition to the Brazilian Amazon, there are another 1.3 million square kilometres of tropical forest in the nations to the west – Venezuela 0.31 million square kilometres, Colombia 0.27 million square kilometres, Ecuador 85,000 square kilometres, and Peru 0.6 million square kilometres.

Despite an extensive colonisation programme during the past two decades the Brazilian Amazon still contains only about 5 per cent of Brazil's 140 million people. An area as large as the Sahara, it is as thinly populated (outside urban area and the main settlement centres) as the Sahara. At the same time, it contributes a mere 5 per cent to Brazil's GNP, and the region's forests contribute only 10 per cent of Brazil's output of industrial timber. In a country with the largest foreign debt of any nation on earth, now more than $110 billion, many Brazilian leaders have been inclined to look upon the Amazon as a valuable asset going to waste, lacking only capital investment, technology and entrepreneurship. In the past few years, however, a number of prominent Brazilians, not only ecologists and economists but also some political figures, have been urging caution in further 'conventional' development of Amazonia in order that maximum long-term benefit be derived from exploitation of the region's vast resources.

The state-by-state distribution of forest cover is set out in Table 15.1. This reflects LANDSAT imagery and other types of remote-sensing surveys during the late 1970s. The data can be considered accurate to the extent that they document outright deforestation – that is, complete and permanent elimination of all forest cover. But because remote-sensing technologies of the late 1970s were not able to differentiate disturbed forest, and because Amazon planners preferred to distinguish only between deforested lands and forest cover of whatever type, the figure of 125,000 square kilometres deforested by 1980 must be looked upon as a *minimum* figure. Very large numbers of small areas, sometimes amounting to only a few dozen hectares, but collectively covering an extensive area, were being cleared by slash-and-burn farmers and other small-scale settlers. Their 'deforestation' could not be picked up by remote-sensing surveys.

The case of Rondônia state, 244,000 square kilometres in size, shows how this additional deforestation can prove significant. Since 1980, there has been extensive deforestation along the southern fringes of the Amazon in Rondônia, also in Acre. The intensive settlement process was working up only initial momentum during the late 1970s, whereas the 1980s have seen forest attrition at accelerating rates (Malingreau and Tucker, 1988). In the mid-1970s, the Brazilian government started to sponsor settlement in Rondônia, and by the early 1980s in Acre as well. The population of Rondônia was only 111,000 persons in 1975, but it soared to well over 1 million by 1986, for an almost tenfold increase in only 11 years. In 1975, only 1,200 square kilometres of forest had been cleared, but by 1982 the amount had grown to well over 10,000 square kilometres, by 1985 to almost 28,000 square kilometres, and by 1987 to around 60,000 square kilometres. Were the exponential rate of increase in deforestation during 1975–85 to be maintained, it would lead to the elimination of half the state's forest by the early 1990s, and to the elimination of the whole expanse by the year 2000.

In addition there has been much disturbance of forest ecosystems in Rondônia, to a degree that impoverishes biotic communities. It occurs primarily through localized clearing that stops short of broad-scale deforestation: road building, logging, mining and small-scale agriculture. By 1985

Table 15.1 LANDSAT surveys of forest clearing in Brazilian Amazonia

State or territory	Area of state or territory (Km²)	Area cleared (Km²)				% of state or territory classified as cleared			
		By 1975	*B 1978*	*By 1980*	*By 1988*	*By 1975*	*By 1978*	*By 1980*	*By 1988*
Amapá	140,276	152.5	170.5	183.7	571.5	0.1	0.1	0.1	0.4
Pará	1,248,042	8,654.0	22,445.3	33,913.8	120,000.0	0.7	0.8	2.7	9.6
Roraima	230,104	55.0	143.8	273.1	3,270.0	0.0	0.1	0.1	1.4
Maranhão	257,451	2,940.8	7,334.0	10,671.1	50,670.0	1.1	2.8	4.1	19.7
Goiás	285,793	3,507.3	10,288.5	11,458.5	33,120.0	1.2	3.6	4.0	11.6
Acre	152,589	1,165.5	2,464.5	4,626.8	19,500.0	0.8	1.6	3.0	12.8
Rondônia	243,044	1,216.5	4,184,5	7,579.3	58,000.0	0.3	1.7	3.1	23.7
Mato Grosso	881,001	10,124.3	28,355.0	53,299.3	208,000.0	1.1	3.2	6.1	23.6
Amazonas	1,567,125	779.5	1,785.8	3,102.2	105,790.0	0.1	0.1	0.2	6.8
Legal Amazonia (total)	5,005,425	28,595.3	77,171.8	125,107.8	598,921.5	0.6	1.5	2.5	12.0

Source: Fearnside (1986); Mahar (1988).

the amount had reached 87,000 square kilometres. Thus by 1987 a total of at least 147,000 square kilometres of forest had been either disturbed or destroyed, making up 60 per cent of the state.

A similar process has been overtaking Acre, where a major new highway has recently been paved, extending as far as the border with Peru. By 1985, at least 5,300 square kilometres of Acre's forest had been cleared, and 30,000 square kilometres disturbed, making a total of 35,300 square kilometres, or 23 per cent of the state (Malingreau and Tucker, 1988).

A similar process of progressive deforestation is starting in Roraima state in northern Amazonia (Fearnside, 1986). How fast is the Brazilian Amazon losing its forest cover overall? As noted, the LANDSAT Monitoring Program estimated in 1980 that at least 124,000 square kilometres of Amazon forests of whatever type, had been eliminated. This amounted to only 2.47 per cent of the Legal Amazon, and only 4.27 per cent of the original forested area. But other observers, notably Fearnside (1986) and Hecht (1985), with their many years of on-ground experience in the Brazilian Amazon, both assert that the figure of 124,000 square kilometres is on the low side. More recently Salati *et al.* (1988), drawing on latest remote-sensing surveys, assert that the amount of Brazilian Amazonia that has been deforested to date amounts to about 10 per cent of the original forested extent, or 290,000 square kilometres, and that deforestation is proceeding at some 20,000 square kilometres per year.

One must remember, moreover, that Amazon deforestation is not entirely a recent phenomenon. The forest of the Bragantina Zone at the easternmost end of the Amazon, an area of some 25,000 square kilometres, was settled by immigrant farmers around the start of this century, and within the space of some three decades it lost virtually all its forest cover, until all that now remains is a grossly denuded landscape with exhausted soils.

As for those parts of the Amazon which lie outside Brazil, latest estimates (Salati *et al.*, 1988), plus the authors' own data, indicate that some 19,000 square kilometres (7 per cent) of Colombian Amazonia have been deforested by 1987; almost 73,000 square kilometres (12 per cent) of Peruvian Amazonia; and almost 7,000 square kilometres (8 per cent) of Ecuadorian Amazonia. Only Venezuelan Amazonia remains little affected; a mere 14,000 square kilometres (4 per cent) have been deforested, this being due to the country's oil revenues that reduce the impulse to extract revenues from such forest exploitation as may be feasible.

This all means that a total of 113,000 square kilometres of non-Brazilian forest has been eliminated in the recent past, making up 9 per cent of what was formerly there. Amalgamating figures for Brazilian and non-Brazilian Amazonia, we find that a total of 313,000 square kilometres has been deforested, or some 7–8 per cent of the original 4.2 million square kilometres (2.9 million square kilometres in Brazil and 1.3 million square kilometres in other countries). Let us further note, moreover, that in all countries, except possibly Venezuela, deforestation rates seem to be accelerating.

A TYPOLOGY OF ECONOMIC VALUES

It is tempting to contrast economic value with other values – ethical concerns, the rights of non-human biota, moral stewardship and so on. The former might appear to relate to utilitarian outputs of the tropical forest – timber, agriculture, etc. – the latter with 'meta-economic' concerns. Work in the last decade or so in environmental economics suggests that this distinction is false, and that at least some of the concerns of conservationists are subsumed in the broad concept of 'total economic value'. In order to assess the consequence of deforestation, then, we must digress briefly to discuss economic value concepts.

While the terminology is still not agreed, environmental economists begin by distinguishing user values from intrinsic values. User values, or user benefits, derive from the actual use of the environment. An angler, wildfowl hunter, fell walker, ornithologist, all use the natural environment and derive benefit from it. Those who like to view the countryside or experience the rain forest, directly or through other media such as photography and film, also 'use' the environment and secure benefit. The values so expressed are economic values in the sense we have defined. Slightly more complex are values expressed through *options* to use the environment – that is, the value of the environment as a potential benefit as opposed to actual present use value. Economists refer to this as *option value*. It is essentially an expression of preference, a willingness to pay, for the preservation of an environment against some probability that the individual will make use of it at a later date. Provided the uncertainty concerning future use is an uncertainty relating to the availability, or 'supply', of the environment, the theory states that this option value is positive (Bishop, (1982)). In this way we obtain the first part of an overall equation for total economic value. This equation says:

Total User Value = Actual Use Value + Option Value

Intrinsic values present more problems. They suggest values which are in the real nature of the thing and unassociated with actual use, or even the option to use the thing. The briefest introspection will confirm that there are such values. A great many people value the remaining stocks of rainforest. Very few of those people value them in order to maintain the option of seeing them for themselves. What they value is the *existence* of the forest, a value unrelated to use – although, to be sure, the vehicle by which they secure the knowledge for that value to exist may well be film or photography or the recounted story. The example can be repeated many thousands of times for other species, threatened or otherwise, and for whole ecosystems such as wetlands, lakes, rivers, mountains and so on.

These *existence values* are certainly fuzzy values. It is not very clear how they are best defined. We can agree with Brookshire and his colleagues (1985) that these values are not related to vicarious benefit – i.e., securing pleasure because others derive a use value. Vicarious benefit belongs in the class of option values, in this case a willingness to pay to preserve the environment for the benefit of others. Nor are existence values what the literature calls 'bequest' values, a willingness to pay to preserve the environment for the benefit of our children and grandchildren. That motive also belongs with option value. Note that if the bequest is for our immediate descendants we shall be fairly confident at guessing the nature of their preferences. If we extend the bequest motive to future generations in general, as many environmentalists would urge us to, we face the difficulty of not knowing their preferences. This kind of uncertainty is different to the uncertainty about availability of the environment in the future which made option value positive. Assuming it is legitimate to include the preferences of as yet unborn individuals, uncertainty about future preferences could make option value negative (Bishop, 1982). We may provisionally state that:

Intrinsic Value = Existence Value

where existence values relate to values expressed by individuals such that those values are unrelated to use of the environment, or future use by the valuer or the valuer on behalf of some future person.

In this way we can write a formula for *total economic value* as:

Total Economic Value = Actual Use Value + Option Value + Existence Value

Within this equation we might also state that:

Option Value = Value in use (by the individual) + Value in Use by Future Individuals (descendants and future generations) + Value in Use by Others (vicarious value to the individual).

The context in which we tend to look for total economic values should also not be forgotten. In many of those contexts, three important features are present. The first is *irreversibility*. If the rainforest is not preserved, it is likely to be eliminated with little or no chance of regeneration. The second is *uncertainty*: the future is not known, and hence there are potential costs if the asset is eliminated and a future choice is forgone. A dominant form of such uncertainty is our ignorance about how ecosystems work: in sacrificing one asset we do not know what else we are likely to lose. The third feature is *uniqueness*. What empirical experiments we have on existence values tend to relate to endangered species and unique scenic views. Economic theory tells us that this combination of attributes will dictate preferences which err on the cautious side of exploitation. That is, preservation will be relatively more favoured in comparison to development. Clearly, all three attributes are of importance for Amazonia.

CONFLICTS OF VALUES

In the Amazonian context, the component values within total economic value are in conflict. Basically, existence and use values appear to be incompatible, and differing use values conflict. Livestock development is thus inconsistent with the environmental existence values, although it is related to purely consumption values in developed economies through the demand for livestock products (see the discussion below on the 'hamburger connection'). Government security values are inconsistent with aboriginal demands for land if the former can be only secured by agricultural colonisation.

When values conflict it is necessary to proceed in stages. The first stage is to establish the overall aim of maximising the *net* benefits of economic development, which can be done only by making every effort to value correctly the differing uses of the forest and by carefully investigating the alternative technologies for use. For example, if mechanisms exist for appropriating existence benefits, then it is important to establish the scale of those benefits so that they can be compared to, say, livestock development. If option and existence values exceed the livestock profits, the appropriate policy is to preserve the forest. The second stage is to investigate the appropriate technologies for use of the forest. Clear-felling for timber results in totally modified configurations of species as habitat changes. The timber benefits of clear-felling may be achievable by selective felling systems, with reduced environmental impact. Even in very narrow, commercial terms, forest clearance for agriculture may simply not pay. To date this is the picture for livestock farming on cleared land.

The whole exercise of making optimal use of a natural capital asset is complex. But we can be sure that it is a procedure that is not even approximated in Amazonia. Political considerations dominate economic value assessment. Yet the mechanisms used to secure political ends are primarily based on fiscal policies which in turn have economic consequences for state and federal budgets, and for economic development as a whole. This suggests that a reappraisal might indicate ways in which the development process can be sustained, and perhaps even improved, without the environmental losses that sustainable development thinking suggests are avoidable and economically costly. A first stage, however, is to obtain some idea of the values at stake in deforestation.

THE CONSEQUENCES OF NATURAL CAPITAL LOSS

As Amazonia's forest disappears, so there is a loss of valuable natural resources associated with forest ecosystems. True, some of these should be renewable, in principle at least. For instance, hardwood stocks can regenerate themselves, whether in natural forest stands (provided the logging is sustainable in its operations) or in deforested areas – provided they have not been given over to permanent alternative human uses.

But certain natural resources are not renewable by their very nature. Primarily this applies to wild species of flora and fauna. The Amazon is believed to harbour at least 30,000 plant species, as compared with only 10,000 species in all of temperate South America. It also contains one in five of all bird species on Earth. On a single tree in the Peruvian Amazon there have recently been discovered 43 ant species more than in the whole of the British Isles.

Species communities are not equally distributed across the entire region. More than 20 centres of biodiversity have been identified (Prance, 1982), accounting for less than 20 per cent of the region. If biodiversity is to be highly valued these localities deserve priority attention from conservationists and developers alike, especially insofar as they usually feature high levels of endemism. To gain an idea of the biotic richness of these areas, consider the 15,000 square kilometre Manu Park in the Peruvian Amazon. Within its forests are at least 8,000 plant species described, or almost half as many as in the 520-fold larger United States; 200 mammal species, or more than in the United States and Canada; and 900 bird species, or more than one in ten of all on Earth.

Another biodiversity case in point is Rondônia/Acre (Lisboa *et al.*, 1987), an area which, as we have seen above, is undergoing ultra-rapid degradation and destruction of its forest cover. We can conservatively estimate that of Amazonia's 30,000 plant species, at least 5,000 occur in Rondônia/Acre, roughly 800 (16 per cent) of them being endemic.

Direct use values arise in agriculture, medicine, industry, and in environmental services. In the field of *agriculture*, for instance, all modern crops need regular infusions of genetic variability to maintain their productivity and to resist new diseases. During the past few decades, genetic resources from the Amazon have helped to safeguard commercial rubber plantations in South-East Asia. In the Amazon forest habitat, where rubber originated, there is a potent fungal parasite known as South American Leaf Blight. The disease has threatened to make its way to South-East Asia, where the blight could cripple the natural rubber industry within just a few seasons. Fortunately, a number of wild species of the *Hevea* (rubber) genus offer resistance to the blight – just as others offer tolerance to cold, or to too much or too little moisture. But the survival of these germplasm resources, valuable as they could be to Brazil were the nation to be able to charge a 'user fee' to South-

East Asian growers, is increasingly threatened by a series of development projects sponsored by the Brazilian government in the Amazon. Of 15 such projects, 11 are located in the original forest habitats of the genus *Hevea*, and at least 8 of them overlap with the blight-resistant populations of the rubber tree.

In the *medicinal* field, it may be noted that the estimated cross-counter sales of drugs and pharmaceuticals manufactured from startpoint plant materials from tropical forests is now in the order of $12 billion a year (Myers, 1984). Of course the startpoint materials contribute only a small part of the finished products – many other inputs are entailed. All the same, the finished product depends absolutely on the critical raw materials from tropical forest plants. A notable instance is the rosy periwinkle, a plant of Madagascar's forests, with alkaloids that have led to the production of two potent drugs for use against Hodgkin's disease, leukemia and other blood cancers. The commercial sales of these drugs are now worth more than $150 million a year in the United States alone, while the economic value to American society can be computed at more than twice as much. According to the National Cancer Institute in the United States (Duke, 1980; Douros and Buffness, 1980), field experience shows that the Amazon could well contain another five plant species with capacity to generate superstar drugs against cancer – and with benefits that would accrue only marginally to Brazilians, Colombians, etc. as compared with the rest of the global community on grounds of both numbers of people and incidence of cancer.

Much the same applies too to *industrial materials*. The *Babassú* palm grows wild in just 3 per cent of the Amazon. It bears fruit with a higher proportion of oil than the coconut palm, while being similarly usable for margarine, shortening, general edibles, fatty acids, toilet soaps, detergents, plastics and liquid fuels (Balick, 1979). So prolific is the palm's oil output that a stand of 500 trees on one hectare could produce 125 barrels of oil per year, plus 5.7 tons of edible protein and 250 barrels of ethanol as by-products. The plant resists domestication, so it can be harvested only in the wild. The species is losing much of its native habitat to settlement projects and encroaching agriculture.

The Amazon forest also generates a number of *environmental services*. Foremost is the hydrological cycle function. According to research with water isotopes (Salati *et al.*, 1988), much of the moisture in Amazonia – between half and four-fifths in Central and Western sectors – remains within the ecosystem. That is, it is constantly transpired by plants into the atmosphere, where it gathers in storm clouds before being precipitated back onto the forest within days. Thus much of the Amazon forest represents a significant source of its own moisture: it does not have to depend on moisture imported from the ocean surrounds. Were Amazonia to be widely deforested, there could be a pronounced decrease in the amount of moisture being evapotranspired into the atmosphere, leading to a decline in rainfall of one-quarter or even more.

A decline of just this scale alone would entrain profound and irreversible ecological changes in many parts of the Amazon region. But more important still, it could trigger a self-reinforcing process of growing desiccation for remaining forest cover, with declining moisture stocks followed by yet more desiccation, and so forth. Eventually the repercussions could extend outside Amazonia, even to southern Brazil with its major agricultural lands.

As Amazonia becomes deforested, we witness the irreversible elimination of many material products and environmental services with pronounced utilitarian value. Moreover, apart from these repercussions of deforestation, there are *humanitarian concerns*. Only five centuries ago the Brazilian Amazon supported 230 native groups, with an estimated minimum of 6 million people – conceivably several times as many. Today there are only half as many such groups, with a total of about 100,000 persons. Aside from the humanitarian issue, there is a utilitarian concern. These forest-dwelling tribes represent a fund of traditional knowledge that is critically relevant to pharmacologists, industrial researchers, agronomists and others who seek to identify key plant raw materials for modern economies. The Chacobo Indians of northern Bolivia utilise a full 80 per cent of the forest trees, shrubs, vines and herbs in their area, using them for foods, clothes, housing and medicines (Boom, 1985). The Gaviões indians of western Brazil are so capable of production, transporting and marketing of forest goods, that in some areas with dense stands of Brazil nut trees, the produce generates more revenue than an equivalent area of pasture devoted to cattle ranching (Bunker, 1981).

The Amazon forests thus offer many outputs apart from the few conventional products, notably hardwood timber, and space for large-scale cattle ranching and small-scale cultivation, that have hitherto been virtually the sole modes of exploitation. Given the way these modes are practised, especially the agricultural forms of exploitation, the deforesters are in effect saying that the many goods and services available through the forests' existence should be reckoned as less valuable than the products of agricultural exploitation. Yet the net value of economic output from agriculture is meagre at best, and is rarely sustainable. That is to say, current exploiters view forest ecosystems as single-output entities, whereas there is much scope for them to be exploited, and sustainably so, through multi-purpose practices. After all, the many non-wood goods of the forests, such as genetic resources and germplasm materials, are low in volume while high in value, hence they can be sustainably harvested while imposing virtually zero disruption on forest ecosystems. Environmental services are supplied through the mere existence of forest ecosystems.

An example of multi-purpose exploitation is to be found in the Ecuador sector of Amazonia, in the form of an experimental project. According to some preliminary figures (Paucar and Gardner, 1981), as much as 85 per cent of animal protein consumed by local people comes from wild animals, notably peccaries, deer, tapirs, pacas and agoutis, among some 40 species of mammals in all. A sustainable harvest of wild meat could, according to exploratory appraisals, amount to 240 kg per square kilometre with a market value of about $1.8 per kg, or a total of almost $440 per square kilometre overall (1980 prices). Were the harvest to be systematised, and expanded to include birds, turtles and fish, the minimum potential value could be increased as much as 10 times. Furthermore, since almost one-third of Amazonian Ecuador consists of swamps and rivers, there is plenty of scope for the harvesting of caimans, at least two per hectare per year; and a caiman measuring 1.5 metres in length is worth $145. Thus a single square kilometre of such habitats could yield $14,550 per year for hides alone, rising to $16,370 when caiman meat is included. In addition, each square kilometre of forest can *renewably* produce 20 primates each year for biomedical research, an individual being worth between $200 and $300. So a sustainable harvest of wild primates could generate a minimum of $4,000 per square kilometre per year.

This all means that a forest tract of 500 square kilometres could, under scientific management, produce a self-renewing crop of wildlife with a potential value of at least $10 million per year, or slightly more than $200 per hectare. These revenues are to be contrasted with the return from commercial logging of only little more than $150 per hectare – and hardwood timber tends to be harvested as a once-and-for-all product, leaving little prospect that a further harvest can be taken within another several decades.

CLIMATE LINKAGES

A further 'output' of Amazonia is the rainforest's capacity to stabilise climate. This has rarely been considered a significant factor in Amazonian development. But evidence is rapidly accumulating that continued broad-scale deforestation could cause substantial dislocations in climatic patterns, notably rainfall regimes – and not just in Amazonia, but in territories further afield.

First, there is a recycling effect. Over half of Amazonia's precipitation derives from moisture circulating within the regional ecosystem, rather than from the Atlantic Ocean (Salati and Vose, 1984). As fast as moisture falls from the skies to earth, it is returned through evapotranspiration to the atmosphere, whereupon it gathers in clouds for a fresh series of thunder-storms. Were a substantial part of Amazonia's forest cover to be removed, the residual forest would be less able to evapotranspire as much moisture as was circulating within the ecosystem before – and this would make for a steadily desiccating ecosystem, with rainforest becoming drier forest, eventually even woodland. There could also be linked effects of albedo change (cleared forestlands reflect more solar energy than the dense foliage of rainforest), with perhaps a further 20 per cent decline in precipitation (Shuttleworth, 1988).

Nor would these repercussions be necessarily confined to Amazonia itself. The climate of extensive territories to the south of Amazonia now appears to depend, in part at least, on the same hydrological cycles of Amazonia. The upshot could be reduced rainfall in Brazil's main farmlands in the South of the country.

In addition, there is a connection to the greenhouse effect that appears to be overtaking the entire planetary ecosystem. Burning of Amazonia to make way for cattle ranches and small-scale cultivation ejects sizeable amounts of carbon dioxide, also other greenhouse gases such as carbon monoxide and nitrous oxide, into the global atmosphere. Of course, Brazil and other Amazonian countries may respond that the greenhouse effect would not (so far as we can discern) have such adverse effect on their climate as on that of temperate regions; so why should they be concerned? But a related linkage could ultimately work the other way round. Political leaders worldwide may come to determine that a greenhouse effect could prove so detrimental to agriculture, coastal settlements and other salient human activities, that they will seek ways to reduce its impact. One of the most efficient ways will be to sequester carbon from the atmosphere through broad-scale reforestation of the humid tropics (fast-growing trees can 'soak up' large amounts of carbon). A prime location for such an effort would surely be Amazonia in the light of the significant capacity of the region to regulate global climate.

These climate linkages raise all manner of complex questions of public policy, both within Amazonian countries and elsewhere. What are some rough costs and benefits of maintaining the forest cover for purposes of climate stabilisation? Who would mainly bear costs and receive benefits? What intergenerational factors are at work? Who can claim 'property rights' in climatic workings, whether local, regional or global? These could eventually prove to be the most pertinent, as well as challenging, questions of Amazonia's 'outputs.'

THE BEEF CONNECTION

There is an expanding appetite for beef around the world, and especially in Japan, nations of the Soviet bloc and the oil-producing nations of the Middle East. In fact, demand for beef is rising more rapidly than for any other food category except fish. To meet this expanding demand for beef with its lucrative profits, Brazil has become an important exporter of beef, with four-fifths of its exports going to Britain, West Germany and Italy among other countries. To this extent, there is external involvement in the beef sector in Brazilian Amazonia. The price paid for beef in a European supermarket does not reflect all the costs that have gone into its production, especially deforestation costs in the Amazon.

There is much experience, moreover, to indicate that cattle raising in the Amazon is a doubtful venture. Stocking rates are generally low, only one head of cattle to 2 hectares immediately after the new pasture is established, and only one animal to every 5 hectares within just five years, due to decline in soil fertility and nutritional grasses. Steers require an average of four years to reach a weight of 450 kg for slaughter, which makes for a productivity rate only one-quarter or less than what it is in US ranchlands. There is also a problem of toxic weeds that invade as the soil becomes compacted.

Ironically, Brazil could almost certainly derive much more animal protein through exploitation of its Amazon river system, which, with over 2,000 known species of fish, (and possibly 3,000 species in all), contains eight times as many species as in the Mississippi river system, and ten times as many as in the whole of Europe. These fish stocks could supply the nation's entire requirements for animal protein, plus a surplus for export. The Amazon fishery overall can be roughly estimated at a minimum of 1 million tons, with a renewable offtake of a quarter of a million tons.

Moreover, the Amazon River floodplains with their rich alluvial deposits cover almost 100,000 square kilometres (roughly the size of England). They could, without using artificial fertiliser, allow an agriculture as rich as any known in the great river civilisations of the past, such as those of Egypt, Mesopotamia and China. Indeed at a generous rate of 10 hectares per farmer (just 2 or 3 hectares are usually enough in this area with its ultra-fertile soils), the floodplains could support 1 million farmers, which, together with their families, could amount to more than 5 million people.

SOME ECONOMIC CAUSES OF ENVIRONMENTAL DEGRADATION

While official policies for the development of Amazonia date from the end of the Second World War, the major 'push' took place after the military take-over of 1964. The main instrument of development has been the Superintendancy for the Development of the Amazon (SUDAM) set up in 1966, and the main focus has been private investment capital through the Fundo para Investimento Privado no Desenvolvimento de Amazônia (FIDAM), renamed and restructured as the Fundo de Investimento da Amazônia (FINAM) in 1975. Development has been heavily subsidised. Subsidies have an economic justification where development generates external benefits that cannot be appropriated by the market. If, for example, the development gains 'trickle down' to the poor, then there can be justification for subsidies in terms of the wider equity objectives. But the SUDAM procedures have this aim only on paper, and SUDAM itself had its origins in attempts to accommodate and further the interests of the rich. Moreover, many of the developments have no commercial rationale in the absence of the subsidies. By and large, then, the subsidies have served neither equity nor efficiency objectives in Amazonia. They have also encouraged and been responsible for a major part of the deforestation that has occurred.

Livestock Development

Browder (1985) estimates that subsidies to investment, rural credit, small farmer colonisation and export products totalled some $2.25 billion (1985 prices) over the two decades 1965–85. The major subsidy item has been that for small farmers, at some $1.5 billion, but this has accounted for perhaps only 20 per cent of all deforestation. Livestock developments have been the major factor behind deforestation, accounting for perhaps three-quarters of deforestation.

Direct Tax Credits

The main motive for deforestation is landownership. In turn, landownership is greatly encouraged by inflation, land being a reasonable hedge against rising and fluctuating prices. This already significant basis for the demand for land is reinforced by tax incentives which make landownership even more desirable. This results in both 'frontier' land purchase, and the transfer of land from small farmers to large farmers and corporations. Given the obviously different tax liabilities of the rich and poor, the incentive system is biased to the rich. Moreover, the incentives become capitalised into land values. The poor become doubly disadvantaged: they secure no benefit from the tax incentive and they cannot buy land as the price is inflated. Land acquisition for the poor is thus by the only remaining means: squatting.

Commentators generally agree that the tax credit scheme operated by SUDAM is a dominant driving force in Amazonian land acquisition (Browder, 1985; Binswanger, 1987; Mahar, 1988). In turn, that scheme encourages livestock development which, as noted above, has been a major factor in deforestation, mainly in Mato Grosso and Goiás. Private corporations can exempt up to 50 per cent of their income tax liability provided the tax savings so secured are invested in FINAM in exchange for FINAM shares. The share purchase is linked to a specific development project in Amazonia. FINAM buys stock in the approved project and transfers this to the private corporation. Effectively, then, the private corporation has its investment costs subsidised. Since the requirement is that it need commit only 25 per cent of the investment cost from its own resources, up to 75 per cent of the cost is financed by the transfer of stock via FINAM. As Binswanger (1987) observes, the private investor will be concerned only with the rate of return on the 25 per cent contribution of its own funds and not with the total social cost of all the funds. Projects may thus yield *negative* rates of return and still be attractive.

This tax credit system accounts for a sizeable part of the investment in livestock schemes in Amazonia, and these accounted for some 42 per cent of all direct tax credits released by SUDAM from 1965 to 1983. Industrial projects absorbed a little more at 45 per cent, and this amount was absorbed by only 59 industrial wood producers (Browder, 1985), but it is livestock that has primarily accounted for deforestation. By late 1985, some 950 livestock projects had been approved by SUDAM (Mahar, 1988), with an average size of some 24,000 hectares. Technically, half the resulting land area of some 8.4 million hectares was supposed to be retained as forest.

The livestock units were largely failures. Gasques and Yokomizo (1985) report that only 16 per cent of anticipated livestock production was realised. Interestingly, however, virtually 100 per cent of tax credits were disbursed, suggesting that the subsidy structure has created its own 'industry', one of collecting the subsidies and then not using them for the stated purpose. This result is a conspicuous example of the outcome of 'rent-seeking' – i.e., of the ways in which economic distortions divert energies and expenditures into non-productive activities, effectively shuffling national income from the less well off to the rich. Moreover, livestock development has not even secured commercial rates of return. Forest soils are notoriously unsuited to sustained livestock production so that, as pasture yields decline, maintenance costs rise, making it cheaper to move to virgin land than to continue with existing holdings. Browder (1985) reports that:

> large scale cattle ranching in the Amazon is inherently unremunerative during the initial five years that correspond to the normal pasture life cycle.

Morover, employment benefits have been minimal, with only one person employed per 250–300 hectares of pasture (Gasques and Yokomizo, 1986).

LAND-TITLING

Deforestation is accelerated by the manner in which land titling laws work. Land claims quickly follow the opening of the frontier through infrastructure development, notably road construction (Mahar, 1988). Claims, however well-founded, are bought and sold. Claims based on SUDAM-approved projects are likely to be settled as priorities. Squatters (*posseiros*) can also quickly establish claims. Usufruct rights can be acquired after one year and one day of effectively using unclaimed public land. Title can be obtained after five years of effective use (a recent change, previously it was ten years). Similarly, squatting on private land can result in title if the owner makes no challenge. Binswanger (1987) notes that, contrary to popular conception, such land acquisition procedures extend to large plots of even 3,000 hectares. Moreover, land acquistion is biased to the large corporation or private ranch because of the initial capital required to gain access to, and market produce from, land beyond the immediate vicinity of highways. Indeed, most deforestation arising from an agricultural conversion is on large private and corporate ranches, suggesting that, in contrast to other countries, the small shifting cultivator is not the main instrument of deforestation in the Amazon. While this is true for the Amazon as a whole, local variations are substantial. Browder (1985) estimates that nearly all deforestation in Rondônia has been due to small farmer colonisation.

The deforestation process is greatly encouraged by the system of titling which tends to confer title on the basis of some multiple of the land cleared for pasture. The bigger the clearance, the bigger the land area likely to be under title. Moreover, clearance is in the interests of the owner since it discourages squatters. Clearance may also be carried out in collaboration with a logging company which takes the timber as payment. The titling system thus reinforces the already considerable drive for landownership in Amazonia, a drive that is characterised by unequal competition between small shifting cultivator and large landowner. Since the stated objective of assisted colonisation is the livelihood of peasants in areas of land scarcity, it is essential to modify the bias in current land allocation schemes. The alternative is continuation of the trend towards urbanisation, with consequent urban unemployment and impoverishment, or rural unemployment arising from landlessness. Given the way in which colonisation 'follows' frontier infrastructure, mainly roads, there is also an indirect mechanism for controlling deforestation through more geographic direction of the road building programme.

The previous brief discussion suggests two fundamental reasons for non-sustainable development of Amazonia: the bias of policy towards the interests of the rich, and the encouragement that land ownership receives from underlying macroeconomic conditions, notably inflation. The cure for the latter lies in the whole structure of macroeconomic policy which is beyond the scope of this chapter. The cure for the former lies in the restructuring of policy instruments in terms of taxation and landtitling. Both, of course, strike at the interests of the wealthy and both therefore face formidable political obstacles.

CAPTURING EXISTENCE VALUES

In reformulating internal policy to achieve optimal use of the Amazonian forest, attention should be given to policy measures which appropriate the external benefits of the forest. Economic theory dictates that the generation of an external benefit warrants charging the beneficiaries for the gains they secure. Brazil has the property rights in Brazilian Amazonia and the means to extract payment through agreements *not* to deforest. Such a prescription does, of course, involve an underlying danger of threat situations, akin to the political use that has already been made of the OPEC oil weapon. But some recent developments already suggest workable compensation measures.

'Debt-for-nature' Swaps

In July 1987, a United States environmental organisation, Conservation International, reached agreement with the Bolivian government for the protection of 1.6 million hectares of forest and grassland in the area east of the Beni Biosphere reserve. The agreement heralded the first of a small series of 'debt-for-nature' swaps. Such swaps hold the potential for appropriating the external benefits, in the form of option and existence values, of neutral assets such as the Amazonian rainforest. Procedures could vary, but the Bolivian case proceeded as follows. Conservation International, using money received from a conservation foundation, bought Bolivian debt on the secondary loan market. Since the international community knows that many developing countries do not have the capacity to repay outstanding levels of debt, it is available at a discount on these markets. Bolivian debt at the time of the swap was trading at about one-sixth of its face value. Conservation International bought $650,000 of nominal debt for $100,000 and then handed the debt over to Bolivia in exchange for the conservation agreement. The Bolivian government additionally set up a $250,000 trust fund in local currency for management of the area. Part of it is to be protected totally and part is to be open for use by the Chimene indians.

Similar deals have been struck for Ecuador in which debt acquired on the secondary loan market is traded in for the issue of non-negotiable local currency bonds to be held by the conservation purchasers. The yields from the bonds are then to be used for conservation purposes. Costa Rica has indicated a similar scheme.

The Bolivian case implies some remarkably cheap conservation, at around 6 US cents per hectare. One could surmise that as more deals are struck the willingness of indebted nations to make further deals will decline and the implicit price per hectare will rise. The 6 cents thus appears to have little relationship at the moment to the scale of existence and option value in the rich nations. Indeed, there are good economic reasons for supposing that debt-for-nature swaps will grossly understate intrinsic values. The benefits of such schemes are mutual: debt can be retired (albeit in tiny amounts if the existing examples of such swaps prove to be typical), and environments preserved. Debt swaps thus initially appear to be good examples of the complementarity of environment and development. Clearly, there are pitfalls. Commitments by an existing government may prove not to be binding on future governments, but this may be overcome by annual, as opposed to once-for-all, payments. Given the uniqueness of Amazonia and its focus of international attention, debt-for-nature swaps, or other international compensation, might prove a fruitful mechanism for capturing just a little of the rich world's concern.

TOWARDS SUSTAINABLE DEVELOPMENT IN AMAZONIA

What should be the way forward for governments of the region? How can they foster sustainable development for their vast tracts of forest? A number of initiatives are open to them.

A basic problem lies with our incapacity to measure the values of all goods and services generated by tropical forests. It is easy, for example, to calculate the amount of hardwood timber cut, and to put a price tag on its commercial value. The same applies to forest land, beef and other products registered through the market place with its sensitive signals of consumers' evaluation. Far more difficult is to appraise the non-use values, and the less tangible use values, such as biodiversity. Were we to appraise all outputs of tropical forests in money terms, it is feasible that we would find the forests can support human welfare in a more bountiful and sustainable fashion than through the narrow focus and ultra-disruptive harvest of timber and beef.

In short, the analytic methodologies based on total economic value concepts need further development and empirical application in the tropical forest case. Economists, politicians, development planners and others are little inclined to keep a close eye on non-wood products and environmental functions, precisely because they do not supply their benefits so speedily and obviously as does beef. What carries weight with planners are recognisable concerns such as cash revenues: there is undue emphasis on what can be counted, to the detriment of what also counts.

For policy-makers to adjust the asymmetry of present evaluation procedures, they need to undertake economic analysis on a scale that encompasses the entire range of goods and services available. True, this is not readily done, especially as concerns environmental services. How to put a dollar value on climatic stability? But cost-benefit analysis could include an 'environmental dimension' by considering the safeguard effect of forest cover with respect to hydrological functions. For sure, this is not easily incorporated into the calculus. Nonetheless it merits anticipatory attention through evaluatory methodologies, such as risk-reduction analysis.

A few extended cost-benefit analyses have been attempted (e.g., Gregersen *et al.*, 1987) but they tend to be limited in both number and scope. They are far from matching up to the nature and scale of the problem. The many fine-grain analyses of, for example, beef demand and supply trends should be complemented by equally rigorous analyses of the concealed costs that arise when Amazon forests are eliminated.

In addition, the analytic methodologies should be geared to social equity as well as economic efficiency. The benefits of beef ranching tend to accrue to only a limited sector of the community, whereas the costs of deforestation fall on the community as a whole – and they extend far further into the future than do the benefits of unsustainable beef ranching.

As far as the Amazon forests' contributions to exceptional-value products that are mainly traded internationally – e.g., plant-bases for drugs – are concerned, it is feasible that some form of 'user charge' could be made. It might then be possible that conservation of threatened forest species with their gene stocks would become a competitive form of land use in forest territories. In the meantime, however, Madagascar receives nothing from the end-product sales of the two anti-cancer drugs based on the rosy periwinkle. The country's forests may well harbour other anti-cancer materials; and thousands of Madagascar's plant species are on the verge of extinction if not eliminated already. Yet the country has zero incentive to preserve the plants in their forest habitats in light of their potential medicinal applications, valuable though they could prove to be.

The answer is to devise an institutional mechanism that would enable tropical forest countries to receive some sort of 'rental fee'. The same applies in the case of other forest outputs that contribute markedly to human welfare in lands far removed from the source country: notable instances include plant genetic resources that support modern agriculture through regular germplasm infusions to upgrade the productivity of crops such as rubber. (We can note too that Brazil benefits significantly from germplasm supplies for its coffee crop, imported from Ethiopia.) An institutional initiative to devise a user charge would be a taxing challenge and, to date it has been considered simply too problematic. But could it prove a tougher proposition than trying to live in a world that has lost its Amazon forests and their wild species?

Perceived within an expanded evaluatory framework, exploitation of Amazon's forests could come to include all manner of activities. Genetic reservoirs should count together with forest; a national park could sometimes prove as valid a use as a paperpulp plantation. Maintenance of climatic stability could be seen as a form of 'use' that ranks alongside timber harvesting. In certain localities, 'use' could even entail outright preservation of forest ecosystems for scientific research, especially applied-biology research as it relates to new modes for man to reap Earth's natural bounty in Amazon forests as the most productive ecosystems anywhere.

In sum, what is needed is an approach that addresses all features of the Amazon landscape – not only the ecological landscape but also the economic, institutional and development landscape as well. If the Amazon forests can be 'developed' in the sense that their entire array of outputs is mobilised in support of human welfare, then their intrinsic all-round worth will be better recognised – and they will be more readily permitted survival in an ever more crowded world. If not, they may not last beyond another half century or so, since their space will be taken for other, and supposedly more valuable, purposes.

It is also clear that government has been the main destroyer of the Amazonian forest, and all for a trivial economic return. Fiscal incentives have played a major role in inducing deforestation and in accelerating the clearance impacts of infrastructure development in the region. Government, in turn, has devised policies which reflect the political and economic power of a limited number of investors. Land-titling has played its role in 'legitimising' deforestation as if it were an improvement to the land instead of an economically inferior use.

While the Amazon forests are vital to development as properly understood, so too is proper development of the Amazon forests vital to their continued existence. They may be accorded development of an enduring kind to the extent that they are enabled to demonstrate their full spectrum of goods and services.

Note

1. The authors are indebted to Edward Barbier for comments on an earlier draft.

References

Balick, M. J. (1979), 'Amazonian Oilpalms of Promise: A Survey', *Economic Botany*, 33(1), pp. 11–28.

Binswanger, H. (1987), 'Fiscal and Legal Incentives with Environmental Effects on the Brazilian Amazon', Report No. ARU 69, Agriculture and Rural Development Department, World Bank (May).

Bishop, R. (1982), 'Option Value: An Exposition and Extension', *Land Economics*, 58(1) (February), pp. 1–15.

Boom, B. M. (1985), 'Amazonian Indians and the Forest Environment', *Nature*, 314, p. 324.

Brookshire, D., Eubanks, L. and Sorge, C. (1985), 'Existence Values and Normative Economics', Department of Economics, University of Wyoming, Laramie (September) (mimeo).

Browder, J. O. (1985), '*Subsidies, Deforestation and the Forest Sector in the Brazilian Amazon*', Report to the World Resources Institute, Washington D.C.

Bunker, S. G. (1981), 'The Impact of Deforestation on Peasant Communities in the Medio Amazonas of Brazil', *Studies in Third World Societies*, 13, pp. 45–60.

Bunker, S. G. (1988), *Underdeveloping the Amazon: Extraction, Unequal Exchange and the Failure of the Modern State*, Urbana: University of Illinois Press.

Douros, J. and Buffness, M. (1980), 'The National Cancer Institute's Natural Products Antineoplastic Development Program', in S. K. Carter and Y. Sakurai (eds), *Recent Results in Cancer Research*, 70, pp. 21–44.

Duke, J. A. (1980), *Neotropical Anti-Cancer plants*, Economic Botany Laboratory, Agricultural Research Service, Beltsville, Maryland.

Fearnside, P. M. (1986), 'Spatial Concentration of Deforestation in the Brazilian Amazon', *Ambio* 15, pp. 74–81.

Gasques, J. and Yokomizo, C. (1985), 'Avaliacão dos Incentivos Fiscais na Amazônia', Instituto de Planejamento Econômico e Social, Brazil (December).

Gasques, J. and Yokomizo, C. (1986), 'Resultados de 20 Anos de Incentivos Fiscais na Agropecuária de Amazônia', X1V *Encontro Nacional de Economia*, ANPEC, vol. 2, quoted in Mahar (1988).

Gregersen, H. M., Brooks, K. M., Dixon, J. A. and Hamilton, L. S. (1987), *Guidelines for Economic Appraisal of Watershed Management Projects*, FAO Conservation Guide, 16, Rome: FAO.

Hecht, S. B. (1985), 'Environment, Development and Politics: Capital Accumulation and the Livestock Sector in Eastern Amazonia', *World Development* 13, pp. 663–84.

Lisboa, P. L. B., Maciel, U. N. and France, G. T. (1987), 'Some Effects of Colonization on the Tropical Flora of Amazonia: A Case Study from Rondônia', *Ciência Hoje*, 6, pp. 48–56.

Mahar, D. (1988), *Government Policies and Deforestation in Brazil's Amazon Region*, Environmental Department, Working Paper, 7, Washington, D.C.: World Bank (June).

Malingreau, J.-P. and Tucker, C. J. (1988), 'Large-Scale Deforestation in the Southern Amazon Basin of Brazil', *Ambio*, 17, pp. 49–55.

Myers, N. (1984), *The Primary Source: Tropical Forests and Our Future*, New York and London: W. W. Norton.

Paucar, A. and Gardiner, A. L. (1981), *Establishment of a Scientific Research Station in the Yasuni National Park of the Republic of Ecuador*, National Forestry Program, Ministry of Agriculture, Quito, Ecuador.

Pearce, D. W. (1986), *Cost Benefit Analysis*, London: Macmillan.

Pearce, D. W. (1988), 'Sustainable Development: Ecology and Economic Progress', Discussion paper, Department of Economics, University College London.

Prance, G. T. (ed.) (1982), *Biological Diversification in the Tropics*, New York: Columbia University Press.

Salati, E., de Oliveira, A. E., Schubart, H. O. R., Novaes, F. C., Dourojeanni, M. J. and Umana, J. C. (1988), 'Changes in the Amazon over the Last 300 Years', in B. Turner (ed.), *Earth Transformed by Human Action*, Proceedings of Conference at Clark University, Worcester, Mass. (October 26–30, 1987).

Salati, E., and Vose, P. P. (1984) 'Amazon Basin: A System in Equilibrium', *Science*, 225, pp. 129–38.

Shuttleworth, W. J. (1988), 'Evaporation from Amazonian Rain Forest', Proceedings of Royal Society of London B (in press).

World Commission on Environment and Development (1987), *Our Common Future*, Oxford: Oxford University Press.

Index